Neuroscience and Philosophy

Neuroscience and Philosophy

Edited by Felipe De Brigard and Walter Sinnott-Armstrong

The MIT Press
Cambridge, Massachusetts
London, England

The MIT Press would like to thank the anonymous peer reviewers who provided comments on drafts of this book. The generous work of academic experts is essential for establishing the authority and quality of our publications. We acknowledge with gratitude the contributions of these otherwise uncredited readers.

This book was set in Stone Serif by Westchester Publishing Services.

Library of Congress Cataloging-in-Publication Data

Names: Brigard, Felipe de, editor. | Sinnott-Armstrong, Walter, 1955– editor.
Title: Neuroscience and philosophy / edited by Felipe De Brigard and
    Walter Sinnott-Armstrong.
Description: Cambridge, Massachusetts : The MIT Press, [2022] |
    Includes bibliographical references and index.
Identifiers: LCCN 2021000758 | ISBN 9780262045438 (paperback)
Subjects: LCSH: Cognitive neuroscience—Philosophy.
Classification: LCC QP360.5 .N4973 2022 | DDC 612.8/233—dc23
LC record available at https://lccn.loc.gov/2021000758

This book is dedicated to all philosophers and all neuroscientists who have built, and continue to build, bridges between the two disciplines.

Contents

Acknowledgments

On December 16, 2013, we received notice from the John Templeton Foundation informing us that our grant proposal to run the Summer Seminars in Neuroscience and Philosophy (SSNAP) at Duke University had been accepted. Since then, this ambitious project benefited from the help of many people.

We would like to thank the Department of Philosophy, the Department of Psychology and Neuroscience, the Duke Institute for Brain Science (DIBS), and the Center for Cognitive Neuroscience at Duke University for their support, as well as all the faculty members from these and other departments who have been instructors at SSNAP. Special thanks to DIBS for offering the infrastructure to carry out this grant, and particularly to Diane Masters and Lynnette Thacker for their support during all these years.

We are also extremely grateful to the project coordinators who, through these years, helped with the organization of the first four SSNAPs: Gregory Stewart, Jacqueline DeRosa, and, especially, Maria Khoudary, the current project coordinator whose help compiling the glossary as well as proofreading, formatting, and providing feedback on the chapters of the current volume has been invaluable.

We also greatly appreciate all of the instructors and speakers, from all over the world, who have come to Duke to support SSNAP and to enlighten and inspire a diverse group of fellows to continue working at the intersection of philosophy and neuroscience. Likewise, we are happy to thank all of the sixty-two fellows who have participated in SSNAP for the past four years. To us, you feel like family, and we continue to be impressed and heartened by the ways in which you move forward both philosophy and neuroscience with your research and teaching.

Finally, we are very grateful to the John Templeton Foundation for their faith in this project, and we especially want to thank Michael Murray, John

Churchill, and Alexander Arnold for their help and support. We also gratefully acknowledge the Templeton World Charity Foundation for supporting the continuation of SSNAP for the next four years.

The book you have in your hands is not only the work of the authors who put together terrific chapters, but also the work of many individuals who made SSNAP possible. We hope that this volume, which will support the instruction of future SSNAP fellows, will help to cement further collaborations between philosophers and neuroscientists working to answer big questions about the human mind.

Felipe De Brigard and Walter Sinnott-Armstrong

September 29, 2020

Introduction

Felipe De Brigard and Walter Sinnott-Armstrong

Philosophers and neuroscientists try to answer many of the same questions: How can consciousness emerge from brain activity? How are perception, memory, and imagination related to what we know? How do brain development and activity influence moral choice, behavior, and character? Are specific decisions based on general intentions? Is brain activity determined or stochastic? Do we control our choices and actions? And so on.

Each field can learn a lot from the other about these and other topics of mutual interest. Unfortunately, researchers in these disciplines rarely work together or even understand each other. Fortunately, recent developments signal the coming end of this misfortune. New methods as well as theories in both neuroscience and philosophy increase openness on both sides and pave the way to mutually beneficial collaborations. Nonetheless, significant impediments remain. We need to understand both the promise and the obstacles in order to make progress.

The Growth of Neuroscience

The past few decades have witnessed an exponential increase in neuroscience research. Whether this growth is due to its brilliant investigators from various disciplines, to its multiple synergistic research methods, or to its numerous promises of profound results, neuroscience is now the flagship among biological and medical sciences.

The White House confirmed this prominent position in April 2013 when it unveiled the Brain Research through Advancing Innovative Neurotechnologies—or BRAIN—Initiative (https://obamawhitehouse.archiv es.gov/BRAIN). This Initiative hopes to enable us not only to understand better the relationship between the mind and the brain but also to "foster science and innovation" and "to inspire the next generation of scientists"—two of the so-called Grand Challenges of the twenty-first century.

Encouraging as this plan might sound, the truth is that we are far from possessing a detailed picture of how the brain works and even farther from understanding how brain functions relate to operations of the human mind. This gap is no secret, as neuroscientists will be the first to admit. Part of the problem lies in the sheer difficulty of developing noninvasive technologies that are powerful and sensitive enough to allow us to measure brain activity in vivo and at different levels of complexity, from molecules to large-scale brain areas. Similarly, neuroscientists face the daunting task of systematizing, analyzing, and modeling enormous amounts of data. Above and beyond these fundamental yet tractable practical difficulties rests a less discussed but equally pivotal challenge: that of understanding how answers about the brain can resolve questions about the mind and vice versa. No matter how much neural data we amass and no matter how sophisticated or computationally powerful our statistical models become, the goal of endeavors such as the BRAIN Initiative will be out of reach unless we have a clear theoretical framework that reveals how the brain relates to the mind. This relation is the bailiwick of philosophers.

A Role for Philosophy

Since Plato's Phaedo, philosophers have repeatedly asked how our minds are related to our bodies. This perennial problem has been reshaped in the context of twenty-first-century neuroscience into the question of how our brains can cause or realize our minds. Many profound and original answers have been proposed by philosophers in recent years, including not only subtle variations on reductionism but also forceful arguments against reductionism, as well as several sophisticated alternatives to reductionism that take neuroscience into account. These philosophical theories could potentially provide the kind of framework that the BRAIN Initiative needs. Thus, neuroscience can gain many benefits from working with philosophers.

Philosophers also stand to gain insights from neuroscience. Any approach to the philosophy of mind needs to be consistent with what we know about how the brain works, but neuroscience also affects many other central issues in philosophy. Epistemologists try to develop theories of knowledge that need to take into account how information is processed in human brains, including the neural systems that enable perception, memory, and reasoning. Philosophers worry about free will and responsibility, and their views can be tested against discoveries in the neuroscience of decision, action, self-control, and mental illness. Again, moral philosophy—including practical ethics, normative theory, and meta-ethics—often hinges on empirical assumptions that can be tested by neuroscience together with psychology. Are moral beliefs based on emotion or reasoning? Do people have moral virtues and vices that remain stable across situations? Are moral judgments unified by a shared substrate in the brain? These questions (and many more) are asked by philosophers, and their answers depend on empirical claims that can and should be tested scientifically. In these ways, philosophers can benefit tremendously by working with neuroscientists.

Barriers to Collaboration

At this juncture, philosophy and neuroscience are poised to work together, learn together, and advance together. However, just as philosophers have come to appreciate the results coming from neuroscience, and just as neuroscientists have come to appreciate the importance of asking foundational questions to grasp the significance of their results better, both fields are keenly aware of the substantial gaps between their vocabularies, research methods, and cultures. These gaps stand in the way of progress.

One barrier is language. Each discipline creates and sometimes insists on its own proprietary vocabulary. Yet, interdisciplinary collaboration requires mutual understanding of each other's technical terms. Disciplines also develop distinctive methodologies, standards, and practices, and they cannot work together without becoming familiar with these central features of their collaborators' fields. One aim of each chapter in this book is to help readers from both fields become familiar and comfortable with each other's vocabularies, methods, standards, and practices.

Successful collaboration also requires mutual respect, which comes from appreciating the value of the other field. Real appreciation, however, cannot come from a single lecture or conference. It takes time. That is why each of the chapters in this book was written by a team that includes at least one philosopher and at least one neuroscientist. These teams had worked together before on research projects and then spent many months writing a chapter together. The authors of each chapter also met for an intense week in which they helped each other to improve their chapters and make them more intelligible to readers who were trained in different disciplines or worked on different topics. This lengthy and intense process of collaboration helped them appreciate each other other's disciplines and led them to respect each other. This mutual appreciation is clearly evident in their chapters. They provide tangible proof that interdisciplinary collaboration between neuroscientists and philosophers is feasible and fruitful (as well as fun!).

The teams of authors of these chapters were drawn from the fellows in Summer Seminars in Neuroscience and Philosophy (SSNAP), which we co-directed in 2016–2018 with generous support from the John Templeton Foundation (for which we are very grateful). Fellows in these seminars studied each other's fields intensely for two weeks, and then most of them collaborated in research projects over the next year or two. Competition for slots in these seminars was fierce, and these authors represent some of the best fellows in that elite group. They are all future leaders in neuroscience and philosophy who will shape interactions between these disciplines for years to come. If you want to know how this interdisciplinary field will develop, they will show and tell you. That is another goal of the chapters in this book.

In addition, this book displays the wide variety of topics that fall under the general heading of neuroscience and philosophy. Although each chapter combines both disciplines, they use them in different ways to make progress on different issues.

Part I: Social Neuroscience and Philosophy

The first group of chapters addresses basic issues about our social and moral lives: how we decide to act and ought to act toward each other, how we understand each other's mental states and selves, and how we deal with pressing social problems regarding crime and mental or brain health.

Moral Judgment

Since ancient Greece, philosophers have debated about which acts are morally right or wrong. Moral controversies have also polarized whole societies, not only recently but throughout history and around the world. Neuroscience by itself cannot solve these problems, but it might be able to help us understand what makes people hold strong but contrary views about many important moral issues. Some skeptics have suggested that knowledge of the neural processes that produce moral disagreements should make us doubt that any moral belief is true or known to be true.

In their chapter, **May, Workman, Haas, and Han** survey what neuroscience has learned so far about these issues. They divide neuroscientific studies of morality into three periods: the first period from 1990 to 2000 emphasized gut feelings, the second period from 2000 to 2010 introduced dual process models using brain imaging and manipulations, and the third period from 2010 to 2020 added computational models and theories of how adults and children learn their moral views. In the end, they argue that neuroscience does not undermine all moral judgment but instead can help us to determine when moral judgment is trustworthy or not.

Empathy

Moral judgments often concern harm to other people. As a result, many philosophers see empathy as central to moral judgment. However, morality is also often said to be impartial, and several critics have argued that empathy can interfere with impartial moral judgments when we empathize only with individuals or groups whom we favor. These debates have been clouded by imprecision and ambiguity, since empathy can have many different meanings.

In their chapter, **Spaulding, Svetlova, and Read** try to untie these Gordian knots by distinguishing automatic versus conscious empathic reactions, cognitive versus affective and motivational aspects of empathy, and self-oriented versus other-oriented concerns. They trace the development through childhood of self-awareness and other-awareness as well as concern for others, sympathy, and mentalizing (theory of mind). Contrary to some recent claims, they argue that these abilities cannot be understood solely in terms of mirror neuron activations, for various kinds of empathy also involve the right temporal parietal junction, precuneus, and superior temporal sulcus, among other brain regions. In the end, they bring this neuroscientific and philosophical theorizing to bear on the issues of how (and whether) empathy is related to morality.

Agency

In addition to substantive moral judgments about actions, we also judge some people to be morally responsible for some actions but not for others.

These judgments depend on various aspects of our actions, such as who does the act or causes its effects, whether the agent knows or could know what was being done, which factors affect what the agent does, and whether the agent controls the act or its consequences. These issues have been investigated in depth by decision neuroscientists, and their findings have far-reaching implications for philosophy of action as well as philosophical theories of free will and moral responsibility.

In their chapter, **Waller and Brager** summarize recent findings about two main issues: the sense of agency (the feeling that I did an action) and the initiation of physical movements (which bears on control of whether and when an agent acts). Then, they go into more detail about the previously unexplored issue of how sleep deprivation affects the brain processes that underlie the sense of agency and the initiation of action. This case study raises questions about whether we should excuse mistakes by agents who are required to make important decisions without adequate sleep—including military personnel, shift workers, college students, and parents of young children. These questions cannot be answered without empirical information from neuroscientists together with normative reflection from philosophers.

Self

We make moral judgments not only about what we and others do in the present, but also about what we and others did in the past—sometimes the distant past. Therefore, we need to ask whether the agent is the same person today as they were when they did the act in the past. If not, the person at present does not seem responsible for what the other person did in the past. Philosophers have addressed this issue of personal identity over time with elaborate thought experiments, but they have rarely used empirical methods—until recently.

In their chapter, **Everett, Skorburg, and Livingston** review psychological studies of how common folk identify people over time when those people have undergone major life changes, such as those that result from severe brain damage. They focus on the perhaps surprising finding that changes in moral characteristics, especially in a negative direction, affect folk judgments about personal identity more than other changes, including memory loss and disruptions in non-moral personality traits. This "moral self" effect has been replicated in many groups, including real families of people with brain damage, and its neural basis is starting to be studied. It raises profound challenges to traditional philosophical theories of moral responsibility as well as personal identity, and it suggests that our notion of personal identity over time is not a descriptive matter of metaphysics alone.

Mental Illness

Mental illnesses provide more striking examples of when people might say that an agent was not herself when she did something wrong. Her illness did it, not her—but when is she "not herself"? That question cannot be answered properly without a deeper understanding of mental illness. Mental illnesses are, of course, related to brain dysfunction, and neuroscientists have made significant progress in understanding and treating several mental illnesses. Neuroscience alone cannot determine when a brain abnormality is a mental illness, disease, disorder, or dysfunction, but it can discover neural changes that lead to the feelings, thoughts, and behaviors that usually constitute mental illnesses.

In their chapter, **Washington, Leone, and Niemi** address these and other issues with respect to obsessive-compulsive disorder (OCD), schizophrenia, and addiction, focusing on deep brain stimulation (DBS) as a treatment for OCD and heterogeneity within schizophrenia and addiction as illuminating for what (if anything) unifies a mental illness and for how brain states are related to mental states. The authors discuss what neuroscience has found out about these three examples, but then they go on to show how neurobiological models of these and other mental illnesses have affected philosophical theories, scientific methods, medical treatments, government policies, court procedures, and public opinions in ways that have been controversial and problematic. These applications illustrate some ways in which neuroscience can be misused and can raise important moral issues.

Crime

People with certain mental disorders are sometimes confined against their will because they are dangerous to themselves and others. Such "involuntary commitment" is based on a prediction of future behavior. Our legal system also incarcerates criminals for longer or shorter periods based on predictions of future crimes when defendants are granted or denied bail, when those found guilty are sentenced to punishments, and when inmates do or do not receive parole. These predictions are often criticized as inaccurate, especially when they are based on subjective assessments by judges or forensic clinicians. Actuarial statistics can be more objective and accurate, but these run into new criticisms because they do not treat people as individuals and often use classifications that are indirect proxies for race and socioeconomic disadvantage. Some neuroscientists have recently attempted to use brain scans to improve the accuracy and fairness of these predictions, but neuroprediction remains controversial.

In their chapter, **Aharoni, Abdulla, Allen, and Nadelhoffer** explain the methods that statisticians and neuroscientists use to predict crime, critically assess some limited successes using these tools, and suggest directions for improving both legal procedures and statistical and neuroscientific methods. They respond to criticisms by neuroscientists as well as ethicists and lawyers, but they conclude optimistically that neuroprediction has the potential to become more accurate and at least as fair as the methods of prediction that courts use today.

Optogenetics

In addition to prediction in law, neuroscience can also be used for treatment in medicine. Pharmacological interventions on brain function have been around for a while and have helped to ease some mental illnesses, but new techniques are quickly coming online. Among the most striking are electrical DBS, transcranial direct current stimulation, gene therapy, and now optogenetic DBS. Each new treatment raises important new ethical issues.

In their chapter, **Felsen and Blumenthal-Barby** explain the basic neurobiology behind optogenetics—how light can affect brain function—along with the potential advantages of this technique over other methods of treatments for various conditions, including retinitis pigmentosa, Parkinson's disease, depression, and OCD. Next, they address ethical concerns involving how to select subjects fairly, how to weigh risks against benefits, how to ensure scientific validity, and how to ensure that users and stakeholders know enough about the treatments to give valid consent. As with several other chapters in this volume, the authors aim not to settle these issues but only to begin an informed discussion.

Part II: Cognitive Neuroscience and Philosophy

The second group of chapters addresses basic issues about our mental lives: how we classify and recall what we experience, how we see and feel objects in the world, how we ponder plans and alternatives, and how our brains make us conscious and create specific mental states.

Touch

Our senses enable us not only to gather information about the world, but also to navigate surroundings successfully. Given the pivotal role our senses play in our mental life, it is unsurprising that philosophers' interest in understanding them is as old as philosophy itself. Indeed, philosophers' views on the reliability of our senses marks a foundational distinction in the history of philosophy. Empiricist philosophers, on the one hand, hold that our sensory interactions with the external world are our main—and sometimes only—source of knowledge, while rationalists, on the other hand, highlight the unreliability of our senses as a reason to distrust them as dependable sources of knowledge.

Historically, both the philosophy and the sciences of the mind have been much more interested in vision than in the other senses. However, in the past few years, there has been growing interest in some of the more neglected senses, such as somatosensation. In their chapter, Cheng and Cataldo survey the history of the psychology and neuroscience of somatosensation, and hone in on a particular kind of somatosensation: touch. Next, they suggest ways in which scientific research on touch can have important implications for traditional and contemporary questions in the philosophy of perception.

Sight

As mentioned before, vision is the most extensively studied of all of our senses, and researchers have approached the study of vision from a wide variety of perspectives. Much of that research has been aided by the development of precise mathematical and computational models that help to account for a number of behavioral findings, as well as to predict performance in a variety of tasks. These models have been used to capture, for instance, reaction times under different conditions, perceptual sensitivity, discrimination among different stimuli, and precision as a function of different visual features.

It is less clear, however, how these different mathematical and computational models relate to a critical feature of our visual experience: its phenomenological dimension. In their chapter, **Denison, Block, and Samaha** survey four commonly used quantitative models of visual processing: signal detection theory, drift diffusion, probabilistic population codes, and sampling. Then, based upon research in the philosophy and the science of visual experience, they discuss possible ways in which these models can map onto different features of the contents of our visual experiences. While they find that no model seems to fit all relevant features of our visual contents equally well, they offer several avenues for continuing to explore the relationship between quantitative models of visual processing and the phenomenology of visual perception.

Consciousness

Research on the neural substrates of our conscious experience has surged in the last few decades. Nevertheless, the question as to how our brains give rise to our conscious, subjective experience remains unanswered. A common approach to investigate the neural substrates of conscious experience is to analyze differences in brain activity associated with stimuli of which participants are consciously aware relative to those of which participants are not consciously aware. The thought here is that such differential activity would reflect the neural mechanisms responsible for the conscious awareness of the stimuli.

Although widely used, this strategy is almost always confounded with perceptual task performance (i.e., the measured degree to which a participant succeeds at achieving the experimental goal), which has been shown to dissociate from conscious awareness. In their chapter, **Morales, Odegaard, and Maniscalco** explore the issue of task performance as a confound in research on conscious perception. Specifically, they critically evaluate the strategy of performance matching, which has been thought to provide a solution to the problem of performance confounds, and suggest a signal detection theoretical model that can help to overcome the challenges associated with this experimental strategy in particular and with experimental design in conscious perception in general.

Memory

We are conscious not only of what occurs in our surroundings but also of what happened in our past. Memory, after all, allows us to bring back to mind previous experiences. It is therefore tempting to think of memory as a cognitive faculty dealing primarily with time. Although how exactly the brain manages to support our recollections is still a matter of debate, most researchers agree that the hippocampus and adjacent medial temporal lobe regions are necessary for encoding and, likely, retrieving memories too. Surprisingly, however, for the first two thirds of the twentieth century, work on the neurophysiology of the so-called hippocampal complex showed it to be strongly associated not so much with time but with space. A little more than forty years ago, the theory of the hippocampus as a cognitive map brought together time and space in the hippocampus in a profoundly influential way.

In their paper, **Robins, Aronowitz, and Stolk** critically revisit the notion of a cognitive map and explore its uses in cognitive neuroscience above and beyond merely being a representation of space. As they show, many researchers use the term "cognitive map" to cover representations of nonspatial information as well. How do cognitive maps scale up from representing space to capturing nonspatial information? As their chapter shows, the answer to this question is complex and has important consequences for understanding not only memory but also several other cognitive processes.

Concepts

In addition to perceptions and memories of particular events, our minds also seem to be furnished with general concepts. Theories of concepts have long traditions in both the philosophy and sciences of the mind. Philosophers have traditionally wondered how abstract concepts fit into a world of particular objects. Cognitive neuroscientists who study concepts often see themselves as investigating the constituents or building blocks of semantic memories. Unfortunately, not only is there lack of clarity as to how precisely to characterize what a concept is, there is also quite a bit of disagreement when it comes to relating the notion of "concept," as employed in cognitive neuroscience, with its uses in psychology and philosophy.

In their chapter, **Leshinkaya and Lambert** explore the nature of this disagreement first by critically examining the way in which concepts are said to be studied in the cognitive neuroscience of long-term semantic memory. Then, they go on to show why this way of understanding "concept" does not dovetail with the way psychologists and philosophers use the term. But the issue is not simply about words. The concern is that if one wants concepts to perform a certain role in our cognitive ontology, then what cognitive neuroscientists investigating concepts are doing now cannot fulfil that role. At the end, the authors offer a suggestion, inspired by recent developments in philosophical discussions of concepts, according to which concepts exhibit "sharedness"—a feature, they suggest, to which cognitive neuroscientists interested in understanding the neural underpinning of concepts should pay attention.

Mind Wandering

As much as we wish to stay focused and undistracted when carrying out a particular task, we often find that our minds wander about, plagued with unrelated thoughts. This phenomenon is frequent, common, and almost always involuntary. For the past two decades, research on mind wandering has blossomed, with several labs now dedicated to studying not only the cognitive characteristics of our minds' tendency to wander away from a task at hand, but also the neural underpinnings of this tendency. However, despite this explosion in scientific interest, the scientific study of mind wandering has been also heavily criticized, in no small part because of its reliance on self-reports.

In their contribution to this volume, **Murray, Irving, and Krasich** walk us through the history and justification of the use of self-reports in the scientific study of mind wandering. In so doing, they build a case as to why self-reports are, in fact, indispensable. Moreover, they critically survey other methods, which allegedly can offer "objective" measures of mind wandering, and they suggest that their validity ultimately depends on self-reports. Finally, they offer both scientific and philosophical arguments as to why self-reports are critical not only for the study of mind wandering but also for our understanding of this phenomenon itself.

Mind and Brain

An overarching, central question in both the philosophy and the sciences of the mind asks what the relationship is between the mind and the brain. Theories about the nature of this relationship abound, but the prospect of them receiving empirical validation has been less obvious. In a sense, cognitive neuroscience was conceived as offering precisely that: the possibility of bridging the psychological and the neural levels, aided in part by the use of novel and revolutionary technologies for exploring the brain in vivo. Many philosophers and neuroscientists think that in order to reach such bridging, we need first to identify our "cognitive ontology"—that is, a taxonomy of the basic psychological kinds that can be correlated with structural and/or functional brain structures. Unfortunately, despite its promise, cognitive neuroscience has yet to deliver a clear consensus on what that cognitive ontology might be.

In their paper, **McCaffrey and Wright** critically examine different attempts to characterize a "correct" cognitive ontology and instead advocate for a kind of pragmatic pluralism according to which different cognitive ontologies would be required for different purposes. The consequences of such pluralism for cognitive neuroscience are indeed profound, but as argued by the authors, there is much promise in relinquishing the idea that there is a single taxonomy of the human mind.

Conclusion

Overall, these chapters display a wide variety of ways in which philosophers and neuroscientists can learn from each other. Of course, this volume does not claim to cover all of the issues that are shared between these disciplines. Nor do any of these chapters claim to be the last word on its topic. Still, we hope that these chapters will inspire philosophers and neuroscientists to work together to the great benefit of both fields.

I Social Neuroscience and Philosophy

1 The Neuroscience of Moral Judgment: Empirical and Philosophical Developments

Joshua May, Clifford I. Workman, Julia Haas, and Hyemin Han

1.1 Introduction

Imagine reading in the news about a country far, far away that won't admit poor, helpless, asylum-seeking refugees. "That's just *wrong*," you think. Yet, your neighbor passionately draws the opposite conclusion. In this way, humans are able to judge actions to be right or wrong and people to be good or evil, even if those criticized (or praised) are "third parties" with no connection to the individual making the judgment. Another interesting feature of moral judgment is that even if it usually motivates one to follow it, the corresponding action often fails to materialize. Many people lie, cheat, and steal and yet feel guilty afterward because they believe such acts are immoral. So, while human moral psychology has many components—including motivation, decision, and action—our focus is on *moral judgment* specifically.

How does this capacity arise in the brain? In this chapter, we identify and weave together some of the major threads of research emerging over the past thirty years, before discussing future directions. We'll explain how the first thread in the neuroscience of moral judgment focused on brain abnormalities, from lesion studies to disorders such as psychopathy. Early research (ca. 1990–2000) concentrated on the amygdala and ventromedial prefrontal cortex (vmPFC), which led to theories emphasizing gut feelings as integral to moral judgment. The second thread (ca. 2000–2010) ushered in new methods, especially brain imaging, brain stimulation, and neurotransmitter manipulations. Moral neuroscience highlighted brain areas associated with complex computation and reasoning, such as the dorsolateral prefrontal cortex (dlPFC) and the temporoparietal junction (TPJ)/posterior superior temporal sulcus (pSTS). Theorists introduced dual-process models to explain how both gut feelings and conscious deliberation influence moral cognition.

More recent trends have drawn more on animal models and computational mechanisms to develop theories of moral learning in terms of reward and valuation. We expect the future will include more work on how brain development from childhood to adolescence relates to moral cognition.

Ultimately, we'll see that neuroscience, when combined with findings from psychology and allied disciplines, helps to address perennial philosophical questions about moral judgment, particularly the possibility and limits of knowing right from wrong. By helping to uncover the neural circuits and neurocognitive mechanisms underlying mature moral cognition, how it normally develops, and how it breaks down in pathology, neuroscience sheds light on the trustworthiness of our moral judgments and the possibility and shape of moral progress. It's by no means easy to bridge the dreaded gap between how we *do* form our moral beliefs and how we *ought* to. Nevertheless, with caution and critical reflection, neuroscience can enrich our understanding of moral judgment and when it works well or poorly.

1.2   First Thread: Gut Feelings

In the nineteenth and twentieth centuries, scientists and physicians inferred brain function primarily by examining patients with brain abnormalities, whether due to freak accidents, genetic anomalies, neurosurgery to treat debilitating seizures, or diseases such as herpes that can damage the nervous system. When lesions disrupted patients' moral capacities, researchers inferred that the affected brain area facilitates the corresponding capacity.

1.2.1   Somatic Markers

The most famous, even if poorly documented, case is that of Phineas Gage. In 1848, while working on a railroad in Vermont, an explosive accident catapulted a three-foot iron rod up through Gage's left cheek and out the top of his head. Remarkably, he survived, but his personality reportedly changed so much that he was virtually unrecognizable as his former self. Unfortunately, there is little reliable corroborating evidence about the details of Gage's case (Macmillan, 2000), but some of the changes were ostensibly apparent in his moral character. He reportedly became more impulsive, vulgar, rash, and even childish.

Although the affected brain area is difficult to pinpoint precisely, Gage seemed to suffer significant damage to the vmPFC, which overlaps with the orbitofrontal cortex behind the eyes ("orbits"). We know much more now about individuals with damage to this area in adulthood. Patients allegedly develop what Antonio Damasio (1994) dubbed "acquired sociopathy." The label is misleading, though, because the psychological profile of patients with vmPFC lesions is hardly similar to that of psychopaths (or what are sometimes referred to colloquially as "sociopaths"). Unlike individuals considered psychopathic, adults with vmPFC lesions are not callous, antisocial, or remorseless, but instead demonstrate a shortage of gut feelings that help guide decisions about what to do in the moment. Physiological measures suggest these patients, relative to controls, show diminished emotional responses when making a wide range of decisions, not just about how to treat others but even how to get points in a card game or where to eat for dinner. Patients with vmPFC damage generally give normal responses to questions about how various hypothetical choices should be resolved, including moral dilemmas (Saver & Damasio, 1991), but they struggle to make decisions about what to do *oneself* in a *particular situation* (Kennett & Fine, 2008). A patient might recognize that it's impolite to talk about gory injuries at dinner, but does that mean the hilarious hiking story is off the table? Damasio attributes this deficit in decision making to an underlying impairment of "somatic markers" that guide everyday decisions about how to behave. Without the relevant gut feelings, patients are able "to know but not to feel," as Damasio (1994, p. 45) puts it.

1.2.2 Psychopathy

A more deviant example of abnormal moral thought and behavior lies in psychopathy. Psychopaths are characteristically callous, remorseless, manipulative, and pompous, and exhibit superficial charm, among other similar vices that often leave injured or indigent victims in their wake (Hare, 1993). Unlike "acquired sociopathy," individuals with psychopathy exhibit abnormal functioning in the vmPFC and amygdala (and their connectivity), along with other paralimbic areas associated with gut feelings (Blair, 2007; Kiehl, 2006). Part of the limbic system, the amygdala is a pair of small almond-shaped nodes deep in the brain (see figure 1.1) and contributes, among other things, to assessing the significance of a stimulus, such as whether it is threatening, which plays a crucial role in learning. Unlike adults who acquire damage to the vmPFC or amygdala (Anderson et al., 1999; Taber-Thomas et al., 2014), psychopathic traits are associated with abnormal structure and functioning in these regions (Kiehl & Sinnott-Armstrong, 2013; Glenn & Raine, 2014), whether due to unfortunate genes (e.g., alleles known to disrupt neurotransmitters) or adverse circumstances (e.g., childhood trauma, neglect, and even lead exposure).

Importantly, individuals with psychopathy not only behave immorally, their understanding of right and wrong seems impaired to some extent. Some, but not all, incarcerated psychopaths exhibit some difficulty distinguishing moral rules from mere conventions (Aharoni, Sinnott-Armstrong, & Kiehl, 2012). Interviews also suggest that some inmates with psychopathy have an inconsistent and tenuous grasp of moral concepts and reasons, particularly when attempting to justify decisions or to use relevant emotion words such as "guilt" (Hare, 1993; Kennett & Fine 2008). Yet, some people with psychopathic traits seem rational—indeed, all too cold and calculating in their apparently skilled manipulation of others—although they do exhibit irrationality too, such as delusions of grandeur, poor attention span, and difficulty learning from punishment (Maibom, 2005; May, 2018). Thus, not only do the vmPFC and amygdala seem to be crucial moral circuits, perhaps emotions are necessary for moral competence (Prinz, 2016).

1.2.3   Post Hoc Rationalization

Also in the 1990s, moral psychologists began emphasizing gut feelings in moral judgment. Imagine being asked whether, hypothetically, it's morally acceptable for someone to clean a toilet with the national flag, to eat a pet that had been run over by a car, or to engage in consensual protected intercourse with an adult sibling. Most people in studies automatically condemn such "harmless taboo violations" without being able to articulate appropriate justifications (Haidt, Koller, & Dias, 1993; Stanley, Yin, & Sinnott-Armstrong, 2019). Indeed, it seems that we often intuitively regard actions as right or wrong first, and only afterward does conscious reasoning concoct a defense of it (Haidt, 2001; Cushman, Young, & Hauser, 2006).

Similar ideas followed studies of split-brain patients, starting around the 1960s. When the corpus callosum is severed, often to treat seizures from epilepsy, the two hemispheres of the brain can no longer communicate with one another. Studies of such split-brain patients suggest that in the absence of crucial information from one side of the brain, patients often confabulate a story to make sense of their behavior (Gazzaniga, 1983). One commissurotomy patient, for instance, was tasked with choosing out of

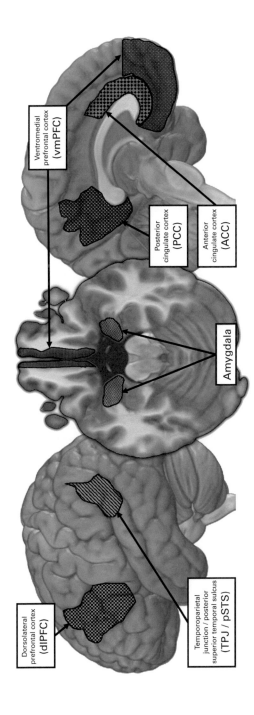

Figure 1.1

The moral brain. Brain areas consistently activated when people make moral, compared to nonmoral, judgments.

Labels in figure:

Ventromedial prefrontal cortex (vmPFC)

Posterior cingulate cortex (PCC)

Anterior cingulate cortex (ACC)

Amygdala

Dorsolateral prefrontal cortex (dlPFC)

Temporoparietal junction / posterior superior temporal sulcus (TPJ / pSTS) a row of eight images which two best relate to the two pictures recently presented to him. The catch was that only one picture was presented to each visual field, and thus each side of the brain could process only one of the two pictures first presented. One half of the patient's brain saw a home covered in snow, while the other half saw a chicken claw. To go with these, the patient almost instinctively chose an image of a snow shovel and an image of a chicken head. However, language abilities appear to be partly lateralized to one side of the brain. So one hemisphere can't readily communicate linguistically what it saw. As a result, the patient articulated a reason that appeared to be concocted just to make sense of his intuitive choice, saying, "You have to clean out the chicken shed with a shovel" (p. 534). Recent work suggests callosotomy patients also provide confabulations in the context of moral judgments (Miller et al., 2010).

1.2.4 Lessons

In light of the above, and other studies in psychology and neuroscience, many theorists in the first thread adopted what we might call "sentimentalist" theories of moral judgment. Proponents asserted that moral attitudes and decisions are generated predominantly by automatic gut feelings, whereas reasoning is largely post hoc rationalization (Haidt, 2001; Nichols, 2004; Prinz, 2016). Now, some of these theorists took the evidence to erode the supposed division between reason and emotion (e.g., Damasio, 1994) but with little emphasis on reasoning, inference, or complex computation underlying gut feelings and moral intuitions.

Sentimentalist theories do reconcile various observations of lesion patients, but several limitations remain. First, the centrality of gut feelings is insufficiently corroborated. While psychological studies initially appeared to support the importance of gut feelings in moral judgment, much of the findings were overblown (Huebner, 2015; May, 2018). One meta-analysis, for example, found limited evidence for the famous effect of incidental disgust priming on moral judgment, which disappeared entirely after controlling for publication bias (Landy & Goodwin, 2015). Moreover, the vmPFC is unlikely a source of gut feelings, but rather a hub wherein such feelings are incorporated with or weighed against other considerations before making a decision (Shenhav & Greene, 2014; Hutcherson et al., 2015). Second, theories focusing on the vmPFC and amygdala are incomplete. Brain damage to rather different areas in the frontal and temporal lobes lead to moral dysfunction as well—for example, in frontotemporal dementia. Moreover, as we'll see, subsequent brain imaging studies confirm that additional brain areas are integral to moral judgment. Indeed, early research on brain abnormalities often studied the patient's social behavior and choice rather than the moral evaluation of other people and their actions. Patients with vmPFC lesions, for example, don't appear to have distinctively moral deficits but rather problems with decision making generally.

1.3   Second Thread: Reasoning

The second thread in moral neuroscience followed the development of neuroimaging technologies, which enabled the live, noninvasive measurement of brain functioning. Of these technologies, functional magnetic resonance imaging (fMRI) has dominated the methodological landscape. Moreover, partly given worries about post hoc rationalization, researchers primarily investigated the neural correlates of moral judgments made in response to particular moral statements (Moll, Eslinger, & de Oliveira-Souza, 2001) or hypothetical scenarios (Greene et al., 2001). By varying the features of hypothetical scenarios, for instance, one can infer which factors shape moral judgments, instead of relying on the factors participants articulate as reasons for their moral judgments, which may be misleading.

1.3.1   Dual Process
The second thread arguably began with an influential article published by Joshua Greene and collaborators (2001), in which participants underwent fMRI scanning while responding to sacrificial dilemmas familiar from longstanding philosophical debates in ethical theory. These hypothetical scenarios pit moral values against each other by describing an opportunity to sacrifice one person in order to save even more lives (typically five). Interestingly, most people consider it morally acceptable to sacrifice one to save five when this can be done in an impersonal way, such as diverting a runaway trolley away from five workers stuck on one track but onto another track with only one stuck worker (Switch scenario). But sacrificing one to save five is deemed unacceptable if the harm must be up close and personal, such as pushing someone in front of a trolley to save the five others (Footbridge scenario).

Such sacrificial dilemmas have long been used by philosophers to distinguish between and evaluate ethical theories. Treating personal harm as immoral, even when it would save more lives, ostensibly reflects character-istically "deontological" judgments that align with moral rules (e.g., don't kill), even if violating them would produce better consequences. Sacrificing one to save five, on the other hand, ostensibly reflects characteristically "util-itarian" (or consequentialist) judgments that privilege the maximization of overall welfare.

Greene's (2014) model adopts the tools of dual-process theory, which posits the operation of competing psychological processes, particularly automatic versus deliberative thinking (e.g., Kahneman, 2011). Applying this to the moral domain, Greene theorizes that "utilitarian" responses to moral dilemmas are driven by controlled, deliberative reasoning, while non-utilitarian ("deontological") responses are driven by automatic, intuitive, emotional heuristics that are relatively insensitive to the consequences of an action. Some of the support for this dual-process model comes from psy-chological experiments but also fMRI. Early on, Greene and collaborators (2001, 2004) reported the engagement of predictably different brain areas when participants respond to personal and impersonal moral dilemmas. Compared to impersonal (and nonmoral) dilemmas, personal dilemmas elicited greater activity in some areas associated with automatic, emotional, and social processing—namely, the vmPFC, amygdala, pSTS, and posterior cingulate cortex (PCC). Responses to impersonal dilemmas yielded greater activity in areas associated with controlled deliberative reasoning—namely, the dlPFC and inferior parietal lobe.

A diverse body of evidence appears to corroborate the dual-process model (Greene, 2014), but let's focus on some of the brain science that goes beyond neuroimaging. Prior lesion studies appear consistent with the mod-el's account of intuitive moral judgments being driven by gut feelings, but what about the claim that characteristically "utilitarian" moral judgments are driven by calculative reasoning? Researchers have found, as the model predicts, more "utilitarian" responses to dilemmas among people with emo-tional deficits, such as psychopaths, people suffering from frontotemporal dementia, and patients with damage to the vmPFC (Koenigs, Young, et al., 2007; Koenigs, Kruepke, et al., 2012; Mendez, Anderson, & Shapira, 2005). Activity in the amygdala correlates negatively with "utilitarian" judgments but positively with adverse feelings in personal moral dilemmas (Shenhav & Greene, 2014). Related to this finding, people with psychopathic traits appear to exhibit lower amygdala activity when responding to personal dilemmas (Glenn, Raine, & Schug, 2009). As a final example, "utilitarian" responses were lower among participants whose brains had been flooded with serotonin, which especially influences the amygdala and vmPFC, among some other regions (Crockett et al., 2010).

Despite the array of corroborating evidence, there are many criticisms of the very dual-process elements of the theory. For example, the personal/impersonal distinction, based largely on reaction time data, was driven by a handful of stimuli from the complete set of about sixty dilemmas judged by participants (McGuire et al., 2009). Furthermore, while different reaction speeds—fast or slow—can reflect intuitive or deliberative processes, this may be due to the particular examples of those types of dilemmas the researchers happened to choose (Krajbich et al., 2015). Indeed, moral dilemmas can be constructed that yield "utilitarian" responses that are intuitive and "deontological" ones that are counterintuitive (Kahane et al., 2012). Another concern is that the automatic versus controlled moral judgments measured with sacrificial dilemmas don't clearly track the relevant moral values or types of moral reasoning. Some apparently "utilitarian" resolutions to personal dilemmas appear to be driven by callousness (e.g., indifference to pushing), not a utilitarian concern for the greater good (Kahane et al., 2015; but see Conway et al., 2018).

Some versions of dual-process theory also treat automatic moral intuitions as relatively simple and inflexible, which understates how they can be shaped by unconscious learning (Railton, 2017). Indeed, some neuroimaging evidence suggests distinct brain areas underlie the resolution of moral dilemmas in terms of factors familiar from moral theory, such as the act/omission distinction and harming as a means versus a side effect (Schaich Borg et al., 2006). What seemed like a simplistic emotional aversion to pushing or prototypically violent acts in personal dilemmas turns out to be driven by complex concerns about how involved an agent is in bringing about harmful outcomes (Mikhail, 2011; Feltz & May, 2017; May, 2018).

1.3.2 Beyond Dilemmas

Some neuroscientists have gone beyond sacrificial dilemmas and dual-process theory when investigating moral cognition (although they do remain fixated largely on harming/helping others). Most of the extant research has extensively studied moral judgments about hypothetical scenarios involving attempted versus accidental harms. Now, intentionality may not be crucial for all moral situations—sleeping with your cousin is deemed impure and morally problematic by many people, even if you're completely oblivious to the family connection (Young & Saxe, 2011). Nevertheless, across cultures, an actor's mental states (intent, belief, knowledge, or lack thereof) influence moral evaluations of harmful acts (Barrett et al., 2016; McNamara et al., 2019).

Liane Young and collaborators have found that increased activation in the TPJ/pSTS is associated with attribution of intent during the evaluation of attempted harms (Young et al., 2007; Young & Saxe, 2008). One study even decoded activity in this region—using multi-voxel pattern analysis—to predict individual differences in moral evaluations of accidental versus intentional harms (Koster-Hale et al., 2013). Moreover, while "no harm, no foul" usually does not apply to attempted murder, disrupting the TPJ with transcranial magnetic stimulation made participants judge less harshly an agent's failed attempt to harm someone by downplaying the agent's malicious intent and focusing instead on the lack of harmful outcomes (Young et al., 2010).

These findings are consistent with theories suggesting that the TPJ—which overlaps with the pSTS—is critical for the domain-general capacity of mental state understanding and empathy (Decety & Lamm, 2007; Young & Dungan, 2012). Unsurprisingly, some evidence even suggests that patients with high-functioning autism have difficulty integrating mental state information into their moral judgments of harm, causing them to judge accidental harms more harshly than neurotypical individuals do (Moran et al., 2011). Similar results have been found with split-brain patients, which coheres with evidence that belief attribution in the TPJ is partly lateralized to the right hemisphere (Miller et al., 2010).

A different neuroimaging paradigm asks participants to judge statements, instead of scenarios, as right or wrong, some of which go beyond harm and even fairness (Moll, Eslinger, et al., 2001; Moll, de Oliveira-Souza, et al., 2002). While in the scanner, participants judged as "right" or "wrong" moral statements (e.g., "The elderly are useless"), nonmoral statements (e.g., "The elderly sleep more at night"), and scrambled statements (e.g., "Sons push use eat work"). The researchers found that moral judgments, relative to judgments about nonmoral statements, elicited greater activity in brain areas now familiar in moral neuroscience: the vmPFC, left temporal pole (highly connected to the vmPFC and amygdala), and TPJ/pSTS.

Studies using electroencephalography further suggest a temporal order over which various neural circuits contribute to the moral evaluation of harmful acts. When forming such moral judgments, participants rapidly computed information about mental states in the TPJ, around sixty milliseconds after viewing a short video depicting accidental or intentional harm (Decety & Cacioppo, 2012). Just a few hundred milliseconds later, the amygdala provided evaluative input to areas of the prefrontal cortex before a moral judgment emerged, whether concerning harmful or helpful acts (Yoder & Decety, 2014). At least when it comes to the domain of harm, the brain appears to compute the positive and negative consequences of the act while weighing up how involved an agent was in bringing them about.

1.3.3 Limitations

One criticism facing much work in the second thread is the overreliance on "reverse inference" to infer the existence of certain mental states from activations in brain areas, when such areas perform multiple psychological functions (see, e.g., Klein, 2010). The amygdala, for example, is associated with motivation, fear, and reward, among other things; greater activity in the amygdala when participants give "deontological" responses to moral dilemmas doesn't necessarily mean participants felt increased motivation as opposed to fear, reward, or perhaps something else entirely by way of functional interactions with other regions at the network level.

Reverse inference can be warranted, however, if the observation of brain activity in some region provides better support for one theory over another (Machery, 2014). Moreover, as we've seen, moral neuroscience does not rely solely on neuroimaging, but also uses other methods that are less susceptible to concerns relating to reverse inference. We have seen some appeal to lesion studies, which go beyond merely correlating brain areas with moral cognition to provide evidence that a region is necessary for moral cognition. Some research is even able to discern which areas are necessary and sufficient for moral cognition by employing noninvasive brain stimulation techniques or psychotropic drugs (or both). Some of the studies cited above used transcranial magnetic stimulation, for instance, to increase (or decrease) neuronal excitation or medications to enhance (or impair) the functioning of neurochemicals such as serotonin. Even the dual-process model, despite being born of fMRI, has been tested against brain lesion data and the manipulation of neurotransmitters, not to mention various psychological experiments.

Another important limitation is that extant theories are woefully incomplete. Much of moral judgment doesn't involve death or bodily harm, let alone dilemmas featuring these moral considerations. Reams of evidence now suggest that across cultures, fundamental moral values include not just harm or care but also fairness, group loyalty, sanctity, and respect for authority (Gilligan, 1982; Haidt et al., 1993; Doğruyol, Alper, & Yilmaz, 2019). Even if these other moral values are ultimately reducible to harm/care, moral neuroscientists have largely ignored them (a few exceptions: Parkinson et al., 2011; Decety, Pape, & Workman, 2018; Workman, Yoder, & Decety, 2020).

1.3.4 Lessons

The second thread in moral neuroscience primarily examined differences in brain activity elicited by moral compared to nonmoral stimuli, or to moral stimuli of one kind compared to another, which were then localized to specific parts of the brain. Combined with lessons from the first thread, a general picture emerges in which at least some core aspects of moral cognition are underpinned by a network of predominantly frontal and temporal regions dedicated to various morally relevant factors recognizable to both commonsense and moral theorizing (for further reviews, see Moll et al., 2005; Greene, 2009; Eres, Louis, & Molenberghs, 2018; Han, 2017; Demaree-Cotton & Kahane, 2018).

A central lesson is that moral cognition is not localized to one brain area or otherwise particularly unified in the brain (Greene, 2009; Young & Dungan, 2012; Parkinson et al., 2011). Instead, our capacity for moral judgment involves a spatially distributed network of areas with various domain-general psychological functions that are also relevant to moral evaluation, such as understanding the consequences of an agent's action, the agent's mental states, how the action was causally brought about, and the social norms it violates.

Another key lesson is that moral judgment is not always driven only by gut feelings or localized just to the vmPFC and amygdala. In contrast with the first thread, we see that some moral judgments involve rapid reasoning, served by areas such as the dlPFC and TPJ. Moreover, even automatic moral intuitions can involve complex computation (Mikhail, 2011). While some moral intuitions are heavily dependent on the amygdala and vmPFC, these are part of a network of brain areas that engage in sophisticated, even if unconscious, learning and inference (Huebner, 2015; Woodward, 2016).

1.4   Third Thread: Learning

The first and second threads in moral neuroscience focused on brain areas and their functions. A third thread focuses on the level of computational analysis in neuroscience. In particular, many proponents of this type of approach draw on reinforcement learning in order to illuminate the nature and workings of moral judgment.

1.4.1   Value, Reward, and Learning

The field of reinforcement learning asks how an agent can learn to optimize its behavior strictly from interactions with its environment. For example, a baby plover is able to leave the nest and feed itself within a few hours of hatching, receiving relatively little assistance from its parents. How is it able to learn and perform these behaviors?

Research in reinforcement learning analyzes idealized versions of this question (Sutton & Barto, 2018). Specifically, it recasts the question in terms of how an agent can learn to maximize its reward and value over time. *Reward* refers to the intrinsic desirability of a given stimulus, whereas *value* refers to the total, expected, future reward associated with a given state. For example, drinking a cup of coffee is intrinsically desirable for many people because it is flavorful and provides caffeine, and so is rewarding. By contrast, grinding some coffee by hand is not rewarding, and may even be annoying, but it is valuable because it subsequently leads to the rewarding state. Many reinforcement learning methods use the notions of reward and value to estimate what it would be good for an agent to do in the long run.

The computational study of reward and value extends into the study of the neurobiological bases of human decision making. An early discovery in computational neuroscience revealed an important correspondence between one of these reinforcement learning methods, known as the temporal-difference learning algorithm, and the firing of dopamine neurons in the mammalian brain (Schultz, Dayan, & Montague, 1997). A substantial body of animal and human behavioral evidence has since suggested that there are at least three different decision systems in the brain: the Pavlovian system, the model-free (or habitual) system, and the model-based (or goal-directed) system (e.g., Glimcher & Fehr, 2014).

The Pavlovian system produces basic stimulus-driven behavioral responses. The term "Pavlovian" frequently leads to confusion (see Rescorla, 1988). In most fields, as well as in everyday usage, the term is usually associated with Pavlov's original experiments with dogs, where he conditioned dogs to salivate at the sound of a bell by repeatedly ringing a bell and then consistently feeding them afterwards. By contrast, in reinforcement learning, "Pavlovian" refers to the relationship between the *unconditioned* stimulus (the food) and the relevant unconditioned response (the salivating). Thought to be evolved and roughly "hardwired," these unconditioned responses include both outcome-specific responses, such as inflexibly licking water, and more open-ended, valence-dependent responses, such as generally approaching something rewarding. Both classes of response are characteristically recalcitrant to changes in outcome, as when chickens will continue to peck at a feeder that will not dispense any seeds over hundreds of trials (Macintosh, 1983; Huys et al., 2011, 2012). The Pavlovian system is supported primarily by the brain stem and subcortical areas of the limbic system—for example, the amygdala, nucleus accumbens, and hypothalamus (Rangel, Camerer, & Montague, 2008).

The model-free system produces instrumental responses by evaluating actions based on their previously learned values in different contexts. For example, a button press that has previously resulted in a reward is a good state–action pair, while a button press that has previously resulted in a punishment is a bad state–action pair. Because the model-free system does not explicitly represent future values, it can be slow to update in the face of changing circumstances. However, unlike the Pavlovian system, the model free system is not "hardwired," and does gradually update. The model-free system is associated with activity in the basal ganglia and the orbital and medial portions of the prefrontal cortex (Yin & Knowlton, 2006).

Finally, the model-based learning system uses a forward-looking model to represent possible actions, outcomes, and associated values. This model is typically represented by a decision tree. Each node in the tree represents a possible choice, where the model-based system "searches" through the decision tree to find the branch with the highest total value. For example, a chess player may represent three upcoming moves in a game of chess, with each possible move further branching into a wide range of subsequent moves. To win, the player tries to represent and choose the best possible sequence of moves overall. The model-based system is primarily associated with activation in the vmPFC (Hare et al., 2008).

1.4.2   Moral Learning

Learning-based approaches to moral judgment are developed using the three decision systems. Which system plays a defining role in moral judgment? Echoing the dual-process theories discussed in the second thread, Cushman (2013, 2015) argues that much of moral cognition depends on a body of objective rules together with the model-free decision system (see also Greene, 2017). When exhibiting the latter process, people often continue to adhere to norms outside of the context in which those norms are in play. For instance, American tourists frequently continue to tip in restaurants abroad, even when there is no relevant norm dictating that they should (Cushman, 2015, p. 59).

Cushman argues that the role of the model-free decision system helps explain participants' diverging responses to the switch and footbridge scenarios. Cushman (2015) reasons that people's tendency to resist harming the single individual in footbridge is "the consequence of negative value assigned intrinsically to an action: direct, physical harm" (p. 59). That is, participants' responses may be underwritten by the model-free decision system: since directly harming others has reliably elicited punishments in the past, this option represents a bad state–action pair and leads people to reject it as an appropriate course of action.

One difficulty with Cushman's general view is that it is in tension with the aforementioned evidence suggesting that all three decision systems trade off and interact to produce our everyday behaviors. Another more specific difficulty comes from the fact that participants' avoidance of harm can just as plausibly be explained by the role of the evolved Pavlovian system as it can by the role of its model-free counterpart. One way to disentangle which system is in effect could be to devise an iterative version of the trolley problem. If participants gradually shifted their views on the matter, we could say that it was the model-free system; if they did not, we could say that it was recalcitrant Pavlovian responding.

In contrast to Cushman's highlighting of the role of only one of the three decision systems in producing moral judgments, Crockett (2013, 2016) argues that all three systems play a role and even interact in the process of producing a single judgment. On Crockett's view, *both* the model-free and Pavlovian systems assign negative values to actions that cause others harm through physical contact. Consequently, when the "votes" of all three systems are tallied up, participants will find it morally acceptable to sacrifice one to save five in impersonal dilemmas but not in personal dilemmas. Hence, even the iterative version of Cushman's view and Crockett's voting explanation provide competing explanations of responses to the trolley problem and so leave open questions for further investigation.

1.4.3   Lessons

Computational approaches to understanding moral judgment complement rather than compete with the first and second threads discussed above. In particular, computational approaches complement strictly behavioral and neuroscientific accounts by illuminating the relationships between the components of moral cognition, using formal mathematical models. Adopting such strategies has the further advantage of enabling researchers to leverage additional bodies of research from computer science and economics.

Notably, this third thread in the neuroscience of morality coheres with the suggestion commonly found in other threads that we use domain-general decision-making systems to make specifically moral judgments. It seems we use the same algorithms and neural mechanisms to make, for example, choices about which car to buy and decisions about which people to blame for moral wrongdoing (see Berns et al., 2012; Shenhav & Greene, 2010; Crockett et al., 2017). This emerging picture is also consistent with findings in the first thread that suggest that breakdowns in general decision making are associated with related breakdowns in moral decision making (Blair, 2007; Mahmut, Homewood, & Stephenson, 2008; Aharoni et al., 2012).

Limitations found in previous threads remain, however, including the ongoing reliance on sacrificial dilemmas. Going forward, computational approaches will need to model more than judgments about dilemmas and go beyond the domain of harm by studying loyalty, care, and other values that arise in everyday circumstances.

1.5   Future Directions: Moral Development

Moral learning is a process that occurs over time and stretches back to the critical period of childhood. We saw in section 1.2 that psychopathy involves dysfunction of at least the amygdala and vmPFC, presumably early in development (Glenn & Raine, 2014; Taber-Thomas et al., 2014), which seems to affect one's moral capacities. In contrast, moral deficits are much less profound in patients who acquire damage to these regions in adulthood because normal brain development affords the acquisition of moral competence. Psychopathy is only one form of moral dysfunction, however, and we should seek a complete understanding of normal moral development that incorporates not only empathy and compassion but also harm (including the weighing of outcomes and the actor's intent) and other moral values (e.g., loyalty). Neuroscientists are increasingly interested in understanding how such elements of moral cognition work and develop in the brains of children and adolescents.

1.5.1 Moral Judgment and Brain Development

As in other areas of neuroscience, we do well to consider brain development in conjunction with relevant psychological theories and evidence. Building on work by Piaget (1932), Kohlberg theorized the development of moral cognition in terms of reasoning and reflection that, once fully developed, employed universal principles that could even be used to question existing conventions in society (Kohlberg, 1984).

An important concern, however, is that this approach only tracks the development of conscious moral reasoning, which could be merely rationalizing moral judgments one has already made intuitively on different grounds (Haidt, 2001). If we seek a theory of moral judgment, not of our (often poor) attempts to justify them verbally, then we need to explain the development of unconscious processes that generate automatic moral intuitions (Cushman et al., 2006).

Taking this approach, researchers have investigated moral development with age-appropriate moral scenarios. Using morality plays with puppets, for example, researchers have found that even infants discriminate and prefer a puppet that helps other characters to achieve, as opposed to hinder, their goals (Hamlin, 2015; Cowell & Decety, 2015). Children as young as four years old begin making moral judgments focused on outcomes, such as whether an action harmed or saved more people (Pellizzoni, Siegal, & Surian, 2010), regardless of whether it was accidental or intentional. The intent of the actor appears to grow increasingly relevant in the next few years of development (Cushman et al., 2013).

Corresponding to the psychological research, studies in developmental neuroscience have found relevant differences in brain structure and function across age groups during moral cognition. One neuroimaging study, for example, found greater activity in the vmPFC in adults, compared to younger participants, when they viewed moral relative to nonmoral transgressions (Decety, Michalska, & Kinzler, 2012). Among older participants, the researchers also observed greater task-based functional connectivity between the vmPFC and amygdala and between the vmPFC and TPJ/pSTS. Consistent with other studies, younger participants' evaluations of a person who caused harm were less sensitive to whether the harm was intentional or accidental. In another study, when both adolescent and adult males evaluated images of moral violations, researchers found greater activity in the TPJ/pSTS and PCC among older participants (Harenski et al., 2012). In their review of these and other studies, Decety and Cowell (2018) concluded that "mature moral cognition" at least requires continued development in brain areas that underlie "aversion to interpersonal harm, empathic concern, and mental state understanding" (p. 160).

Further research is needed, but studies in developmental neuroscience thus far fit well with the moral circuits identified in adulthood. Central players include the limbic regions (particularly the amygdala), portions of the prefrontal cortex (especially the vmPFC), and relevant areas of the temporal lobe (namely, the STS, including its posterior/TPJ). Brain activity in these moral circuits changes over the course of development, and such changes are associated with key elements of moral cognition, particularly: assigning value to outcomes such as harm, representation of the actor's knowledge or intentions, and retrieval of relevant social information. However, again, morality involves more than harm or even fairness (Gilligan, 1982; Haidt et al., 1993; Doğruyol et al., 2019). Further developmental neuroscience research should study more than simplistic depictions of harm, altruism, or compassion and make sure their findings generalize to, say, fraud, torture, betrayal, and filial duties.

1.5.2 Integrative Approaches

The neuroscience of moral development also suggests an interesting overlap between regions that support moral cognition and regions that support thinking about the self. Meta-analyses of neuroimaging studies reveal that several moral circuits—for example, the vmPFC and PCC—overlap with the default mode network (Bzdok et al., 2012; Eres et al., 2018; Sevinc & Spreng,

2014; Han, 2017). In many studies, participants are asked to evaluate other people and their actions. So it's striking to find such extensive overlap with self-related regions.

One explanation for this is that participants often make moral judgments in response to emotionally charged stories with actors who intend to cause harm, which naturally recruits brain areas that contribute to understanding narratives, theory of mind, and distinguishing self from other. However, another explanation is that while moral judgment and motivation are distinguishable, they are intimately connected, especially in normal development. Extensive interview studies do suggest that the integration of moral values and one's self-concept occurs throughout adolescence and into adulthood (Damon, 1984). Moreover, we've seen that psychopathy affects moral cognition by causing dysfunction in areas associated not only with conscious reasoning or social knowledge but also with emotion and motivation.

Thus, it may be that a brain develops normal moral judgment only through proper development of a suite of connected moral capacities, including emotions, motivation, and identity. An analogy may help. Suppose that in educated adults the ability to solve algebraic equations is localized, more or less, to the prefrontal cortex. It doesn't follow that a child can learn algebra so long as the prefrontal cortex is functioning properly. If other areas are dysfunctional during development—even including swaths of the motor or primary visual cortex—one may be unable to develop mathematical capacity properly, even if this is later grounded in only one of the many brain areas necessary for initial development.

One approach would be to integrate moral judgment, motivation, and behavior via moral identity, or the degree to which moral values are central to one's self-concept (Aquino & Reed II, 2002). Experimental evidence suggests that people are more likely to behave according to their moral judgments if they regard moral values as both central to themselves and more important than nonmoral values (Reed, Aquino, & Levy, 2007; Winterich et al., 2013). Qualitative studies corroborate the idea that strong moral identity is required for sustained commitment to moral behavior (Colby & Damon, 1992). Psychopathy may even involve weak moral identity, since people with psychopathic tendencies have been shown to report weaker moral identities (Glenn et al., 2010).

Another integrative approach, which is rather mainstream in moral education (Han, 2014), is neo-Kohlbergian. Unlike classical Kohlbergian theory, which focused only on moral judgment and reasoning, the Four-Component Model incorporates additional aspects of moral functioning (Bebeau, 2002)—namely, moral motivation, character, and sensitivity. In this model, moral development and maintenance involves orchestrating these four components to cooperate with each other (Rest & Narvaez, 1994), which are associated with interactions among various limbic and frontal regions (Narvaez & Vaydich, 2008).

We can perhaps situate integrative developmental theories within integrative models of the neurobiology of mature moral judgment. The Event-Feature-Emotion framework (Moll et al., 2005), for example, identifies a spatially distributed network of frontal, temporal, and subcortical brain areas involved not just in the moral evaluation of others but also in moral emotion and motivation. Such frameworks cohere with the meta-analysis of neuroimaging studies suggesting that moral circuits significantly overlap with self-related psychological processing (Han, 2017). Thus, when it comes to the development of moral cognition and its improvement in adulthood, it is wise to consider the integration of otherwise dissociable moral capacities, including both moral judgment and motivation (May, 2018).

1.6   Conclusion: Philosophical Implications

The neuroscience of moral judgment is still fairly young. There is no doubt that conclusions about moral judgment on the basis of neurobiology should be drawn with caution. Nevertheless, in this final section, we aim to show how combining brain science with philosophical analysis can aid our understanding of moral judgment, particularly by elucidating concrete mechanisms and corroborating or disconfirming theories developed in other areas of cognitive science (Prinz, 2016; Demaree-Cotton & Kahane, 2018).

Indeed, the neuroscience of ethics is already greatly improving due to philosophy and science continuously informing one another. We've already seen how decades of ethical theorizing about the trolley problem, for instance, have shaped experimental paradigms. In this section, however, let's conclude by briefly drawing out how the advances in neuroscience discussed above can contribute to debates in moral philosophy.

1.6.1   Reason versus Emotion in Ethics

The dichotomy between reason and emotion stretches back to antiquity. But an improved understanding of the brain has, arguably more than psychological science, questioned the dichotomy (Huebner, 2015; Wood-ward, 2016). Brain areas associated with prototypical emotions, such as the vmPFC and amygdala, are also necessary for complex learning and infer-ence, even if largely automatic and unconscious. Even psychopaths, often painted as the archetype of emotionless moral monsters, have serious defi-cits in learning and inference. Moreover, even if our various moral judg-ments about trolley problems, harmless taboo violations, and the like are often automatic, they are nonetheless acquired through sophisticated learn-ing mechanisms that are responsive to morally relevant reasons (Railton, 2017; Stanley et al., 2019). Indeed, normal moral judgment often involves gut feelings being attuned to relevant experience and made consistent with our web of moral beliefs (May & Kumar, 2018).

Blurring the line between reason and emotion may seem to render the corresponding philosophical disputes meaningless, but that's too fast. If emo-tions are required for moral judgment only because affect is integral to rea-soning generally, then moral judgment isn't special in requiring emotional processes, which is a core tenant of sentimentalism. Instead, what seems vin-dicated is a core thesis of rationalism: that moral cognition involves domain-general capacities for learning and reasoning, just like nonmoral cognition. Rather than obliterate the philosophical dispute, the evidence may support sophisticated forms of rationalism (May, 2018), despite early proclamations that the science preferentially supports sentimentalism.

1.6.2 Moral Knowledge (or Lack Thereof)

A consensus already seems to be emerging that, under the skull's hood, moral cognition is a complex affair, not just among ethicists and the intel-ligentsia but also among ordinary people. Moral judgment is not merely a matter of expressing one's emotions, divorced from reasoning. Moralizing is informed at least by one's own experiences, as well as knowledge from one's society and ancestors, in the form of cultural norms and evolved predispo-sitions. Yet, moral beliefs are not fixed after maturation. Even staunchly held moral attitudes, such as opposition to same-sex marriage, can rapidly change in response to a greater understanding of others and consistency reasoning (Campbell & Kumar, 2012).

However, even if most moral cognition involves learning and inference, these may be too biased and unreliable to yield moral knowledge or justi-fied belief. By helping to uncover the causal sources of moral cognition, neuroscience can aid in either debunking or vindicating certain kinds of moral beliefs (Greene, 2017; Kumar, 2017), although sweeping conclusions about all moral cognition are likely to falter (May, 2018; Kumar & May, 2019). Of course, neuroscience alone can't settle philosophical issues without making normative assumptions (Berker, 2009), but together they can advance debates in ethics.

1.6.3 Moral Education and Progress

Understanding the workings of mature moral judgment, as well as its development, also promises to illuminate how we can improve the acquisition of moral knowledge and perhaps even speed up moral progress. Extant evidence already suggests that mature moral judgment requires the proper development of an interwoven tapestry of moral capacities, including appropriate reasoning, sentiments, motivations, learning mechanisms, and self-conception.

Of course, neuroscience alone is unlikely to demonstrate how to improve our moral selves. But such findings *can* suggest useful directions for moral psychology and moral education, especially when applied to a particular social context. For example, given the association between morality and identity at the neural level, Han and colleagues (2017) predicted and found that stories of closely related moral exemplars, such as peers, more effectively promoted moral elevation and emulation than stories of distant exemplars, such as historical figures (see also Han et al., 2018). Or consider more newfangled proposals for moral improvement, such as indiscriminately amplifying moral emotions—whether through pills, brain stimulation, or lacing the water with oxytocin (Earp, Douglas, & Savulescu, 2017). The neuroscience of moral judgment already speaks against such tactless tactics. Of course, devastating a person's moral capacities may be as simple as disrupting moral circuits in childhood—unfortunately, it's generally easier to harm than to benefit (Persson & Savulescu, 2012). But distinguishing right from wrong is an incredibly complex process that requires the coordinated orchestration of a diverse range of brain areas and neurotransmitters. Novel neurobiological methods for moral improvement will certainly require finesse.

Acknowledgments

For helpful feedback on this chapter, we thank the editors and other contributors to this volume, as well as Vanessa Bentley, Jean Decety, Andrea Glenn, and Andrew Morgan. Special thanks to Felipe De Brigard and Walter

Sinnott-Armstrong for directing the Summer Seminars in Neuroscience and Philosophy and to the John Templeton Foundation for funding them. Authorship: J. H. wrote section 1.4; H. H. wrote section 1.5; J. M. and C. W. coauthored the remaining sections and edited the entire manuscript.

References

Aharoni, E., Sinnott-Armstrong, W., & Kiehl, K. A. (2012). Can psychopathic offenders discern moral wrongs? A new look at the moral/conventional distinction. *Journal of Abnormal Psychology, 121*(2), 484–497.

Anderson, S. W., Bechara, A., Damasio, H., Tranel, D., & Damasio, A. R. (1999). Impairment of social and moral behavior related to early damage in human prefrontal cortex. *Nature Neuroscience, 2*(11), 1032–1037.

Aquino, K., & Reed II, A. (2002). The self-importance of moral identity. *Journal of Personality and Social Psychology, 83*(6), 1423–1440.

Barrett, H. C., Bolyanatz, A., Crittendend, A. N., Fessler, D. M. T., Fitzpatrick, S., Gurven, M., . . . Laurence, S. (2016). Small-scale societies exhibit fundamental variation in the role of intentions in moral judgment. *Proceedings of the National Academy of Sciences of the United States of America, 113*(17), 4688–4693.

Bebeau, M. J. (2002). The defining issues test and the four component model: Contributions to professional education. *Journal of Moral Education, 31*, 271–295.

Berker, S. (2009). The normative insignificance of neuroscience. *Philosophy and Public Affairs, 37*(4), 293–329.

Berns, G. S., Bell, E., Capra, C. M., Prietula, M. J., Moore, S., Anderson, B., . . . & Atran, S. (2012). The price of your soul: Neural evidence for the non-utilitarian representation of sacred values. *Philosophical Transactions of the Royal Society B: Biological Sciences, 367*(1589), 754–762.

Blair, R. J. R. (2007). The amygdala and ventromedial prefrontal cortex in morality and psychopathy. *Trends in Cognitive Sciences, 11*(9), 387–392.

Bzdok, D., Schilbach, L., Vogeley, K., Schneider, K., Laird, A. R., Langner, R., & Eickhoff, S. B. (2012). Parsing the neural correlates of moral cognition: ALE meta-analysis on morality, theory of mind, and empathy. *Brain Structure and Function, 217*, 783–796.

Campbell, R., & Kumar, V. (2012). Moral reasoning on the ground. *Ethics, 122*(2), 273–312.

Colby, A., & Damon, W. (1992). *Some do care: Contemporary lives of moral commitment.* New York: Free Press.

Conway, P., Goldstein-Greenwood, J., Polacek, D., & Greene, J. D. (2018). Sacrificial utilitarian judgments do reflect concern for the greater good: Clarification via process dissociation and the judgments of philosophers. *Cognition, 179*, 241–265.

Cowell, J. M., & Decety, J. (2015). Precursors to morality in development as a complex interplay between neural, socioenvironmental, and behavioral facets. *Proceedings of the National Academy of Sciences of the United States of America, 112*(41), 12657–12662.

Crockett, M. J. (2013). Models of morality. *Trends in Cognitive Sciences, 17*(8), 363–366.

Crockett, M. J. (2016). How formal models can illuminate mechanisms of moral judgment and decision making. *Current Directions in Psychological Science, 25*(2), 85–90.

Crockett, M. J., Clark, L., Hauser, M. D., & Robbins, T. W. (2010). Serotonin selectively influences moral judgment and behavior through effects on harm aversion. *Proceedings of the National Academy of Sciences of the United States of America, 107*(40), 17433–17438.

Crockett, M. J., Siegel, J. Z., Kurth-Nelson, Z., Dayan, P., & Dolan, R. J. (2017). Moral transgressions corrupt neural representations of value. *Nature Neuroscience, 20*(6), 879.

Cushman, F. A. (2013). Action, outcome, and value: A dual-system framework for morality. *Personality and Social Psychology Review, 17*(3), 273–292.

Cushman, F. A. (2015). From moral concern to moral constraint. *Current Opinion in Behavioral Sciences, 3*, 58–62.

Cushman, F. A., Sheketoff, R., Wharton, S., & Carey, S. (2013). The development of intent-based moral judgment. *Cognition, 127*(1), 6–21.

Cushman, F. A., Young, L. L., & Hauser, M. D. (2006). The role of conscious reasoning and intuition in moral judgment: Testing three principles of harm, *Psychological Science, 17*(12), 1082–1089.

Damasio, A. R. (1994). *Descartes' error: Emotion, reason, and the human brain.* New York: Avon Books.

Damon, W. (1984). Self-understanding and moral development from childhood to adolescence. In W. M. Kurtines & J. L. Gewirtz (Eds.), *Morality, moral behavior and moral development* (pp. 109–127). New York: John Wiley.

Decety, J., & Caioppo, S. (2012). The speed of morality: A high-density electrical neuroimaging study. *Journal of Neurophysiology, 108*(11), 3068–3072.

Decety, J., & Cowell, J. M. (2018). Interpersonal harm aversion as a necessary foundation for morality: A developmental neuroscience perspective. *Development and Psychopathology, 30*(1), 153–164.

Decety, J., & Lamm, C. (2007). The role of the right temporoparietal junction in social interaction: How low-level computational processes contribute to meta-cognition. *The Neuroscientist, 13*(6), 580–593.

Decety, J., Michalska, K. J., & Kinzler, K. D. (2012). The contribution of emotion and cognition to moral sensitivity: A neurodevelopmental study. *Cerebral Cortex, 22*(1), 209–220.

Decety, J., Pape, R., & Workman, C. I. (2018). A multilevel social neuroscience perspective on radicalization and terrorism. *Social Neuroscience, 13*(5), 511–529.

Demaree-Cotton, J., & Kahane, G. (2018). The neuroscience of moral judgment. In A. Z. Zimmerman, K. Jones, & M. Timmons (Eds.), *The Routledge handbook of moral epistemology* (pp. 84–104). New York: Routledge.

Doğruyol, B., Alper, S., & Yilmaz, O. (2019). The five-factor model of the moral foundations theory is stable across WEIRD and non-WEIRD cultures. *Personality and Individual Differences, 151*, 109547.

Earp, B. D., Douglas, T., & Savulescu, J. (2017). Moral neuroenhancement. In L. S. Johnson & K. S. Rommelfanger (Eds.), *The Routledge handbook of neuroethics* (pp. 166–184). New York: Routledge.

Eres, R., Louis, W. R., & Molenberghs, P. (2018). Common and distinct neural networks involved in fMRI studies investigating morality: An ALE meta-analysis. *Social Neuroscience, 13*, 384–398.

Feltz, A., & May, J. (2017). The means/side-effect distinction in moral cognition: A meta-analysis. *Cognition, 166*, 314–327.

Gazzaniga, M. S. (1983). Right hemisphere language following brain bisection. *American Psychologist, 38*(5), 525–537.

Gilligan, C. (1982). *In a different voice: Psychological theory and women's development.* Cambridge, MA: Harvard University Press.

Glenn, A. L., Koleva, S., Iyer, R., Graham, J., & Ditto, P. H. (2010). Moral identity in psychopathy. *Judgment and Decision Making, 5*(7), 497–505.

Glenn, A. L., & Raine, A. (2014). *Psychopathy.* New York: New York University Press.

Glenn, A. L., Raine, A., & Schug, R. A. (2009). The neural correlates of moral decision-making in psychopathy. *Molecular Psychiatry, 14*(1), 5–6.

Glimcher, P. W., & Fehr, E. (Eds.). (2014). *Neuroeconomics: Decision making and the brain.* New York: Academic Press.

Greene, J. D. (2009). The cognitive neuroscience of moral judgment. In M. S. Gazzaniga (Ed.), *The cognitive neurosciences* (pp. 987–999). Cambridge, MA: MIT Press.

Greene, J. D. (2014). Beyond point-and-shoot morality. *Ethics, 124*(4), 695–726.

Greene, J. D. (2017). The rat-a-gorical imperative: Moral intuition and the limits of affective learning. *Cognition, 167*, 66–77.

Greene, J. D., Nystrom, L. E., Engell, A. D., Darley, J. M., & Cohen, J. D. (2004). The neural bases of cognitive conflict and control in moral judgment. *Neuron, 44*(2), 389–400.

Greene, J. D., Sommerville, R. B., Nystrom, L. E., Darley, J. M., & Cohen, J. D. (2001). An fMRI investigation of emotional engagement in moral judgment. *Science, 293*(5537), 2105–2108.

Haidt, J. (2001). The emotional dog and its rational tail. *Psychological Review, 108*(4), 814–834.

Haidt, J., Koller, S. H., & Dias, M. G. (1993). Affect, culture, and morality, or is it wrong to eat your dog. *Journal of Personality and Social Psychology, 65*(4), 613–628.

Hamlin, J. K. (2015). The infantile origins of our moral brains. In J. Decety & T. Wheatley (Eds.), *The moral brain* (pp. 105–122). Cambridge, MA: MIT Press.

Han, H. (2014). Analysing theoretical frameworks of moral education through Lakatos's philosophy of science. *Journal of Moral Education, 43*(1), 32–53.

Han, H. (2017). Neural correlates of moral sensitivity and moral judgment associated with brain circuitries of selfhood: A meta-analysis. *Journal of Moral Education, 46*(2), 97–113.

Han, H., Kim, J., Jeong, C., & Cohen, G. L. (2017). Attainable and relevant moral exemplars are more effective than extraordinary exemplars in promoting voluntary service engagement. *Frontiers in Psychology, 8*, 283.

Han, H., Workman, C., Dawson, K. J., & May, J. (2018). *Which stories of moral exemplars best provoke moral behavior?* Presentation at the 44th Annual Conference of Association for Moral Education, Barcelona, Spain.

Hare, R. D. (1993). *Without conscience: The disturbing world of the psychopaths among us.* New York: Guilford Press.

Hare, T. A., O'Doherty, J., Camerer, C. F., Schultz, W., & Rangel, A. (2008). Dissociating the role of the orbitofrontal cortex and the striatum in the computation of goal values and prediction errors. *Journal of Neuroscience, 28*(22), 5623–5630.

Harenski, C. L., Harenski, K. A., Shane, M. S., & Kiehl, K. A. (2012). Neural development of mentalizing in moral judgment from adolescence to adulthood. *Developmental Cognitive Neuroscience, 2*, 162–173.

Huebner, B. (2015). Do emotions play a constitutive role in moral cognition? *Topoi, 34*(2), 427–440.

Hutcherson, C. A., Montaser-Kouhsari, L., Woodward, J., & Rangel, A. (2015). Emotional and utilitarian appraisals of moral dilemmas are encoded in separate areas and integrated in ventromedial prefrontal cortex. *Journal of Neuroscience, 35*(36), 12593–12605.

Huys, Q. J., Cools, R., Gölzer, M., Friedel, E., Heinz, A., Dolan, R. J., & Dayan, P. (2011). Disentangling the roles of approach, activation and valence in instrumental and Pavlovian responding. *PLoS Computational Biology, 7*(4), e1002028.

Huys, Q. J., Eshel, N., O'Nions, E., Sheridan, L., Dayan, P., & Roiser, J. P. (2012). Bonsai trees in your head: How the Pavlovian system sculpts goal-directed choices by pruning decision trees. *PLoS Computational Biology, 8*(3), e1002410.

Kahane, G., Everett, J. A. C., Earp, B. D., Farias, M., & Savulescu, J. (2015). "Utilitarian" judgments in sacrificial moral dilemmas do not reflect impartial concern for the greater good. *Cognition, 134*(C), 193–209.

Kahane, G., Wiech, K., Shackel, N., Farias, M., Savulescu, J., & Tracey, I. (2012). The neural basis of intuitive and counterintuitive moral judgment. *Social Cognitive and Affective Neuroscience, 7*(4), 393–402.

Kahneman, D. (2011). *Thinking, fast and slow*. New York: Farrar, Straus, and Giroux.

Kennett, J., & Fine, C. (2008). Internalism and the evidence from psychopaths and "acquired sociopaths." In W. Sinnott-Armstrong (Ed.), *Moral psychology* (Vol. 3, pp. 173–190). Cambridge, MA: MIT Press.

Kiehl, K. A. (2006). A cognitive neuroscience perspective on psychopathy: Evidence for paralimbic system dysfunction. *Psychiatry Research, 142*(2–3), 107–128.

Kiehl, K., & Sinnott-Armstrong, W. (Eds.). (2013). *Handbook of psychopathy and law*. New York: Oxford University Press.

Klein, C. (2010). Philosophical issues in neuroimaging. *Philosophy Compass, 5*(2), 186–198.

Koenigs, M., Kruepke, M., Zeier, J., & Newman, J. P. (2012). Utilitarian moral judgment in psychopathy. *Social Cognitive and Affective Neuroscience, 7*(6), 708–714.

Koenigs, M., Young, L. L., Adolphs, R., Tranel, D., Cushman, F. A., Hauser, M. D., & Damasio, A. R. (2007). Damage to the prefrontal cortex increases utilitarian moral judgements. *Nature, 446*(7138), 908–911.

Kohlberg, L. (1984). *The psychology of moral development: The nature and validity of moral stages*. San Francisco: Harper & Row.

Koster-Hale, J., Saxe, R., Dungan, J., & Young, L. L. (2013). Decoding moral judgments from neural representations of intentions. *Proceedings of the National Academy of Sciences of the United States of America, 110*(14), 5648–5653.

Krajbich, I., Bartling, B., Hare, T., & Fehr, E. (2015). Rethinking fast and slow based on a critique of reaction-time reverse inference. *Nature Communications, 6*(1), 7455.

Kumar, V. (2017). Moral vindications. *Cognition, 167*, 124–134.

Kumar, V., & May, J. (2019). How to debunk moral beliefs. In J. Suikkanen & A. Kauppinen (Eds.), *Methodology and moral philosophy* (pp. 25–48). New York: Routledge.

Landy, J. F., & Goodwin, G. P. (2015). Does incidental disgust amplify moral judgment? A meta-analytic review of experimental evidence. *Perspectives on Psychological Science, 10*(4), 518–536.

Machery, E. (2014). In defense of reverse inference. *The British Journal for the Philosophy of Science, 65*(2), 251–267.

Macintosh, N. J. (1983). *Conditioning and associative learning*. New York: Oxford University Press.

Macmillan, M. (2000). *An odd kind of fame: Stories of Phineas Gage*. Cambridge, MA: MIT Press.

Mahmut, M. K., Homewood, J., & Stevenson, R. J. (2008). The characteristics of noncriminals with high psychopathy traits: Are they similar to criminal psychopaths? *Journal of Research in Personality, 42*(3), 679–692.

Maibom, H. L. (2005). Moral unreason: The case of psychopathy. *Mind and Language, 20*(2), 237–257.

May, J. (2018). *Regard for reason in the moral mind*. Oxford: Oxford University Press.

May, J., & Kumar, V. (2018). Moral reasoning and emotion. In A. Zimmerman, K. Jones, & M. Timmons (Eds.), *The Routledge handbook of moral epistemology* (pp. 139–156). New York: Routledge.

McGuire, J., Langdon, R., Coltheart, M., & Mackenzie, C. (2009). A reanalysis of the personal/impersonal distinction in moral psychology research. *Journal of Experimental Social Psychology, 45*(3), 577–580.

McNamara, R. A., Willard, A. K., Norenzayan, A., & Henrich, J. (2019). Weighing outcome vs. intent across societies: How cultural models of mind shape moral reasoning. *Cognition, 182*, 95–108.

Mendez, M. F., Anderson, E., & Shapira, J. S. (2005). An investigation of moral judgement in frontotemporal dementia. *Cognitive and Behavioral Neurology, 18*(4), 193–197.

Mikhail, J. (2011). *Elements of moral cognition*. Cambridge: Cambridge University Press.

Miller, M. B., Sinnott-Armstrong, W., Young, L. L., King, D., Paggi, A., Fabri, M., . . . Gazzaniga, M. S. (2010). Abnormal moral reasoning in complete and partial callosotomy patients. *Neuropsychologia, 48*(7), 2215–2220.

Moll, J., Eslinger, P. J., & de Oliveira-Souza, R. (2001). Frontopolar and anterior temporal cortex activation in a moral judgment task: Preliminary functional MRI results in normal subjects. *Arquivos de Neuro-Psiquiatria, 59*(3B), 657–664.

Moll, J., de Oliveira-Souza, R., Bramati, I. E., & Grafman, J. (2002). Functional networks in emotional moral and nonmoral social judgments. *NeuroImage, 16*(3), 696–703.

Moll, J., Zahn, R., de Oliveira-Souza, R., Krueger, F., & Grafman, J. (2005). The neural basis of human moral cognition. *Nature Reviews Neuroscience, 6*(10), 799–809.

Moran, J. M., Young, L. L., Saxe, R., Lee, S. M., O'Young, D., Mavros, P. L., & Gabrieli, J. D. (2011). Impaired theory of mind for moral judgment in high-functioning autism. *Proceedings of the National Academy of Sciences of the United States of America, 108*(7), 2688–2692.

Narvaez, D., & Vaydich, J. L. (2008). Moral development and behaviour under the spotlight of the neurobiological sciences. *Journal of Moral Education, 37*(3), 289–312.

Nichols, S. (2004). *Sentimental rules*. New York: Oxford University Press.

Parkinson, C., Sinnott-Armstrong, W., Koralus, P. E., Mendelovici, A., McGeer, V., & Wheatley, T. (2011). Is morality unified? Evidence that distinct neural systems underlie moral judgments of harm, dishonesty, and disgust. *Journal of Cognitive Neuroscience, 23*(10), 3162–3180.

Pellizzoni, S., Siegal, M., & Surian, L. (2010). The contact principle and utilitarian moral judgments in young children. *Developmental Science, 13*(2), 265–270.

Persson, I., & Savulescu, J. (2012). *Unfit for the future: The need for moral enhancement.* Oxford: Oxford University Press.

Piaget, J. (1932). *The moral judgment of the child.* New York: Free Press.

Prinz, J. J. (2016). Sentimentalism and the moral brain. In S. M. Liao (Ed.), *Moral brains: The neuroscience of morality* (pp. 45–73). New York: Oxford University Press.

Railton, P. (2017). Moral learning: Conceptual foundations and normative relevance. *Cognition, 167*, 172–190.

Rangel, A., Camerer, C., & Montague, P. R. (2008). A framework for studying the neurobiology of value-based decision making. *Nature Reviews Neuroscience, 9*(7), 545–556.

Reed, A., Aquino, K., & Levy, E. (2007). Moral identity and judgments of charitable behaviors. *Journal of Marketing, 71*(1), 178–193.

Rescorla, R. A. (1988). Pavlovian conditioning: It's not what you think it is. *American Psychologist, 43*(3), 151–160.

Rest, J. R., & Narvaez, D. (1994). *Moral development in the professions: Psychology and applied ethics.* Hillsdale, NJ: Lawrence Erlbaum Associates.

Saver, J. L., & Damasio, A. R. (1991). Preserved access and processing of social knowledge in a patient with acquired sociopathy due to ventromedial frontal damage. *Neuropsychologia, 29*(12), 1241–1249.

Schaich Borg, J., Hynes, C., Van Horn, J., Grafton, S., & Sinnott-Armstrong, W. (2006). Consequences, action, and intention as factors in moral judgments: An fMRI investigation. *Journal of Cognitive Neuroscience, 18*(5), 803–817.

Schultz, W., Dayan, P., & Montague, P. R. (1997). A neural substrate of prediction and reward. *Science, 275*(5306), 1593–1599.

Sevinc, G., & Spreng, R. N. (2014). Contextual and perceptual brain processes underlying moral cognition: A quantitative meta-analysis of moral reasoning and moral emotions. *PLoS One, 9*, e87427.

Shenhav, A., & Greene, J. D. (2010). Moral judgments recruit domain-general valuation mechanisms to integrate representations of probability and magnitude. *Neuron, 67*(4), 667–677.

Shenhav, A., & Greene, J. D. (2014). Integrative moral judgment: Dissociating the roles of the amygdala and ventromedial prefrontal cortex. *Journal of Neuroscience, 34*(13), 4741–4749.

Stanley, M. L., Yin, S., & Sinnott-Armstrong, W. (2019). A reason-based explanation for moral dumbfounding. *Judgment and Decision Making, 14*(2), 120–129.

Sutton, R. S., & Barto, A. G. (2018). *Reinforcement learning: An introduction.* Cambridge, MA: MIT Press.

Taber-Thomas, B. C., Asp, E. W., Koenigs, M., Sutterer, M., Anderson, S. W., & Tranel, D. (2014). Arrested development: Early prefrontal lesions impair the maturation of moral judgement. *Brain, 137*(4), 1254–1261.

Winterich, K. P., Aquino, K., Mittal, V., & Swartz, R. (2013). When moral identity symbolization motivates prosocial behavior: The role of recognition and moral identity internalization. *Journal of Applied Psychology, 98*(5), 759–770.

Woodward, J. (2016). Emotion versus cognition in moral decision-making. In S. M. Liao (Ed.), *Moral brains: The neuroscience of morality* (pp. 87–117). New York: Oxford University Press.

Workman, C. I., Yoder, K. J., & Decety, J. (2020). The dark side of morality: Neural mechanisms underpinning moral convictions and support for violence. *AJOB Neuroscience, 11*(4), 269–284.

Yin, H. H., & Knowlton, B. J. (2006). The role of the basal ganglia in habit formation. *Nature Reviews Neuroscience, 7*(6), 464.

Yoder, K. J., & Decety, J. (2014). Spatiotemporal neural dynamics of moral judgment: A high-density ERP study. *Neuropsychologia, 60*, 39–45.

Young, L. L., Camprodon, J. A., Hauser, M. D., Pascual-Leone, A., & Saxe, R. (2010). Disruption of the right temporoparietal junction with transcranial magnetic stimulation reduces the role of beliefs in moral judgments. *Proceedings of the National Academy of Sciences of the United States of America, 107*(15), 6753–6758.

Young, L. L., Cushman, F. A., Hauser, M. D., & Saxe, R. (2007). The neural basis of the interaction between theory of mind and moral judgment. *Proceedings of the National Academy of Sciences of the United States of America, 104*(20), 8235–8240.

Young, L. L., & Dungan, J. (2012). Where in the brain is morality? Everywhere and maybe nowhere. *Social Neuroscience, 7*(1), 1–10.

Young, L. L., & Saxe, R. (2008). The neural basis of belief encoding and integration in moral judgment. *NeuroImage, 40*(4), 1912–1920.

Young, L. L., & Saxe, R. (2011). When ignorance is no excuse: Different roles for intent across moral domains. *Cognition, 120*(2), 202–214.

2 The Nature of Empathy

Shannon Spaulding, Rita Svetlova, and Hannah Read

2.1 The Development of Empathy

The questions that motivate much of developmental research on empathy are: What cognitive, social, and motivational abilities does one need to have to be able to empathize with others? How—and in what sequence—do these abilities emerge in ontogeny? What drives individual differences in empathic responsiveness? Some important conceptual distinctions are drawn between automatic (involuntary, "contagious") empathic reactions versus conscious, motivated reactions; between affective and cognitive aspects of empathy; and between self-oriented reactions (empathizer's personal distress) and other-oriented reactions (concern for the other, or "sympathy"). An important variable contributing to these distinctions is the degree of overlap between the perceived internal state of the target of empathy and that of the empathizer. Recent advances in neuroscience have informed many of these questions.

2.1.1 Emotional Contagion

A common theme among various conceptions of empathy is that empathizing involves feeling what another agent feels. The phenomenon of "affective sharing" (congruency of affect in the empathizer and the target) has been of interest to both neuroscientists and developmental psychologists as a possible starting point for empathic responding. Indeed, human infants seem to be biologically predisposed to share the affective states of others. From the moment they are born, newborns cry when they hear others' distress signals (Sagi & Hoffman, 1976; Decety & Meyer, 2008; Hoffman, 1984; Geangu et al., 2010; Hoffman, 2001). Newborns' contagious crying is the earliest developmental precursor of empathy.[1]

A particularly interesting question for empathy researchers is whether such "emotional contagion" involves any, even if rudimentary, understanding that the source of distress lies within another person. Such understanding requires two things: (1) distinguishing between self and other and (2) recognizing that one is distressed *because* another is distressed. The general agreement in the developmental literature is that even if infants in the first few months of life possess some sense of self, it does not imply any self-consciousness or self-awareness. Rather, it is an implicit sense of self as an agentive entity in the environment: the "ecological self" (Neisser, 1991). Only later, over the course of the first and second years of life, do the reciprocal social interactions with others provide the child with what Rochat (1995) calls a "social mirror," laying the ground for objective self-awareness— understanding of the self not just as an agent but as an object of others' perception (Kagan, 1981; Moore et al., 2007).

How does objective self-awareness emerge and develop? Already in the first few months of life, human infants begin to engage in face-to-face "protoconversations" with caregivers that involve the mutual regulation of emotional engagement (Trevarthen, 1979). Precursors of self-understanding and other understanding are thought to develop gradually in the course of these interactions as the infant monitors the other's behavior as well as the other's reactions to his or her own behavior—a process that by the end of the first year gives rise to a new "interpersonal self" (Moore et al., 2007; Neisser, 1991). The tendency to match others' affective states may be a part of this process, as matching the affect of a caregiver can elicit a shared affective experience that further fosters reciprocal interactions.

2.1.2 Self versus Other

By the time of their first birthday, children acquire several new social cognitive abilities that point to their nascent understanding of themselves and others as independent psychological agents: they are able to follow the attention of others by looking where others are looking (joint attention), use others' reactions to novel objects to form their own attitudes toward these objects (social referencing), and attempt to imitate what others are doing with novel objects (imitative learning; Carpenter et al., 1998). By the middle of the second year, a new kind of self-understanding becomes evident with the emergence of mirror self-recognition (Brooks-Gunn & Lewis, 1984), explicit references to oneself by name, use of personal pronouns, and self-conscious emotions in response to the attention of others (Bullock & Lütkenhaus, 1990; Kagan, 1981; Lewis & Ramsay, 2004). These are all signs of objective self- and other-awareness—a phenomenon that, according to a number of scholars, is a fundamental prerequisite for the transition from automatic emotional reactivity to empathic responsiveness (Decety & Meyer, 2008; Thompson, Barresi, & Moore, 1997; Nichols, Svetlova, & Brownell, 2009; Moore et al., 2007; Zahn-Waxler et al., 1992; Kagan, 1981).

Several experimental studies have examined associations between self–other awareness and empathy-related responding in young children. Most of them employ variations of the classic mirror self-recognition rouge task (Amsterdam, 1972), as well as references to self, in order to index self-understanding. The outcome measures include empathic responses to an adult whose toy had broken (Bischof-Köhler, 1991), whose toy had been taken (Johnson, 1982), or who expressed pain upon hurting herself, as well as maternal reports of the child's prosocial responsiveness (Zahn-Waxler et al., 1992). Significant relations were found between 18- and 24-month-olds' mirror self-recognition and empathy toward distressed adults (Bischof-Köhler, 1991; Johnson, 1982). Nichols and colleagues (2009) additionally showed that 12- to 24-month-old toddlers with more advanced self- and other-awareness (controlling for age) were more empathically responsive to a distressed peer.

2.1.3 Concern for Others

Empathy-related responding presents several challenges for someone who has just started to form a concept of self and others. To react with empathy, the child must know that another's distress is unique to that person, even if the child herself is also distressed. The child must also know that the other person may need something different from what the child herself needs when she is upset. Ultimately, in order to act on the other's distress helpfully, one must be able to inhibit one's own distress and focus on the other. Developmentally, reactions of other-oriented concern (sometimes also called sympathy; Eisenberg, 2000) emerge later than self-focused reactions of "contagious" personal distress.

Age-related developments revealed by studies of empathic and prosocial responsiveness in young children are quite consistent. In their study of the development of concern for others, Zahn-Waxler and colleagues (1992) found that rates of expressions of empathy and sympathy toward a distressed adult increased significantly between 13 and 24 months of age and that reactions of personal distress were gradually replaced by more constructive, action-oriented patterns. Comparing responses of 12-, 18-, and 24-month-olds to a distressed peer, Nichols, Svetlova, and Brownell (2015) found that 12-month-olds were neither interested nor concerned about the crying peer, 18- and 24-month-olds showed higher levels of social interest in the peer, and 24-month-olds showed greater empathy, sympathy and prosocial responsiveness toward the distressed peer than any of the younger ages. In a study of helping in one- and two-year-olds, Svetlova, Nichols, and Brownell (2010) found that whereas one-year-olds readily helped an adult in action-based situations that required inferring the adult experimenter's goals, they had great difficulty knowing how to help when the situation involved the adult's negative emotions. Two-year-olds were more responsive and required less communicative cues from adults in both action-based and emotion-based situations, suggesting that by that age, children are able to read the emotional cues of others and generate appropriate responses related to empathy and sympathy. Experiments by Denham (1986) and Vaish, Carpenter, and Tomasello (2009) explored affective perspective-taking abilities in toddlers. They found that by two years of age, toddlers begin to be able to make appropriate judgments about others' emotional reactions, even without explicit emotional cues.

The first year of life is thus the time when humans begin to feel what others feels (emotional contagion); the second year is a period of great advances in self–other differentiation and the time when empathy develops and sympathy first emerges. However, the transition is far from being accomplished. In fact, as noted by Zahn-Waxler and colleagues (1992), "individuals clearly struggle between concern for the needs of self and the welfare of others throughout the life cycle" (p. 133). The process of developing more mature forms of empathy-related responding goes in waves and circles, and even though we as adults will probably never return to the experience of neonatal contagious crying, we may continue to experience the rest of the range of empathy-related reactions, including emotional contagion and personal distress. However, our (adult) range of reactions is undoubtedly more diverse and flexible than that of a two-year-old. How does the development of empathy-related responding continue, and what shapes it, between early childhood and adulthood?

Current research shows that different components of empathy-related responding may follow distinct developmental pathways. Indeed, an agent's ability to recognize and share a target's affective experience may be present in its close-to-mature form starting very early in life when certain interpersonal and situational conditions are met. From the experimental evidence reviewed earlier, it appears that already by two years of age, as long as the child has enough information about the other person's internal state (e.g., if the emotion of the other is clearly manifested by a facial expression and/or vocalization, and the type of emotion is familiar to the child), he or she is able to empathize.[2]

2.1.4 Theory of Mind (Mentalizing)

When the situation is more complex and the other's state is less obvious or requires a modulated contextual appraisal, empathy requires "mentalizing." Mentalizing, also known as theory of mind and mindreading, consists in attributing mental states to a target (Wimmer & Perner, 1983; Premack & Woodruff, 1978). What mentalizing contributes to empathy is the awareness that others' internal states may differ from our own (Frith & Frith, 1999). There are two main theories on how mentalizing works: theory theory and the simulation theory. We will describe these below, but it is important to note that most theorists do not think there is only one way to mentalize. Theorizing and simulating are two legitimate ways to infer a target's emotion (or mental state more generally). The question is: Under which conditions do we use each strategy?

One strategy for understanding a target's mental states when they are not already obvious is to consider what one knows about folk psychology in general and this situation in particular, and to draw an inference about what the target's mental states are and how he or she is likely to behave. This process of theorizing about others' mental states may be conscious and explicit or non-conscious and tacit. "Theories" may include models, heuristics, and a body of assumptions. A general and modern way of characterizing theory theory is in terms of an *information-rich* inference to the best explanation (Baker, Saxe, & Tenenbaum, 2011; Carruthers, 2006; Fodor, 1983; Scholl & Leslie, 1999; Wellman, 2015).[3]

An alternative way to figure out a target's mental states is for one to imagine what one would think, feel, and do in the target's situation, and attribute the result of the mental simulation to the target (Davies & Stone, 1995b).[4] Just like theorizing, mental simulation may be explicit and conscious or tacit and non-conscious. Whereas theorizing involves inferring others' mental states on the basis of a rich body of information, simulation is meant to be an *information-poor* process. It does not require access to large bodies of information about folk psychology. Simulation simply requires an ability to put oneself mentally in a target's position and figure out what one would feel, think, and do. In effect, one simply redeploys one's own cognitive mechanisms for the purpose of understanding the other person's perspective.

One important mark of sophisticated mentalizing ability is the capacity to understand that a target's beliefs guide his or her actions, even when those beliefs conflict with reality and/or with one's own beliefs (Wellman, Cross, & Watson, 2001). When children are capable of demonstrating this ability is a matter of intense debate (Apperly & Butterfill, 2009; Carruthers, 2016; Heyes, 2014; Rakoczy, 2012; Baillargeon, Scott, & He, 2010). We can say with a fair degree of confidence that there are various transitions that occur over the first five years of life, and it appears to be challenging for children on the younger end of this range to reason about others' complex mental states, especially when the situational context is unclear and/or complicated.

One of the most significant transitions occurs between three and four years old, and it is marked by the passing of what is known as the false-belief task, also known as the change of location task. This task has been the most widely used test of children's mentalizing ability (Wimmer & Perner, 1983). In a classic version of the false-belief task, subjects watch Sally, a puppet, place a toy in a basket and leave the scene. Next, they watch Ann, another puppet, come in and move the toy to the cupboard and then leave the scene. Finally, they watch Sally return to the scene, and then they are asked a series of questions to ensure that they remember what happened and know where the toy actually is. The key question in this version of the task is: Where will Sally look for her toy? A child counts as passing this task if she (1) remembers the sequence of events, (2) tells the experimenter where the toy actually is, and (3) answers that Sally will look for the toy in the basket. Almost all children tested can do (1) and (2). The test condition is (3). In order to succeed on (3), children must separate what they know about the location of the toy from what Sally thinks about the location of the toy. They must understand that Sally thinks the toy is in one location (the basket), and will look for it there, but actually the toy is in a different location (the cupboard). Several versions of the false-belief task exist, and while the precise age of passing it varies between children and between task versions (Clements & Perner, 1994; Wellman et al., 2001; Garnham

& Ruffman, 2001), children younger than three or four years typically fail, whereas older children typically pass.[5]

If children younger than three or four years of age have difficulty sorting out conflicting information about their own and another's mental states, it ought to be challenging for them to deal with others' emotional states if they are significantly different from their own states and/or difficult to read from the situation. In fact, research shows that at the early school age, similarity between the perceiver and the victim of distress is still an important factor in empathy. For example, in a study in which six- and seven-year-old boys and girls observed slide-sequence stories about boys and girls in different emotional states, they had a stronger vicarious emotional response to children of their own sex (Feshbach & Roe, 1968). Mentalizing abilities continue to be a source of variability in empathic and prosocial responsiveness, even in early adolescence. For instance, children who are better at reconciling conflicting emotional cues also exhibit more helping behaviors (Carlo et al., 1991). The transition between childhood and adolescence is also marked by an important development that is believed to be related to a combination of self-concept development, growth of perspective-taking abilities, and moral reasoning: the ability to empathize with a generalized group of others (Fabes et al., 1999).

2.1.5 Sympathy

Sympathy—feeling concern for a target—seems to develop later than empathy. Sympathy is another domain in which top-down processes, including mentalizing, perspective-taking, and emotion regulation, are implicated. Unlike empathy, sympathy requires the perceiver to evaluate the perceived affective state of the other and to form a motivational attitude toward it. It also involves effortful control—the ability to inhibit a predominant response—in this case, the reaction of personal distress (Eisenberg & Eggum, 2009; Rothbart, 2007). While personal distress is one of the earliest reactions to others' distress that children show, effortful control is a quite late-developing phenomenon. Neurologically, it relies on the prefrontal cortex (PFC)—the latest-developing brain structure, whose myelination continues throughout adolescence and into early adulthood (Dahl, 2004).

The ways in which the abilities of mentalizing and emotion regulation interact in predicting patterns in children's empathy-related responding have been studied extensively by Eisenberg and Fabes (1991), Eisenberg and colleagues (1994), and Eisenberg, Spinrad, and Sadovsky (2006). One of their studies compared adults' and second- to fifth-graders' empathic reactions to others' distress, as well as their tendency to help the victims. The measures included self-reported reactions to a video of a woman and her two injured children who had been in a car accident, as well as the subjects' facial expressions and heart rates recorded while they were watching the video. The subjects were then given the opportunity to offer help to the characters in the video. For both the self-reports and the physiological measures, the researchers looked separately at expressions of self-focused personal distress versus sympathetic concern about the others. Consistent with the earlier findings of Batson and colleagues (1983) and Batson, Fultz, and Schoenrade (1987), in adults, self-reported sympathy and facial expressions of sympathy predicted helping. Children's self-reports of personal distress and sympathy were not related to their prosocial behavior (possibly suggesting children's immature abilities to self-reflect or to report their experiences accurately). However, physiological measures, in particular children's facial display of personal distress, were negatively related to their helping behavior. Other developmental studies, using various measures of empathy-related responsiveness, have confirmed that children's increased sympathy is positively correlated with other-oriented behavior. For a review, see Eisenberg (2000).

Putting all these data together yields a specific developmental trajectory. Emotional contagion is present at birth. Personal distress is online very early in development as well. The capacity to empathize with others when their emotions are simple and the context is straightforward develops sometime in the second year of life and continues to mature as children's mentalizing abilities further develop. The capacity for sympathy is relatively late developing in comparison to the other empathy-related responses because sympathy involves effortful control, which continues to develop throughout childhood and adolescence.

2.2   Mechanisms of Empathy

In philosophy and the cognitive sciences, much of the discussion of empathy concerns two subsystems of empathy: neural resonance and mentalizing. We shall explain each subsystem below, but it is important to note at the outset that it is likely that both subsystems play a role in various empathic phenomena. Thus, the issue is what each subsystem contributes to empathy rather than whether either one does.

2.2.1   Neural Resonance

One mechanism of empathy is neural resonance, which is underwritten by so-called mirror neurons. Mirror neurons are neurons that activate when a subject acts, emotes, or experiences a certain sensation and also when a subject observes a target acting, emoting, or experiencing a certain sensation. For example, a host of neurons in the premotor cortex and posterior parietal cortex fires when I grasp an object, and this same host of neurons fires when I observe another person grasping an object (Rizzolatti & Craighero, 2004). There are similar mirror neuron systems for experiencing and observing certain emotions. When I experience disgust and when I observe another person experiencing disgust, the same collection of neurons in the insula fires (Calder et al., 2000; Wicker et al., 2003). Similar findings hold for the experience and observation of fear (Adolphs et al., 1994), anger (Lawrence et al., 2002), pain (Singer et al., 2004), and touch (Keysers & Perrett, 2004). In each of these cases, groups of neurons are endogenously activated when the subject acts, emotes, or feels a certain way, and these same groups of neurons are exogenously activated (at an attenuated level) when the subject observes another acting, emoting, or feeling in those same ways.

Mirror neurons, which are present in both humans and other primates, are hypothesized to be an evolutionarily old mechanism underlying the tendency to mimic and synchronize facial expressions, postures, and movements automatically with those of others (Di Pellegrino et al., 1992; Rizzolatti & Craighero, 2004; Gallese, 2003). This ability has been linked to empathy in several areas of cognitive science. In social psychology, where this phenomenon is called the chameleon effect, evidence suggests that the tendency to imitate others' bodily expressions automatically is correlated with self-reported empathy (Chartrand & Bargh, 1999). In social cognitive neuroscience, some argue that the action mirror neuron system is connected to the emotion-related limbic system via the insula, an area involved in self-representation (Carr et al., 2003). According to this account, the insula serves as a relay from action representation to emotional reaction, and "empathy may occur via a mechanism of action representation that shapes emotional contents" (Carr et al., 2003, p. 5497). In other words, the hypothesis is that we come to understand others' feelings by first perceiving, processing, and mimicking their actions or appearances.

It is quite plausible that neural resonance is connected to various empathy-related concepts. Most clearly, neural resonance seems to be a mechanism of emotional contagion, that is, feeling an emotion as a result of perceiving another person feel that emotion. Of course, this may not be the only mechanism, but plausibly, it is one such mechanism (see Debes, 2017, for a similar argument). As a result of being a mechanism of emotional contagion, it is plausible that neural resonance plays a role in personal distress, as described in the previous section. However, there are many routes to a subject feeling personally distressed by a target's emotional experience. I may feel afraid for myself when I see an angry person, which would meet the definition of personal distress but clearly would not involve neural resonance in the relevant sense. Thus, there is only a tenuous connection between neural resonance and this empathy-related concept.

Neural resonance may also be connected to other empathic responses. For example, recognizing that the target is having a certain affective experience and, as a result, having a congruent affective experience (when the recognition occurs via mental simulation, this is sometimes called affective empathy, and when it occurs via theoretical inference it is sometimes called cognitive empathy), and recognizing that a target is having a certain affective experience and, as a result, feeling concern for the target (i.e., sympathy).

The connection between neural resonance and these empathic responses is not at all straightforward. Cognitive empathy, affective empathy, and sympathy involve more than just affective sharing. All three responses require recognizing that the target is having a certain emotion. The cognitive element of these empathic responses goes beyond mere affective sharing, and thus neural resonance clearly is not *sufficient* for these empathic responses (Debes, 2010; Spaulding, 2012, 2013; Debes, 2017).

However, the issue of whether neural resonance is *necessary* for cognitive and affective empathy and sympathy is nuanced. On the one hand, there is strong evidence that subjects who have suffered damage to mirror neuron areas of the brain (the amygdala, insula, or anterior cingulate cortex [ACC]) have difficulty inferring others' emotions as a result (Adolphs & Tranel, 2003; Calder et al., 2000; Jackson, Meltzoff, & Decety, 2005; Wicker et al., 2003). For example, subjects who suffer damage to the amygdala (purportedly part of the fear mirror neuron circuit) are deficient specifically in experiencing and attributing fear to others. In a study with a very small sample size of only two subjects with bilateral damage to the amygdala, researchers determined that both subjects were selectively impaired in their ability to experience fear and to recognize fear in isolated, static facial expressions.

This double dissociation is prima facie evidence that neural resonance is somehow involved in some inferences about others' emotions. On the other hand, these subjects were not impaired in their ability to recognize fear in richer visual stimuli that also contain non-facial expression cues, for example static or dynamic body postures (Atkinson, Heberlein, & Adolphs, 2007; Adolphs et al., 2005). Subjects with damage to mirror neuron areas for fear retain the ability to attribute fear based on body language, semantic knowledge, and situations that typically evoke fear. Moreover, when one of the subjects was instructed to pay attention to the eye region of the target's face, her deficit in recognizing facial expressions of fear was eliminated.[6] This suggests that the cognitive deficits result from inputs to mirror neuron systems rather than problems with the mirror neuron systems themselves. Similar deficit patterns are found for the experience and recognition of disgust (Calder et al., 2000) and anger (Lawrence et al., 2002).

Putting all of this information together yields a complicated picture. Mirror neuron activation is clearly not causally sufficient for sympathy and cognitive and affective empathy, as we argued above. The evidence just canvassed suggests that mirror neuron activation may be necessary for certain narrowly constrained aspects of empathic responding. In particular, mirror neuron activation may be necessary for reading others' emotions from their static facial expressions. Static facial expressions are just one not particularly naturalistic source of information about others' emotions. In more ecologically typical environments, we use various sources of information about others' emotions, and mirror neuron activation does not play such an important role in this richer context (Debes, 2010; Spaulding, 2012, 2013).[7] Nevertheless, even if neural resonance is not a universal part of all routes to empathy, in virtue of the close connection between neural resonance and emotional contagion, it is seems to be a building block of some, possibly the most phylogenetically ancient routes. It is also believed to be a part of the early precursors of empathy in human ontogeny (Hoffman, 2000; Decety & Svetlova, 2012).

2.2.2 Mentalizing

A second mechanism of empathy is mentalizing. In the developmental section, we discussed two varieties of mentalizing: theorizing and simulating. Theorizing consists in relying on a broad body of folk psychological information to draw an inference as to the best explanation about a target's mental states. Simulating, in contrast, involves mentally imagining oneself in a target's situation and determining how one would think, feel, and act in that situation and then attributing that to the target.

What are the neural underpinnings of theorizing and simulating? Of the brain regions shown to be involved in the evaluation of others' emotions, the temporoparietal junction (TPJ) is particularly important. In addition to being widely regarded as an important neural realizer of mentalizing (Samson et al., 2004; Saxe & Wexler, 2005), the TPJ is commonly recruited in tasks involving self–other distinction. Specifically, it seems to play an important role in the distinction between self-produced actions and actions generated by others (Blakemore & Frith, 2003). The TPJ is an association cortex with reciprocal connections to the PFC and the temporal lobes, which may contribute to its role in self–other differentiation. More importantly for the study of empathy-related responding, the right TPJ is specifically recruited when participants imagine how other people would feel in distressing situations, but not when they imagine the analogous situations for themselves (Jackson et al., 2005, Lamm, Batson, & Decety, 2007). These data suggest that the right TPJ is especially important for theorizing about others' emotions and mental states more generally.

With respect to simulating, a meta-analytic study by Lee and Siegle (2009) examined common and distinct brain regions involved in explicit evaluation of one's own and others' emotions across thirty-seven neuroimaging studies. The reviewed studies mainly asked participants to evaluate and report the feelings of another person (based on vignettes or photographs), the participant's own feelings, or both. The meta-analysis showed that in these paradigms, the insula and rostral ACC were specifically associated with evaluation of one's own emotion, whereas the precuneus, superior temporal sulcus (STS), and the right TPJ were specifically associated with evaluation of others' emotions. This confirms the idea that the right TPJ—along with the STS and precuneus—are neural mechanisms of theorizing. Brain regions that were involved in explicitly evaluating both one's own and others' emotions included the amygdala, lateral PFC (lPFC), and dorsomedial PFC (dmPFC). These data indicate that simulating others' emotions is realized in a different network of brain regions than theorizing. In particular, the dmPFC and lPFC are important for simulating others and oneself (though see De Brigard et al., 2015, for more fine-grained analysis of the brain regions involved in mentally simulating different sorts of targets).

2.2.3   Which Kind of Empathy?

With the neural realizers of mentalizing and simulating others' emotions on the table, we can now examine how these mechanisms are related to the various empathy-related concepts described above. Consider cognitive and affective empathy and sympathy first. Both theory theory and simulation theory aim to explain how we understand what others' emotions are, which is a necessary aspect of sympathy, cognitive empathy, and affective empathy. For that reason, it seems that mentalizing (of either the theoretical or simulationist sort) is necessary for empathy and sympathy. In particular, theorizing about a target's emotional state is necessary for cognitive empathy. Clearly, however, theorizing about a target's emotions is not sufficient for the affective sharing part of empathy. On the basis of what I know about you, your situation, and folk psychology in general, I may infer that you are experiencing a certain emotion, but that does not entail that I will feel a congruent emotion in response. Similarly, just because I infer a target's emotion in the way theory theory posits does not entail that I will be sympathetic. An intelligent psychopath may be adept at inferring others' emotions but feel no empathy or sympathy for others. Thus, the connection between theory theory and sympathy and empathy is indirect. Theory theorists will need an additional element to explain why sometimes we feel congruent emotions and concern for a target and why sometimes we do not.

In contrast, simulation theory holds that the way in which one infers what a target feels is by trying to feel what the target is feeling (at least in an attenuated way). Thus, simulation theory offers a much more direct explanation of empathy, in particular affective empathy, which is a direct outcome of mentally simulating a target's emotions. We imagine what a target feels and, through this imagining, feel a congruent emotion ourselves. As noted above, many theorists argue that we rely on both theorizing and simulating in understanding others' mental states (Carruthers, 1996; Goldman, 2006; Spaulding, 2012, 2017; Nichols & Stich, 2003). Simulation theory's explanation of sympathy is a bit less obvious. The sympathetic response is not part of the mental simulation. So, the simulation theorist would need to posit an additional subsequent element to explain why it is that the subject feels a complementary emotion.

While theory theory has no direct explanation of emotional contagion, proponents of simulation theory have argued that neural resonance is a mechanism of low-level simulation (see above). If neural resonance is indeed a mechanism of simulational mentalizing,[8] then simulation theory has a clear explanation of emotional contagion. The explanation is similar to the explanation of affective empathy: we feel what a target feels because upon perceiving a target's emotional expressions, our brain activates as if we were experiencing that emotion. The difference between simulation theory's story of affective empathy and emotional contagion is that only the former involves the subject recognizing that the target feels a certain emotion and feeling a congruent emotion *because* of this recognition.

Artificiality of empathy experiments (very simplified stimuli/tasks) and differences in tasks (passive viewing stimuli vs. active inference) have led to the impression that these subsystems are dissociable (Zaki, 2014). While simplification of experiments is important, at least initially, to eliminate confounding factors in the research, now it is important to make stimuli in experiments more ecologically valid and to make sure experiments are comparing similar types of tasks. More recently, theorists have moved beyond questions of which subsystem is responsible for empathy—it is widely recognized that both subsystems contribute to empathy—to questions about when and how each neural mechanism activates empathy (Decety & Meyer, 2008; Eisenberg & Eggum, 2009; Decety, 2010; Zaki & Ochsner, 2012; Lombardo et al., 2010; Zaki et al., 2007).

2.3   The Concept of Empathy

As we have seen, tracking the development of different dimensions of empathy and teasing apart the mechanisms that underlie each are useful to see how these aspects of empathy can occur together or separately and why empathy can therefore appear to take so many different forms. These different forms provide various ways to share the mental lives of others, but what makes them all empathy?

Debates about the concept of empathy are notoriously messy. On the one hand, some, including notable critics of empathy Paul Bloom (2017) and Jesse Prinz (2011a, 2011b), take empathy to involve feeling what one thinks another person is feeling. At the same time, others, such as Nel Noddings (2010) and Lawrence Blum (2011, 2018), take it to involve both shared feelings as well as motivation to respond appropriately to the target. Different still, some, such as Amy Coplan (2011), argue that empathy is primarily a matter of imagining what things must be like for another person in her situation, or "simulating" the target's "situated psychological state," as she puts it (p. 40). Yet, debates about what empathy really is have proven intractable.

We propose that empathy might be best understood as a complex multi-dimensional response. In particular, empathy is comprised of affective, cognitive, and motivational dimensions (Read, 2019). Consider first empathy's affective dimension. At its most basic, this simply involves feeling what another person feels, or sharing another person's affective mental state (Slote, 2007; Sreenivasan, 2020). This very simple affect sharing can be mediated by varying degrees of awareness that one's empathic feeling originated with the target. On the one hand, one might have very little or no such awareness—for instance, you are sad, and I simply catch your sadness. Following Eisenberg and Eggum (2009), we have called this "emotional contagion," and as discussed above, it is something even infants appear to be capable of experiencing. On the other hand, more mature forms of affect sharing might involve feeling what another person feels *for her sake*, as Maibom (2017) puts it. In fact, as mentioned above, some evidence suggests that the degree to which we have more or less other-oriented empathy is at least partly a function of the degree to which we have a clear sense of the differences between self and other (Nichols et al., 2009), as well as the degree to which one's empathy has some cognitive dimension (Maibom, 2012).

The cognitive dimension of empathy involves understanding what another person is thinking and feeling and why—sometimes referred to as mentalizing (Zaki, 2014), theory of mind (Leslie, 1987), or even a kind of cognitive empathy in its own right (Spaulding, 2017). This understanding can come in degrees. For instance, I might understand that you are scared with or without understanding what you are scared of. As discussed above, this cognitive dimension of empathy might be achieved by at least two different means or mechanisms. We can simulate the target's mental state(s) ourselves (simulation theory) or infer the target's mental state(s) based on observed behavior and our folk psychology theories of the relation between mental states and behavior (theory theory). These days, it is most common to endorse a hybrid account according to which empathy's cognitive dimensions is achieved by means of some combination of these different mechanisms (Decety, 2010; Eisenberg & Eggum, 2009; Zaki & Ochsner, 2012).

Perspective taking, which involves simulation, is often invoked as a key route to empathy's cognitive dimension. This can be a matter of imagining either what it would be like to be in another person's situation or what things must be like for another person in her situation—what Batson (2011) refers to as imagine-self versus imagine-other perspective taking. In both cases of perspective taking, one simulates or tries on the target's mental state, imagining what it would one would think and feel in her situation (Coplan, 2011; Goldman, 2006; Gordon, 1995). For instance, I might see someone rushing through the rain without an umbrella and imagine what it would be like for me, now on my morning commute, to be caught in the rain without an umbrella. Alternatively, I might imagine what things must be like for my friend who has recently had a new baby and is struggling to balance her new and existing responsibilities and relationships.

Finally, empathy might also have a motivational dimension by virtue of which the empathizer experiences some motivation to help or respond otherwise appropriately to the target. Blum (2011) characterizes this motivational dimension of empathy in terms of one having an "at least minimal *ceteris paribus* motivation" (p. 173), to respond appropriately to the target, while for Noddings (2010), it as a matter of being moved by the target's situation—that is, being moved to help the target achieve some goal or respond otherwise usefully and sensitively to her. And while empathy's affective dimension might sometimes be motivating in its own right, it isn't always. For example, I might be motivated to help someone as a result of sharing her fear, but doing so might also paralyze me. It is thus conceptually and empirically useful to distinguish the motivational and affective dimensions of empathy.

Empathy's different dimensions and the various forms that it can take, depending on the nature and degree of prominence of these, might be distinguished from related concepts, including empathic concern, sympathy, and personal distress. While these related responses are commonly thought to involve some affectivity, it is typically not thought to be a shared affective experience, unlike empathy. For instance, on Batson's (2011) account, "empathic concern" is a response to the target's welfare. As such, the affective dimensions of empathic concern are conceptually tied to the target's situation and may or may not correspond to the target's own affective experience with respect to her situation. For example, Hoffman (2001) describes a case of empathic anger felt for someone who has been attacked, while the victim herself feels something different, such as fear, sadness, or disappointment. Something similar is often thought to be true of sympathy, the affective components of which are more commonly associated with a target's welfare as opposed to her state of mind (Bloom, 2017; Darwall, 1998; Singer & Klimecki, 2014; Slote, 2010; Maibom, 2009). Empathy might thus be distinguished from these neighboring concepts loosely in terms of *feeling with* another person as opposed to *feeling for* her.

These various conceptual distinctions notwithstanding, it is important to note that there is a great deal of debate about these different concepts and how they relate to one another. For example, Batson (2011) includes empathy under the broad heading of "empathic concern," which includes sympathy. Similarly, for Blum (2011), empathy and sympathy are species of the genus "fellow-feeling." On the other hand, others argue for a much sharper distinction between these different responses (Slote, 2010; Bloom, 2017; Darwall, 1998; Maibom, 2009). Nonetheless, we maintain that distinguishing between empathy and related responses as well as empathy's different dimensions along the lines proposed here is crucial, given the goal of studying the developmental and causal relations between them.

2.4 Future Directions

We propose that fruitful future research might be conducted broadly along the following three lines.

2.4.1 The Development of Empathy

While a significant amount of research has addressed the development of empathy's individual dimensions, as discussed above, relatively little attention has been paid to the way in which development of these different dimensions might mutually reinforce or promote one another, along with the conditions under which this becomes possible. For instance, to what extent, or under what conditions, does the development of the ability to match others' affective states promote or contribute to the development of the ability to differentiate between self and other, and vice versa? Future research along these and related lines would greatly enhance existing developmental studies of empathy and its potential contribution to a number of prosocial tendencies and behaviors.

Additionally, research addressing the developmental relation between empathy's different dimensions could contribute to vindicating empathy's role in the moral life, contra claims by its most vocal critics (Bloom, 2017; Prinz, 2011a, 2011b). According to these critics, affect-matching forms of empathy with little or no cognitive dimension are subject to a number of limitations related to bias and motivation. These limitations render dubious any claims regarding their moral importance. Yet, if it turned out that even rudimentary affect matching promotes the development of empathy's other dimensions as well as more sophisticated, multidimensional forms of empathy, then contra critics, even simple affect matching may in fact occupy an important role in the moral life, particularly with respect to moral development.

2.4.2 Empathy as a Motivated Response

Empathy is, as Zaki (2018) notes, a motivated response, and as such can be developed and exercised across many different lines. Factors that seem to bear on our motivation to empathize in different contexts include our relationship with the target, including whether the target is an ingroup member, family member, friend, and so on (Cikara et al., 2014; Chiao & Mathur, 2010); geographic proximity and similarity with the target (Hoffman, 2001); the target's perceived cuteness (Batson et al., 1995); as well as the perceived emotional, cognitive, and material cost of empathizing (Batson, 2011). Some evidence also suggests that having and highlighting a goal for empathy, as well as the extent to which one's group endorses norms regarding empathy for diverse others, promotes individual willingness and ability to empathize (Weisz & Zaki, 2017). And a person's beliefs about empathy's malleability—or the extent to which it is possible to muster empathy, even when one is not immediately inclined to empathize— seem to affect willingness and ability to empathize (Schumann, Zaki, & Dweck, 2014). Further research might thus continue to examine these and other factors that bear on motivation to empathize.

Related research might also aim to determine the most promising means of intervening on motivation to empathize with the help of public institutions such as schools. Creating structured opportunities for people to practice empathy and develop the relevant empathy-related abilities is likely crucial to ensuring that empathy is motivated in a host of challenging cases where we are prone to avoid it. For instance, cooperative learning models, such as the jigsaw classroom, have proven highly successful in promoting empathic abilities across various racial and socioeconomic group divisions

(Aronson, 2002; Tropp & Saxena, 2018). Future research might thus consider additional ways in which interventions aimed at promoting empathic abilities could be supported by, or implemented in, the context of various public institutions.

2.4.3 Empathy and Morality

A third direction for future study regards the role of empathy in morality. This topic has received a great deal of attention in recent years from empathy's proponents and critics alike. On the one hand, proponents of empathy argue that it plays a crucial role in moral judgment making (Slote, 2007), moral motivation (Noddings, 2010), moral development (Hoffman, 2001), and even the promotion of close personal relationships (Betzler, 2019). On the other hand, critics have pointed out empathy's susceptibility to a number of morally troubling limitations. Most notably, it is often extended to racial (Gutsell & Inzlicht, 2010; Xu et al., 2009), political (Zaki & Cikara, 2015), or social (Cikara, Bruneau, & Saxe, 2011) ingroup members at the expense of outgroup members, as well as those who are relatable to us and even geographically nearby—what Hoffman (2001) refers to as empathy's "familiarity" and "here and now" biases (p. 197).

Yet, across nearly all debates, the emphasis has been on the potential moral benefits of empathy for those in need or those toward whom we are already in some close relationship, for example friendship or romantic partnership. And while some, such as L. A. Paul (2017) have proposed that empathy might be extended to more demographically diverse targets as a means to promote tolerance, how exactly this is so remains to be explained.

Future research in both neuroscience and philosophy might consider the feasibility and potential moral benefits of empathy for non-needy targets. Questions of potential interest include: Is empathy a useful tool for promoting positive relationships between people who are opposed to one another for moral, political, or social reasons? How might empathy's limitations due to bias and ingroup preferences be overcome in these cases such that it is usefully employed? Research addressing questions such as these would prove especially useful given recently high levels of so-called affective polarization (Iyengar & Westwood, 2015) in much of the Western world, as well as other persistent, seemingly intractable domestic and international conflicts.

Notes

1. Though the idea that newborns experience emotional contagion is widely shared in developmental psychology, not everyone agrees. See Ruffman, Lorimer, and Scarf (2017) for a critical analysis of studies purporting to demonstrate emotional contagion. Ruffman and colleagues offer a deflationary account of newborns crying in response to other newborns crying in terms of the aversive acoustic properties of neonate crying.

2. In the neuroscientific literature, this route to empathy is called bottom-up processing (Decety & Meyer, 2008; Singer & Lamm, 2009), although some also use this term to refers to a lower-level perceptual mechanism, e.g., Preston and De Waal (2002).

3. For a historical overview of theory theory, see the collected volumes by Davies and Stone (1995a) and Carruthers and Smith (1996).

4. The basic idea of mental simulation is simple and intuitive, but the details of how this happens are quite nuanced (Spaulding, 2012, 2015).

5. In addition to the classic false-belief task described in the main text, there are now less demanding non-linguistic versions of these tasks that children pass at much younger ages, some before twelve months old. For a recent overview, see Scott and Baillargeon (2017) and Baillargeon, Buttelmann, and Southgate (2018). See Apperly and Butterfill (2009), Heyes (2014), and Rakoczy (2012) for a skeptical take on these experiments.

6. This finding reveals an ambiguity in the data. It could be that damage to a mirror neuron system for a certain emotion results in an impaired ability to recognize that emotion. Or it could be that the lesions damage mechanisms that relay input to mirror neuron systems. However, perhaps it is wise not to overgeneralize from a study on two subjects.

7. Elsewhere, Spaulding (2012) argues that mirror neurons are a contributory (or enabling) cause of low-level mindreading (i.e., rapid, non-conscious attribution of mental states). That is, though it is neither nomologically necessary nor sufficient for low-level mindreading, it nevertheless contributes to bringing about (or enables) mindreading. This hypothesis implies that the activation of mirror neurons precedes some episodes of mindreading, and that interventions on mirror neuron systems will affect mindreading. However, attributing emotions, intentions, and other mental states is possible without the contribution of mirror neurons, at least in adults.

8. See Spaulding (2012) for an argument that mirror neurons are *not* in fact evidence for simulation theory.

References

Adolphs, R., Gosselin, F., Buchanan, T.W., Tranel, D., Schyns, P., & Damasio, A.R. (2005). A mechanism for impaired fear recognition after amygdala damage. *Nature, 433*(7021), 68.

Adolphs, R., & Tranel, D. (2003). Amygdala damage impairs emotion recognition from scenes only when they contain facial expressions. *Neuropsychologia, 41*(10), 1281–1289.

Adolphs, R., Tranel, D., Damasio, H., & Damasio, A. (1994). Impaired recognition of emotion in facial expressions following bilateral damage to the human amygdala. *Nature, 372*(6507), 669–672.

Amsterdam, B. (1972). Mirror self-image reactions before age two. *Developmental Psychobiology, 5*(4), 297–305.

Apperly, I. A., & Butterfill, S. A. (2009). Do humans have two systems to track beliefs and belief-like states. *Psychological Review, 116*(4), 953.

Aronson, E. (2002). Building empathy, compassion, and achievement in the jigsaw classroom. In *Improving academic achievement: Impact of psychological factors on education* (pp. 209–225). New York: Academic Press.

Atkinson, A. P., Heberlein, A. S., & Adolphs, R. (2007). Spared ability to recognise fear from static and moving whole-body cues following bilateral amygdala damage. *Neuropsychologia, 45*(12), 2772–2782.

Baillargeon, R., Buttelmann, D., & Southgate, V. (2018). Interpreting failed replications of early false-belief findings: Methodological and theoretical considerations. *Cognitive Development, 46*, 112–124.

Baillargeon, R., Scott, R., & He, Z. (2010). False-belief understanding in infants. *Trends in Cognitive Sciences, 14*(3), 110–118.

Baker, C., Saxe, R., & Tenenbaum, J. (2011). *Bayesian theory of mind: Modeling joint belief-desire attribution.* Proceedings of the 33rd Annual Meeting of the Cognitive Science Society, Boston, MA.

Batson, C. D. (2011). *Altruism in humans.* New York: Oxford University Press.

Batson, C. D., Fultz, J., & Schoenrade, P. A. (1987). Distress and empathy: Two qualitatively distinct vicarious emotions with different motivational consequences. *Journal of Personality, 55*(1), 19–39.

Batson, C. D., Klein, T. R., Highberger, L., & Shaw, L. L. (1995). Immorality from empathy-induced altruism: When compassion and justice conflict. *Journal of Personality and Social Psychology, 68*(6), 1042.

Batson, C. D., O'Quin, K., Fultz, J., Vanderplas, M., & Isen, A. M. (1983). Influence of self-reported distress and empathy on egoistic versus altruistic motivation to help. *Journal of Personality and Social Psychology, 45*(3), 706.

Betzler, M. (2019). The relational value of empathy. *International Journal of Philosophical Studies, 27*(2), 136–161.

Bischof-Köhler, D. (1991). The development of empathy in infants. In M. E. Lamb & H. Keller (Eds.), *Infant development: Perspectives from German-speaking countries* (pp. 245–273). Mahwah, NJ: Lawrence Erlbaum Associates.

Blakemore, S.-J., & Frith, C. (2003). Self-awareness and action. *Current Opinion in Neurobiology, 13*(2), 219–224.

Bloom, P. (2017). *Against empathy: The case for rational compassion.* London: Random House.

Blum, L. (2011). Empathy and empirical psychology: A critique of Shaun Nichols's neo-sentimentalism. In C. Bagnoli (Ed.), *Morality and emotions* (pp. 170–193). Oxford: Oxford University Press.

Blum, L. (2018). A moral account of empathy and fellow feeling. In N. Roughley & T. Schramme (Eds.), *Forms of fellow feeling: Empathy, sympathy, concern and moral agency* (p. 142). Cambridge: Cambridge University Press.

Brooks-Gunn, J., & Lewis, M. (1984). The development of early visual self-recognition. *Developmental Review, 4*(3), 215–239.

Bullock, M., & Lütkenhaus, P. (1990). Who am I? Self-understanding in toddlers. *Merrill-Palmer Quarterly, 36*(2), 217–238.

Calder, A. J., Keane, J., Manes, F., Antoun, N., & Young, A. W. (2000). Impaired recognition and experience of disgust following brain injury. *Nature Neuroscience, 3*(11), 1077–1078.

Carlo, G., Knight, G. P., Eisenberg, N., & Rotenberg, K. J. (1991). Cognitive processes and prosocial behaviors among children: The role of affective attributions and reconciliations. *Developmental Psychology, 27*(3), 456.

Carpenter, M., Nagell, K., Tomasello, M., Butterworth, G., & Moore, C. (1998). Social cognition, joint attention, and communicative competence from 9 to 15 months of age. *Monographs of the Society for Research in Child Development, 63*(4), i–vi, 143.

Carr, L., Iacoboni, M., Dubeau, M. C., Mazziotta, J. C., & Lenzi, G. L. (2003). Neural mechanisms of empathy in humans: A relay from neural systems for imitation to limbic areas. *Proceedings of the National Academy of Sciences, 100*(9), 5497–5502.

Carruthers, P. (1996). Simulation and self-knowledge: A defence of theory-theory. In P. Carruthers & P. K. Smith (Eds.), *Theories of theories of mind* (pp. 22–38). Cambridge: Cambridge University Press.

Carruthers, P. (2006). *The architecture of the mind.* Oxford: Oxford University Press.

Carruthers, P. (2016). Two systems for mindreading? *Review of Philosophy and Psychology, 7*(1), 141–162.

Carruthers, P., & Smith, P. K. (Eds.). (1996). *Theories of theories of mind.* Cambridge: Cambridge University Press.

Chartrand, T. L., & Bargh, J. A. (1999). The chameleon effect: The perception–behavior link and social interaction. *Journal of personality and social psychology, 76*(6), 893.

Chiao, J. Y., & Mathur, V. A. (2010). Intergroup empathy: How does race affect empathic neural responses? *Current Biology, 20*(11), R478–R480.

Cikara, M., Bruneau, E. G., & Saxe, R. R. (2011). Us and them: Intergroup failures of empathy. *Current Directions in Psychological Science, 20*(3), 149–153.

Cikara, M., Bruneau, E., Van Bavel, J. J., & Saxe, R. (2014). Their pain gives us pleasure: How intergroup dynamics shape empathic failures and counter-empathic responses. *Journal of Experimental Social Psychology, 55*, 110–125.

Clements, W. A., & Perner, J. (1994). Implicit understanding of belief. *Cognitive Development, 9*(4), 377–395.

Coplan, A. (2011). Will the real empathy please stand up? A case for a narrow conceptualization. *The Southern Journal of Philosophy, 49*, 40–65.

Dahl, R. E. (2004). Adolescent brain development: A period of vulnerabilities and opportunities. Keynote address. *Annals of the New York Academy of Sciences, 1021*(1), 1–22.

Darwall, S. (1998). Empathy, sympathy, care. *Philosophical Studies, 89*(2), 261–282.

Davies, M., & Stone, T. (1995a). *Folk psychology: The theory of mind debate.* Oxford: Blackwell.

Davies, M., & Stone, T. (1995b). *Mental simulation: Evaluations and applications.* Oxford: Blackwell.

De Brigard, F., Spreng, R. N., Mitchell, J. P., & Schacter, D. L. (2015). Neural activity associated with self, other, and object-based counterfactual thinking. *Neuroimage, 109*, 12–26.

Debes, R. (2010). Which empathy? Limitations in the mirrored "understanding" of emotion." *Synthese, 175*(2), 219–239.

Debes, R. (2017). Empathy and mirror neurons. In H. L. Maibom (Ed.), *The Routledge handbook of philosophy of empathy* (pp. 54–63). New York: Routledge.

Decety, J. (2010). To what extent is the experience of empathy mediated by shared neural circuits? *Emotion Review, 2*(3), 204–207.

Decety, J., & Meyer, M. (2008). From emotion resonance to empathic understanding: A social developmental neuroscience account. *Development and Psychopathology, 20*(4), 1053–1080.

Decety, J., & Svetlova, M. (2012). Putting together phylogenetic and ontogenetic perspectives on empathy. *Developmental Cognitive Neuroscience, 2*(1), 1–24.

Denham, S. A. (1986). Social cognition, prosocial behavior, and emotion in preschoolers: Contextual validation. *Child Development, 57*(1), 194–201.

Di Pellegrino, G., Fadiga, L., Fogassi, L., Gallese, V., & Rizzolatti, G. (1992). Under-standing motor events: A neurophysiological study. *Experimental Brain Research, 91*(1), 176–180.

Eisenberg, N. (2000). Emotion, regulation, and moral development. *Annual Review of Psychology, 51*(1), 665–697.

Eisenberg, N., & Eggum, N. D. (2009). Empathic responding: Sympathy and personal distress. *The Social Neuroscience of Empathy, 6*, 71–83.

Eisenberg, N., & Fabes, R. A. (1991). Prosocial behavior and empathy: A multi-method developmental perspective. In M. S. Clark (Ed.), *Review of personality and social psychology* (Vol. 12, pp. 34–61). London: Sage.

Eisenberg, N., Fabes, R. A., Murphy, B., Karbon, M., Maszk, P., Smith, M., . . . Suh, K. (1994). The relations of emotionality and regulation to dispositional and situational empathy-related responding. *Journal of Personality and Social Psychology, 66*(4), 776.

Eisenberg, N., Spinrad, T. L., & Sadovsky, A. (2006). Empathy-related responding in children. In M. Killen & J. G. Smetana (Eds.), *Handbook of moral development* (pp. 517–549), Mahwah, NJ: Lawrence Erlbaum Associates.

Fabes, R. A., Eisenberg, N., Jones, S., Smith, M., Guthrie, I., . . . Friedman, J. (1999). Regulation, emotionality, and preschoolers' socially competent peer interactions. *Child Development, 70*(2), 432–442.

Feshbach, N. D., & Roe, K. (1968). Empathy in six- and seven-year-olds. *Child Development, 39*(1), 133–145.

Fodor, J. A. (1983). *The modularity of mind.* Cambridge, MA: MIT Press.

Frith, C. D., & Frith, U. (1999). Interacting minds—a biological basis. *Science, 286*(5445), 1692–1695.

Gallese, V. (2003). The roots of empathy: The shared manifold hypothesis and the neural basis of intersubjectivity. *Psychopathology, 36*(4), 171–180.

Garnham, W. A., & Ruffman, T. (2001). Doesn't see, doesn't know: Is anticipatory looking really related to understanding or belief? *Developmental Science, 4*(1), 94–100.

Geangu, E., Benga, O., Stahl, D., & Striano, T. (2010). Contagious crying beyond the first days of life. *Infant Behavior and Development, 33*(3), 279–288.

Goldman, A. I. (2006). *Simulating minds: The philosophy, psychology, and neuroscience of mindreading.* New York: Oxford University Press.

Gordon, R. M. (1995). Sympathy, simulation, and the impartial spectator. *Ethics, 105*(4), 727–742.

Gutsell, J. N., & Inzlicht, M. (2010). Empathy constrained: Prejudice predicts reduced mental simulation of actions during observation of outgroups. *Journal of Experimental Social Psychology, 46*(5), 841–845.

Heyes, C. (2014). False belief in infancy: A fresh look. *Developmental Science, 17*(5), 647–659.

Hoffman, M. L. (1984). Interaction of affect and cognition in empathy. In C. Izard, J. Kagan, & R. Zajonc (Eds.), *Emotions, cognition, and behavior* (pp. 103–131). New York: Cambridge University Press.

Hoffman, M. L. (2000). *Empathy and moral development: Implications for caring and justice.* Cambridge: Cambridge University Press.

Hoffman, M. L. (2001). *Empathy and moral development: Implications for caring and justice.* Cambridge: Cambridge University Press.

Iyengar, S., & Westwood, S. J. (2015). Fear and loathing across party lines: New evidence on group polarization. *American Journal of Political Science, 59*(3), 690–707.

Jackson, P. L., Meltzoff, A. N., & Decety, J. (2005). How do we perceive the pain of others? A window into the neural processes involved in empathy. *NeuroImage, 24*(3), 771–779.

Johnson, D. B. (1982). Altruistic behavior and the development of the self in infants. *Merrill-Palmer Quarterly, 28*(3), 379–388.

Kagan, J. (1981). *The second year: The emergence of self-awareness.* Cambridge, MA: Harvard University Press.

Keysers, C., & Perrett, D. I. (2004). Demystifying social cognition: A Hebbian perspective. *Trends in Cognitive Sciences, 8*(11), 501–507.

Lamm, C., Batson, C. D., & Decety, J. (2007). The neural substrate of human empathy: Effects of perspective-taking and cognitive appraisal. *Journal of Cognitive Neuroscience, 19*(1), 42–58.

Lawrence, A. D., Calder, A. J., McGowan, S. W., & Grasby, P. M. (2002). Selective disruption of the recognition of facial expressions of anger. *Neuroreport, 13*(6), 881–884.

Lee, K. H., & Siegle, G. J. (2009). Common and distinct brain networks underlying explicit emotional evaluation: A meta-analytic study. *Social Cognitive and Affective Neuroscience, 7*(5), 521–534.

Leslie, A. M. (1987). Pretense and representation: The origins of "theory of mind." *Psychological Review, 94*(4), 412.

Lewis, M., & Ramsay, D. (2004). Development of self-recognition, personal pronoun use, and pretend play during the 2nd year. *Child Development, 75*(6), 1821–1831.

Lombardo, M. V., Chakrabarti, B., Bullmore, E. T., Wheelwright, S. J., Sadek, S. A., . . . Baron-Cohen, S. (2010). Shared neural circuits for mentalizing about the self and others. *Journal of Cognitive Neuroscience, 22*(7), 1623–1635.

Maibom, H. L. (2009). Feeling for others: Empathy, sympathy, and morality. *Inquiry, 52*(5), 483–499.

Maibom, H. L. (2012). The many faces of empathy and their relation to prosocial action and aggression inhibition. *Wiley Interdisciplinary Reviews: Cognitive Science, 3*(2), 253–263.

Maibom, H. L. (2017). Affective empathy. In H. L. Maibom (Ed.), *The Routledge handbook of philosophy of empathy* (pp. 22–32). New York: Routledge.

Moore, C., Mealiea, J., Garon, N., & Povinelli, D. J. (2007). The development of body self-awareness. *Infancy, 11*(2), 157–174.

Neisser, U. (1991). Two perceptually given aspects of the self and their development. *Developmental Review, 11*(3), 197–209.

Nichols, S., & Stich, S. (2003). *Mindreading: An integrated account of pretence, self-awareness, and understanding other minds.* Oxford Cognitive Science Series. Oxford: Oxford University Press.

Nichols, S. R., Svetlova, M., & Brownell, C. A. (2009). The role of social understanding and empathic disposition in young children's responsiveness to distress in parents and peers. *Cognition, Brain, Behavior: An Interdisciplinary Journal, 13*(4), 449.

Nichols, S. R., Svetlova, M., & Brownell, C. A. (2015). Toddlers' responses to infants' negative emotions. *Infancy, 20*(1), 70–97.

Noddings, N. (2010). Complexity in caring and empathy. *Abstracta, 6*(2), 6–12.

Paul, L. A. (2017). First personal modes of presentation and the structure of empathy. *Inquiry, 60*(3), 189–207.

Premack, D., & Woodruff, G. (1978). Does the chimpanzee have a theory of mind? *Behavioral and Brain Sciences, 1*(4), 515–526.

Preston, S. D., & De Waal, F. B. M. (2002). Empathy: Its ultimate and proximate bases. *Behavioral and Brain Sciences, 25*(1), 1–20.

Prinz, J. (2011a). Against empathy. *The Southern Journal of Philosophy, 49*, 214–233.

Prinz, J. (2011b). Is empathy necessary for morality. *Empathy: Philosophical and Psychological Perspectives, 1*, 211–229.

Rakoczy, H. (2012). Do infants have a theory of mind? *British Journal of Developmental Psychology, 30*(1), 59–74.

Read, H. (2019). A typology of empathy and its many moral forms. *Philosophy Compass, 14*(10), e12623.

Rizzolatti, G., & Craighero, L. (2004). The mirror-neuron system. *Annual Review of Neuroscience, 27*(1), 169–192.

Rochat, P. (1995). Early objectification of the self. In P. Rochat (Ed.), *Advances in psychology, 112. The self in infancy: Theory and research* (pp. 53–71). Amsterdam: Elsevier.

Rothbart, M. K. (2007). Temperament, development, and personality. *Current Directions in Psychological Science, 16*(4), 207–212.

Ruffman, T., Lorimer, B., & Scarf, D. (2017). Do infants really experience emotional contagion? *Child Development Perspectives, 11*(4), 270–274.

Sagi, A., & Hoffman, M. L. (1976). Empathic distress in the newborn. *Developmental Psychology, 12*(2), 175.

Samson, D., Apperly, I. A., Chiavarino, C., & Humphreys, G. W. (2004). Left temporoparietal junction is necessary for representing someone else's belief. *Nature Neuroscience, 7*(5), 499.

Saxe, R., & Wexler, A. (2005). Making sense of another mind: The role of the right temporo-parietal junction. *Neuropsychologia, 43*(10), 1391–1399.

Scholl, B. J., & Leslie, A. M. (1999). Modularity, development and "theory of mind." *Mind and Language, 14*(1), 131–153.

Schumann, K., Zaki, J., & Dweck, C. S. (2014). Addressing the empathy deficit: Beliefs about the malleability of empathy predict effortful responses when empathy is challenging. *Journal of Personality and Social Psychology, 107*(3), 475.

Scott, R. M., & Baillargeon, R. (2017). Early false-belief understanding. *Trends in Cognitive Sciences, 21*(4), 237–249.

Singer, T., & Klimecki, O. M. (2014). Empathy and compassion. *Current Biology, 24*(18), R875–R878.

Singer, T., & Lamm, C. (2009). The social neuroscience of empathy. *Annals of the New York Academy of Sciences, 1156*(1), 81–96.

Singer, T., Seymour, B., O'Doherty, J., Kaube, H., Dolan, R. J., & Frith, C. D. (2004). Empathy for pain involves the affective but not sensory components of pain. *Science, 303*(5661), 1157–1162.

Slote, M. (2007). *The ethics of care and empathy.* New York: Routledge.

Slote, M. (2010). *Moral sentimentalism.* New York: Oxford University Press.

Spaulding, S. (2012). Mirror neurons are not evidence for the simulation theory. *Synthese, 189*(3), 515–534.

Spaulding, S. (2013). Mirror neurons and social cognition. *Mind and Language, 28*(2), 233–257.

Spaulding, S. (2016). Simulation theory. In A. Kind (Ed.), *The Routledge handbook of philosophy of imagination* (pp. 262–273). New York: Routledge.

Spaulding, S. (2017). Cognitive empathy. In H. L. Maibom (Ed.), *The Routledge handbook of philosophy of empathy* (pp. 13–21). New York: Routledge.

Sreenivasan, G. (2020). *Emotion and virtue*. Princeton: NJ: Princeton University Press.

Svetlova, M., Nichols, S. R., & Brownell, C. A. (2010). Toddlers' prosocial behavior: From instrumental to empathic to altruistic helping. *Child Development, 81*(6), 1814–1827.

Thompson, C., Barresi, J., & Moore, C. (1997). The development of future-oriented prudence and altruism in preschoolers. *Cognitive Development, 12*(2), 199–212.

Trevarthen, C. (1979). Communication and cooperation in early infancy: A description of primary intersubjectivity. *Before Speech: The Beginning of Interpersonal Communication, 1*, 530–571.

Tropp, L. R., & Saxena, S. (2018). Re-weaving the social fabric through integrated schools: How intergroup contact prepares youth to thrive in a multicultural society. Brief No. 13. https://eric.ed.gov/?id=ED603699

Vaish, A., Carpenter, M., & Tomasello, M. (2009). Sympathy through affective perspective taking and its relation to prosocial behavior in toddlers. *Developmental Psychology, 45*(2), 534.

Weisz, E., & Zaki, J. (2017). Empathy building interventions: A review of existing work and suggestions for future directions. In E. M. Seppälä, E. Simon-Thomas, S. L. Brown, & M. C. Worline (Eds.), *The Oxford handbook of compassion science* (pp. 205–217). New York: Oxford University Press.

Wellman, H. M. (2015). *Making minds: How theory of mind develops*. New York: Oxford University Press.

Wellman, H. M., Cross, D., & Watson, J. (2001). Meta analysis of theory of mind development: The truth about false belief. *Child Development, 72*(3), 655–684.

Wicker, B., Keysers, C., Plailly, J., Royet, J. P., Gallese, V., & Rizzolatti, G. (2003). Both of us disgusted in my insula: The common neural basis of seeing and feeling disgust. *Neuron, 40*(3), 655–664.

Wimmer, H., & Perner, J. (1983). Beliefs about beliefs: Representation and constraining function of wrong beliefs in young children's understanding of deception. *Cognition, 13*(1), 103–128.

Xu, X., Zuo, X., Wang, X., & Han, S. (2009). Do you feel my pain? Racial group membership modulates empathic neural responses. *Journal of Neuroscience, 29*(26), 8525–8529.

Zahn-Waxler, C., Radke-Yarrow, M., Wagner, E., & Chapman, M. (1992). Development of concern for others. *Developmental Psychology, 28*(1), 126.

Zaki, J. (2014). Empathy: A motivated account. *Psychological Bulletin, 140*(6), 1608.

Zaki, J. (2018). Empathy is a moral force. In K. Gray & J. Graham (Eds.), *Atlas of moral psychology* (pp. 49–58). New York: Guilford Press.

Zaki, J., & Cikara, M. (2015). Addressing empathic failures. *Current Directions in Psychological Science, 24*(6), 471–476.

Zaki, J., & Ochsner, K. N. (2012). The neuroscience of empathy: Progress, pitfalls and promise. *Nature Neuroscience, 15*(5), 675.

Zaki, J., Ochsner, K. N., Hanelin, J., Wager, T. D., & Mackey, S. C. (2007). Different circuits for different pain: Patterns of functional connectivity reveal distinct networks for processing pain in self and others. *Social Neuroscience, 2*(3–4), 276–291.

3   I Did That! Biomarkers of Volitional and Free Agency

Robyn Repko Waller and Maj. Allison Brager

Late at night, I opened a second bottle of wine. Did I do that? How can I tell whether I was the agent or whether my conscious choice caused me to drink more? What did cause me to drink more if not my conscious will? Was I too tired to think straight? This chapter attempts to answer questions such as these.

In section 3.1, we sketch a philosophical account of the significance and nature of agency, from both metaphysical and phenomenological points of view. In section 3.2, we discuss how the neuroscience of agency—how agents in fact exercise their agency—can bear upon philosophy of agency—accounts of what capacities or conditions would enable or prohibit free agency. Here, we review established and recent neuroscientific findings on agential control and the experience of agency. In section 3.3, we seek to establish a new avenue in interdisciplinary research on agency. Here, we outline how sleep deprivation can impact sense of agency ("I did that!"). In drawing this connection, we propose new research that bridges sleep research with well-established cognitive neuroscientific paradigms concerning voluntary action. Finally, we consider the importance of such new avenues for philosophical theories of free will and moral responsibility.

3.1   Metaphysics and Phenomenology of Agency

Sense of agency is an inextricable feature of human life. When we decide what to do—for instance, decide to pick up that cup of tea—we typically unreflectively believe that our decision or choice is up to us. When we move about in the world—for example, cycling along the roadside—we typically unreflectively believe that it is up to us when and how we move. Here, "up to us" indicates a kind of control over our decisions and actions. Intentional decisions and actions are goal-directed activities over which the agent has some control.[1] We humans as social beings take it that, typically, other humans too have this kind of control over their decisions and actions. Moreover, we take it that humans extend this control not only over innocuous matters, such as sipping tea, but also over morally weighty ones, such as whether and how to care for and treat our dependents, affiliates, and distant others.

If we believe that at least some of our decisions and actions, especially our intentional bodily movements and their consequences, are up to us, we may believe that we are freely deciding or acting, at least at times. Accounts of free will differ widely, but most free will theorists take free will to be the ability to act freely. An agent who has free will has the appropriate kind of control over her decisions and actions to be an apt target of responsibility practices (see, e.g., Fischer & Ravizza, 1998). If one is an apt target of our moral responsibility practices, then one qualifies as the type of agent deserving of moral blame (for a morally bad action) and moral praise (for a morally good action; see, e.g., Fischer & Tognazzini, 2011; Watson, 1996). Here, by "responsibility practices," we will understand these practices broadly to include not only moral practices of blaming or praising an agent, but also legal practices of punishment. Others take the control typically associated with responsibility to be valuable in its own right as grounding our human nature as self-governing agents.[2]

Exercising one's free will involves control, or influence, over oneself. But what exactly must an agent control in order to exercise free will? One answer is that free will requires control over one's *actions* (and perhaps the consequences of those actions). An agent exercises control over her action insofar as whether she does the action depends on her internal state of mind. This sense of free will is sometimes referred to as *free action*. Here, by "action," we mean overt action, such as a bodily movement and the effects of that movement. This makes sense. When we hold people responsible—for example, blame them, punish them, feel gratitude toward them—we tend to hold them responsible for their conduct, or actions, and the repercussions of those actions. So, on (at least) a backward-looking notion of responsibility, we care about whether the agent had control over her actions. Why?

For instance, suppose on a crowded sidewalk Jan shoves into a group of pedestrians, causing an elderly pedestrian in the group to fall onto a crosswalk and sustain a serious injury. If we come to find out that Jan had a seizure and her "shoving" was the result of that unexpected seizure, it would be strange and inappropriate to feel (or express) blame or anger or indignation at Jan. Likewise, we may take a similar view of the incident if we come to find out that Jan herself was pushed from behind by a crowd crush, thereby causing her to shove into the pedestrians. Although the incident was unfortunate for all involved, Jan's bodily movement and the resulting shove were not her own or "up to her"—she, the agent, did not authorize or causally initiate the action, she did not desire or intend to shove into others or recklessly endanger them. She did not control her action.

Now, suppose instead that circumstances were different: Jan is late to a Broadway play, one for which she has paid significant money. She doesn't want to miss the curtain's going up and the doors closing for seating. She runs down the crowded sidewalk and, knowingly, straight into a group of pedestrians. The elderly pedestrian falls and sustains a serious injury. Here, of course, the group of pedestrians and especially the injured party should be permitted to, and likely would, express blame, anger, resentment (from the injured party), and indignation (from the spectators) at Jan for her reckless action and the resulting injury. To achieve her goal of not missing the opening curtain, Jan intentionally ran down the crowded sidewalk, knowing that she might well knock into others and that someone might get injured (even if Jan didn't intend for others to get injured). Hence, her shoving of the pedestrians and the elderly pedestrian's injury are the result of how she intentionally moved her body, an exercise of control over her actions. Moreover, the action expresses something about Jan as an agent: she was reckless, knowingly taking the risk of injuring others. In this way, the action was up to Jan.

So, it seems that the agent's mental states—goals, desires, perceptions—at the time of action, in addition to how she moves her body, play a central role in whether and to what extent she controls her actions and so acts freely. But now one may reasonably ask the following: Even if an agent *acts freely*, how much control does the agent have over her state of mind? In particular, are agents' decisions really up to them? That is, regardless of whether an agent performs a free action in the sense of her overt movements, we can still ask if the agent *decided freely*.[3] We can make progress on these questions of control over one's actions and decisions via the empirical work on the neuroscience of intentional bodily movement.

3.2 Neuroscience of Intentional Action

We noted that the kinds of actions that are "up to us," so to speak, are those that come (at least partly) from the agent—and not wholly external sources—and further issue from the agent's state of mind—her goals, desires, beliefs, and perceptions. There is a strong conceptual tie between actions (and consequences) that are up to us and actions for which agents are held responsible. For this reason, we don't find it appropriate to blame Jan for injuring the elderly pedestrian in the event that Jan too was shoved from behind by the crowd. Her body moving as a result of a shove is a *passive and involuntary* movement caused by external forces. Furthermore, not all movements that are internally generated qualify as voluntary—think of Jan's seizure-induced push into the crowd. Although the push into the crowd is the product of her (internal) neural and physiological processes, it isn't produced in the right way—from her volitional states and capacities—to count as a voluntary action. It is only when we find out that Jan, the agent herself, initiated her shoving into the crowd that Jan is seen as having acted voluntarily and with the kind of control and mind-set that makes her worthy of blame for the injury that results. But how, then, do we sort out the neural mechanisms that underscore voluntary "up to us" movements as opposed to involuntary ones?

Fittingly, the neuroscience of agency provides a rich source of evidence concerning the mechanisms underlying voluntary actions, especially endogenously generated ones. Endogenously generated actions are actions for which the cue for acting does not come from the experimenter or other external stimuli but rather from within the participant—for example, within a designated window, a participant decides whether or when to execute an overt movement such as pressing a button. When we as agents execute endogenously generated movements, we experience ourselves as in control of those movements and their immediate effects in the world. We typically believe that it is indeed up to us when and how we act, at least some of the time. Moreover, we not only believe that what we do is the result of our decisions, but that what we do is frequently the result of decisions we are aware of. Hence, one commonsense assumption about endogenously generated actions, in line with widespread belief, is that the agent consciously "wills" or decides how and when to act, and that this decision initiates the process leading to action. Call the former phenomenology of control "sense of agency" and the latter assumption "the efficacy of conscious intentions." In the next section, we will discuss two broad neuroscientific paradigms that inform our understanding of the neural processes that subserve these aspects of agency. The intentional binding paradigm, detailed first here, provides an implicit measure of sense of agency. The Libet paradigm, outlined second, investigates the efficacy of conscious intentions in relation to the neural preparatory processes leading to action.

3.2.1 Intentional Binding as Implicit Sense of Agency

The most intuitive method in understanding sense of agency would be to solicit self-reports of the degree to which people see themselves instead of another as the agent who caused an event (e.g., a 100-point scale from "I allowed [event] to happen" to "I intended [event]"; Wegner & Wheatley, 1999). However, research on explicit sense of agency ("I did that") has uncovered a cognitive bias impacting such judgments (Wegner & Wheatley, 1999). Participants who are asked in a social setting whether an outcome was the result of their agency or that of another person tend to misattribute unrelated events to their own agency, especially if the event is a positive one. Further, individuals exhibit reduced explicit sense of agency for negative outcomes than for positive ones (see, e.g., Bradley, 1978; Greenberg, Pyszczynski, & Solomon, 1992; Bandura, 1982, 2001).

In contrast with explicit self-report, implicit measures of sense of agency, including intentional binding, attempt to avoid this self-serving bias (for a review of sense of agency, see Haggard, 2017). Intentional binding is understood in terms of how much the time interval between perceived time of action and perceived time of outcome (of that action) shifts in conditions of voluntary action (the active condition) compared to conditions of involuntary movement (the passive condition).

In the active experimental condition, participants are asked to execute an endogenously generated action, such as a button press, at a time of their choosing. A tone sounds 250 milliseconds after the button is pressed, and the actual time of their action is recorded. Participants watch a modified clockface as they complete the task. Participants report in separate blocks either the time of the button press or the time of the tone. In the passive experimental condition, using transcranial magnetic stimulation (TMS), experimenters induce a finger twitch in the participants with a tone sounding 250 milliseconds later. Similar to the active condition, participants report in separate blocks either the time of the finger twitch or the time of the tone. In comparison, in baseline conditions, participants either execute an action (with no resultant tone) or listen for a tone (with no action) and report the time of the action or the tone. Intentional binding refers, then, to the phenomenon whereby the perceived time of action and tone shift closer to each other—bind together—when agents perform voluntary actions (relative to baseline time judgments; Haggard, Clark, & Kalogeras, 2002; Caspar et al., 2015; Haggard & Clark, 2003; see figure 3.1). Hence, intentional binding marks a difference in sense of agency between actions and non-actions, as well as among kinds of actions.

It should be noted that the designation of this temporal binding as action specific—indicative of an agent's perception of her own voluntary action—is not univocal. In particular, there is an active debate as to the scope of intentional binding. Does the shift together in perceived timing of events accompany just actions and actional outcomes or, rather, more broadly the timing of physical causes with their sensory effects? In support of the latter wider scope, some studies are suggestive of temporal binding as a marker of observed physical causation more broadly (Buehner & Humphreys, 2009; Buehner, 2012, 2015; Kirsch, Kunde, & Herbort, 2019; Suzuki et al., 2019).

In favor of the narrower-scope view of intentional binding as a marker of perceived self-agency, several studies link intentional binding specifically to the motor system and to explicit judgment of agency. Such narrower-scope proponents argue that the kinds of experimental task features that enhance or attenuate intentional binding provide evidence for this action-specific domain of intentional binding (Bonicalzi & Haggard, 2019; Haggard, 2019). For instance, intentional binding is enhanced—that is, the perceived time of the action and the perceived time of the outcome are closer together—to the degree that participants have more courses of action available from which to choose. In one study, the intentional binding effect for voluntary actions was strongest when participants chose from seven button response options, and a three-button response choice evidenced enhanced intentional binding over the one-button option (e.g., push the prescribed response button when you feel like it; Barlas & Obhi, 2013). This finding of open versus instructed action was replicated and extended to support that not only the availability of alternative options but also the perceived valence of the option can impact intentional binding. Participants who are given a "free" choice of options evidence the strongest intentional binding effect when

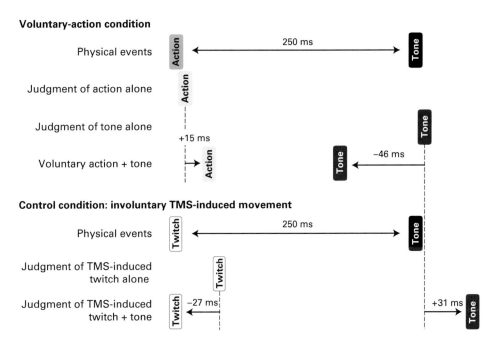

Figure 3.1
This figure illustrates the intentional binding effect. Participants watch a sped-up
analog clock. In the experimental voluntary action condition, participants execute
an endogenously generated movement, a button press, at a time of their choosing.
A tone sounds 250 milliseconds after the button is pressed. Using the clock, in separate blocks, participants judge either the time of the action or the time of the tone. In
the baseline conditions, participants execute a button press with no resultant tone or
hear a tone with no preceding movement. Participants judge the time of the stand-
alone event. In the involuntary TMS movement condition, experimenters induce a
finger twitch in the participants with TMS, with a tone sounding 250 milliseconds
late. Like the voluntary condition, in separate blocks, participants judge the time of
the time of the twitch or the time of the tone. In voluntary active conditions, the
time of the movement and the time of the tone bind together in perceptual temporal
space compared to baseline and involuntary conditions.

the outcome of action is perceived as pleasant (here operationalized as the
pleasantness of the tone). Further, the availability of alternatives and the
valence of the outcome influence the degree to which participates explicitly feel that they are in control (Barlas, Hockley, & Obhi, 2018). The emotional valence, positive or negative, of an outcome vocal tone more generally
affects intentional binding for voluntary actions, with enhanced intentional

binding for positive valence tones and attenuated intentional binding for negative tones (Yoshie & Haggard, 2013, 2017; but see Moreton, Callan, & Hughes, 2017, for conflicting findings). However, importantly, the valence of an outcome for a third-party only, and not the agent, does not impact the agent's implicit sense of agency in terms of intentional binding (Haggard, 2019).[4] In line with these results, Borhani, Beck, and Haggard (2017) found that more painful (temperature) stimulus outcomes for the agent attenuated intentional binding, but this attenuation was weakened for free choice (trials with options) and active (vs. passive) trials. Hence, having options and the value of one's options seem to play a role in generating a stronger (implicit) sense of agency over one's actions and outcomes.

Interestingly, social and emotional factors that impact the likelihood of acting (Latane & Darley, 1968) and the exercise of self-control also modulate both explicit and implicit sense of agency. Recent work suggests that explicit judgments of control are impacted by the presence of others. Participants report reduced feelings of control over consequences of their actions when acting in the company of others who did not act but could have performed the action in place of the agent (Beyer et al., 2017). Implicit sense of agency in terms of intentional binding is attenuated in the lab in a number of situations that impact both exercised and experienced agential control in real life. Participants evidence a weakened temporal binding between voluntary actions and their outcomes when acting under orders in a coercive paradigm (Caspar et al., 2016), when induced to feel fearful or angry (Christensen et al., 2019), and when acting under the influence of alcohol (De Pirro et al., 2019).

We have surveyed how intentional binding can capture aspects of our experience of control and how this experience varies, but what is the neural basis of this experience? Brain studies demonstrate that the intentional binding effect is produced or affected by activity in the frontal and parietal cortex. Noninvasive continuous theta burst stimulation (cTBS) of the pre-supplementary motor area (pre-SMA) using TMS—a procedure that inhibits locally targeted neural activity—disrupts intentional binding compared to control cTBS of other areas (Moore et al., 2010). Furthermore, single-pulse TMS over the left inferior parietal lobe disrupts sense of agency for outcomes of voluntary actions when the participants choose among action options. Here, Chambon, Moore, and Haggard (2015) argue that this disruption to implicit sense of agency in a choice paradigm may reflect the disruption of the functional network involving the dorsolateral prefrontal cortex (dlPFC) and the inferior parietal lobe subserving premovement action selection.

This research on sense of agency still does not give us a full picture of how an action can be up to the agent. As Bonicalzi and Haggard (2019) aptly highlight, self-initiated actions—such as the kind at issue in the free will and responsibility literature—are not characterized simply by the presence of subjective aspects of agency such as explicit and implicit sense of control. Rather, to appreciate fully how an action can be up to the agent, we will need grounding in the kinds of internal activity that are causal contributors and/or constitutive of agent-authored action. Accordingly, the next section will review the neuroscience of action initiation for voluntary (i.e., endogenously generated) actions.

3.2.2 Libet's Paradigm and the Initiation of Intentional Action

The commonsense model, as discussed above, takes it that the kinds of actions that we classify as up to us are not just those that are internally generated, but those that, in addition, are *agent initiated*. This claim in the metaphysics of free action suggests that for our movements to be up to us, the agent should be the causal initiator or at least a prominent causal contributor to the preparation for movement and subsequent movement. For self-initiated actions, the agent consciously wills or decides when and how to act, and that volition or decision initiates and (at least in part) causes the preparatory processes leading to movement. For instance, when I pick up my cup of coffee for a sip, it seems to me that first I, the agent, consciously decide to have a sip of coffee now, and then and only then do I "hand off" control to my neural action preparation circuits to bring it about that I pick up the coffee cup. The empirical question thus became: When does the conscious intention to move appear in relation to neural preparation to move and subsequent movement?

This question has been addressed by a series of experiments. Using electro-encephalography (EEG), Kornhuber and Deecke (1965) discovered that when participants make simple movements in the lab, these movements are preceded by a slow negative buildup of precentral and parietal cortical activity bilaterally, termed the "Bereitschaftpotential" or "readiness potential" (RP). The RP is taken to be indicative of preparation for action and so a motor-specific neural signature. Libet, Wright, and Gleason (1982) extended these findings to demonstrate that the RP preceded specifically endogenously generated simple movements (e.g., wrist or finger flexing). How and when does the agent's conscious intention to move come on the scene for such movements?

Libet and colleagues (1983) asked participants to perform a series of endogenously generated spontaneous (not preplanned) wrist flexes as they felt the desire or intention to do so.[5] The beginning of each trial was signaled by a beep. Post flexing, participants reported the time of their awareness of their intention to flex in each trial by indicating where a moving spot was on a modified clockface when they were first aware of their intention to flex. This modified clock—now referred to as a Libet clock—makes a complete revolution every 2.56 seconds.[6] Averaging over forty trials, the average time when the RP started to ramp up for these trials was about a half second (517 milliseconds) before recorded movement. In contrast, the average reported time of first awareness of intention to move, labeled "W," occurred about 206 milliseconds prior to movement ("M"). Libet and colleagues found, then, that the RP preceded the reported time of conscious intention to move by around a third of a second (see figure 3.2).

Libet and co-authors (1983) and Libet (1985) took these findings to indicate that unconscious brain processes decide (form an intention) to flex, with conscious intentions to flex temporally later. On the basis of his results, some theorists—including Libet himself (e.g., 1999)—have concluded that our conscious intentions never causally initiate our overt actions (see also

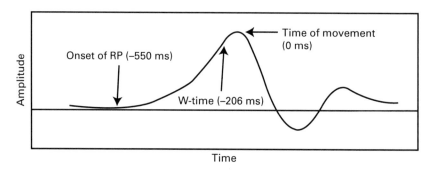

Figure 3.2
Libet's experimental results. The diagram shows the results of Libet and colleagues' (1983) experiment. The readiness potential (RP) onset occurs approximately 550 milliseconds prior to time of movement, labeled here as "0," and approximately 300 milliseconds prior to the average reported time of first awareness of intention to act, labeled here as "W."

Wegner, 2002). Libet meant this account to cover *all* overt actions, including complex movements. Notably, Libet did not test the latency of the RP and W time in the lab for complex actions. However, Libet reasoned that if an RP is present for all overt movements and W appears after the onset of the RP in his studies, then RPs occur prior to conscious intentions and initiate action for all intentional actions, even complex ones (Libet, 2005, p. 560). This result is, notably, in tension with the commonsense model of agency where the agent via a conscious intention initiates and/or causes the preparation for action.[7]

In the years following the study by Libet and colleagues (1983), Libet-style neuroscientific investigation of free agency has become a booming experimental enterprise. Others have replicated and extended the basic finding that brain activity associated with action preparation precedes W time (including, e.g., an fMRI-based version [Soon et al., 2008] and a single-cell recording version [Fried, Mukamel, & Kreiman, 2011]). To cover all of the careful work, empirically and philosophically, on the broadly construed Libet studies here would not be feasible.[8] Nonetheless, we will briefly survey parts of this literature via an assessment of one pressing line of argument for free will skepticism relying on Libet-like studies—what we call the *"unconscious initiation threat to free agency."*[9]

Recall the claims by Libet and others that conscious intentions do not initiate the preparatory process leading to action; rather, unconscious brain processes do.[10] Call this the *empirical* premise. Supplement this empirical premise with the following prescriptive *metaphysical* premise: conscious intentions, not unconscious brain activity, must initiate actions for actions to be free.[11] From a commonsense perspective, it seems that my free agency is threatened if my unconscious brain always settles what I am to do before I am aware of my decision or consciously weigh in via deliberation.[12] That kind of internally generated action isn't enough for the action to be up to me the agent. Put together, these two premises yield the conclusion that agents do not in fact act freely.

The empirical premise above is dependent on two substantive assumptions. First, the neural activity captured by the RP is indicative of neural preparation for a committed particular movement. Second, Libet's interpretation of the findings rests on taking W, the averaged reported time of first awareness of intention to act, to be a reliable experimental measure of the time of a conscious intention to act. Otherwise, we cannot be sure that W occurring after the onset of the RP in the experimental timeline yields any significance for the timing of the conscious intentions in producing action.[13] Here, we will briefly review the evidence for and against these assumptions, and so for and against the unconscious initiation threat to free agency.

Let us first evaluate the claim that the RP is the neural signature of a (perhaps initially unconscious) practical intention. If the RP encodes an intention to move, then one might posit that there ought to be a relation between the RP qua intention to move and agents' awareness of an intention to move. Since participants in the Libet paradigm report the timing of their awareness of their intention to move, W time, one might expect a significant relationship between the time of onset, or initial ramp-up, of the RP and W. However, the evidence for covariance of RP and W is weak at best. While Jo and colleagues (2015) found that the RP co-varies with W time for a particular sub-population, committed meditators,[14] others have failed to find a significant relationship between RP and W for general populations of agents (Haggard & Eimer, 1999; Schlegel et al., 2013; Alexander, Schlegel, Sinnott-Armstrong, Roskies, Tse, et al., 2015). Intriguingly, there is some evidence, albeit not univocal, that the timing of the lateralized RP (LRP) is correlated with W time (Schlegel et al., 2013; Alexander, Schlegel, Sinnott-Armstrong, Roskies, Tse, et al., 2015). When an agent makes a unilateral movement (e.g., left movement or right movement), the LRP represents the difference in electrical potential to the left and right of the vertex prior to that movement.

Even if the timing of the RP ramp-up and reported time of conscious intention are not correlated, one might still expect that RP is sufficient for the *presence* of a conscious intention to act. That is, if an agent exhibits an RP prior to voluntary movement and the RP is related to an intention to act, then the agent will have *some* conscious intention related to her so acting (at some time prior to movement). Even this assumption, taken for granted in earlier Libet-inspired work, has been thrown into question with new empirical evidence supporting similar RP and LRP profiles for both aware and hypnotized participants (Schlegel et al., 2015). Here, the hypnotized participants lack awareness of an intention to act but still show a RP ramp-up prior to action. Recent experimental redesigns and analysis, though, have revived the case for the RP representing intention to move. Using an online—premovement—measure, Parés-Pujolràs and colleagues (2019) argue for a significant relation between the RP and conscious intention to act. Parés-Pujolràs and colleagues (2019) found RP-like ramp-up prior to the button press to report in-progress preparation to move.[15] This on-line report of conscious intentions offers a promising new prospective measure of reported time of conscious intention, especially as recent work has supported that the traditional retrospective off-line measure, W time, is vulnerable to experimental task order effects and so may be worryingly unreliable (Dominik et al., 2017; Sanford et al., 2020). This work may renew worries that W is a good operationalization of awareness of intention to move. Moreover, the finding that externally triggered movements fail to produce an RP adds to the support for the RP's link to settledness of action, in particular endogenously generated action (Jahanshahi et al., 1995).

A weaker interpretation of the Libet results is that even if the RP is not the biomarker of intention to act, it may well still be the biomarker of preparation to act or a bias toward a particular action (Mele, 2009, 2012; Shepherd, 2015). That is, the RP does not signify commitment to an action plan, but rather general readiness or, alternatively, inclination to an action plan. Several studies have challenged this motor-domain interpretation of the RP, finding that the analogue of the RP is present when participants complete non-movement-based cued cognitive decision tasks (e.g., mental addition and subtraction, imagined movement; Alexander, Schlegel, Sinnott-Armstrong, Roskies, Tse, et al., 2015; Alexander, Schlegel, Sinnott-Armstrong, Roskies, Wheatley, et al., 2016; Raś et al., 2019). In contrast, the analogue of the LRP is present for only motor tasks. In support of a motor functional role, Travers and colleagues (2020) argue that while RP-like events were ubiquitous in the EEG recordings preceding both self-initiated and no movement tasks, such RP-like events are likely to be "false-positives" generated by distinct neural mechanisms than the RPs generated by voluntary motor preparation. Here, we might agree that regardless of its motor-specific or more general function, the RP is a neural signature of initiation of voluntary (i.e., endogenously generated) action where we can broaden the tent of voluntary as inclusive of not just overt bodily movement but also purely mental endogenously generated activity, such as mental math and imagining.[16]

Taken together, the threat of unconscious initiation of action from the Libet results seems underwhelming. We at best have good grounds for taking the RP to be neural preparation for voluntary actions. However, on that count, the Libet task appears to remain a good neuroscientific framework for understanding voluntary agency. In particular, the Libet paradigm is solidly established for investigating the neural underpinnings of action initiation for simple self-authored actions. In this vein, in section 3.3, we will begin to sketch how the Libet paradigm can shed light on agency-related changes during conditions of sleep deprivation. Citing known results on agency and decision making during sleep reduction, we will construct a testable theory of the changes in the amplitude and latency of the RP prior to endogenously generated movement in sleep-deprived and non-sleep-deprived individuals. Further, relying on the above cited literature on intentional binding, we will piece together how sense of agency is impacted when agents are subject to reduced sleep. In this way, we aim to inspire new avenues of understanding on how sleep-deprived agents, a sizable contingent of the population, may exercise altered control of their actions and further experience their agency in a deviant fashion.

3.3   Effects of Sleep Deprivation on Reasoning and Agency

False confessions—suspects legally admitting to committing actions that they did not perform—is a popular trope of crime shows and sometimes, more depressingly, real-life criminal cases. Confessing to a crime is, in a sense, admitting ownership of the bodily actions and consequences that constitute the criminal act. Yes, I did that.

Are such false confessions more likely when suspects are sleep deprived? Frenda and colleagues (2016) kept participants awake for twenty-four hours. Those participants who self-reported high levels of sleepiness in the study were four-and-a-half times more likely than those who reported low levels of sleepiness to sign a confession that they had hit the escape key and thereby deleted the study data (when they had, in fact, not done so). Perhaps, though, as the researchers themselves note, these participants were not so much retrospectively reporting their genuine belief that they hit the escape key, but instead suffering from a reduced capacity to inhibit behavioral impulses and to resist suggestive influence.

Sleep deprivation might have similar effects on many other kinds of action ascription. We will discuss how sleep deprivation warps sense of agency in terms of intentional binding. We will then consider how sleep deprivation affects action control and action initiation, and how these effects bear upon action in a moral context. These two ways in which neurobiological regulation of responsible agency unravel with sleep deprivation are of prime importance for the sizable subpopulation of agents who suffer from insufficient sleep yet must perform in high-stakes contexts, such as military personnel, shift workers, college students, psychiatric populations, parents of young children, and others.

3.3.1   Sleep Deprivation

Sleep deprivation is multifaceted. Here, we will focus on *acute sleep deprivation* and *chronic sleep restriction*. Acute sleep deprivation describes a situation or environment in which wakefulness extends over twenty-four hours (Goel, Basner, et al., 2013). Executive function is impaired in a time-dependent manner across twenty-four hours of forced wakefulness (Lo et al., 2012). Because some individuals show resiliency to twenty-four hours of total sleep deprivation (TSD), researchers have found that resiliency reaches a tipping point after forty-eight hours of TSD and beyond (McCauley et al., 2013). Under this paradigm, wakefulness has been extended for sixty-two hours on average (Goel, Basner, et al., 2013), but it has extended as far as ninety-six hours (Goel, Basner, et al., 2013).

Executive function is also severely impaired by chronic restricted sleep (Goel, Abe, et al., 2014). Chronic sleep restriction describes a situation or environment in which an individual is awake for more than eighty percent of a twenty-four-hour period across more than five days (Goel, Abe, et al., 2014). Five hours of sleep for seven days is enough to put executive function in a "danger zone" if subjects report that they need seven to eight hours of sleep a night in order to perform normally (measured by self-report).

In this section, we apply these sleep paradigms first to the Libet paradigm for initiation of action and then to the intentional binding paradigm for sense of agency. Finally, we look at applications of these sleep paradigms to situations in which the agent faces a morally significant decision with real-world relevance. Although empirical work connecting sleep paradigms and the two neuroscience agency paradigms has yet to be conducted, we survey related results and sketch what we might expect to find if we were to conduct the Libet and intentional binding paradigms with acute sleep-deprived and chronic sleep-restricted agents.

3.3.2   Sleep Deprivation and Agency: What We Know
and Future Directions

Studies support compromised cognitive performance under insufficient sleep and also suggest that the functional neuroanatomy of this cognitive

disruption overlaps with the functional neuroanatomy of task competence for intentional binding and the Libet task. Specifically, these studies
show a decline in performance of tasks requiring: (1) cognitive flexibility—
deployment of mental resourcefulness in order to adapt to a dynamic environment (Honn et al., 2018); (2) flexible adaption—attentional control for
"where" and "what" (Whitney et al., 2017); and (3) response inhibition (Zhao
et al., 2019). In these studies, the dlPFC and pre-SMA were shown to underlie compromised functional connectivity manifest from insufficient sleep
(Simon et al., 2017; Yang et al., 2018). As detailed above, both the dlPFC and
pre-SMA are critical in order to perform tasks under the Libet and intentional
binding paradigms. For instance, the early RP, the biomarker of initiation of
agency in the Libet paradigm, is found bilaterally in the pre-SMA (Fried
et al., 1991). Further, although sleep is thought to be a brain-centric process,
this hypothesis has been challenged by the discovery of a sleep regulatory
pathway in skeletal muscle (Ehlen, Brager, et al., 2017). This finding bears
special relevance to the neuroscience of overt bodily actions, as considered
in section 3.2, and suggests that the effects of sleep deprivation on action
production go well beyond the cognitive pathways we review below.

How does sleep impact the neural activity of these areas to compro-
mise cognitive performance? Slow-wave brain activity (SWA), measurable
by EEG, is the basis for the recuperative value of sleep (Ehlen, Jefferson,
et al., 2013). SWA include delta waves (0.5–4 Hz), characteristic of sleep
and dream states, as well as theta waves (4–8 Hz), characteristic of states
of drowsiness. In contrast, busy or active wakeful states are characterized
by higher-frequency gamma (38–42 Hz) or beta waves (12–38 Hz), the former especially for instances of concentration (Abhang, Gawali, & Mehrotra,
2016). If a person is subject to restricted sleep, the temporal distribution
of the EEG architecture changes such that SWA intrudes into the waking
EEG (Vyazovskiy et al., 2011), in particular locally for the dlPFC and neighboring areas (Ehlen, Jefferson, et al., 2013). Specifically across extended
wakefulness, the EEG is redistributed to include more theta compared to
higher-frequency activity. Across recovery sleep following extended wakefulness, the EEG is redistributed to include more delta compared to higher-
frequency activity (Ehlen, Jefferson, et al., 2013).

The Libet paradigm also relies on brain activity measurable via EEG.
So, we could use it to measure how action initiation, understood as the
RP, is likely to change under sleep deprivation. Combining the slow-wave

intrusion into extended wakefulness and the finding that the early RP is generated in the pre-SMA, bilaterally, we should expect that agents subject to acute sleep deprivation and perhaps chronic sleep restriction will have more positive (i.e., smaller amplitude) and later latency RPs than agents who are not sleep deprived. This pattern would indicate weaker and delayed intentions to act or preparation for action, depending on how the RP is interpreted. Further, we might also expect that agents who are sleep deprived on average have delayed W times, or reported times of first awareness of their intention to move, compared to non-sleep-deprived agents. However, this finding by itself would not necessarily indicate that awareness of one's intentions is impaired under lack of sleep. Whether this effect is specific to agency would depend on whether agents with insufficient sleep also evidence a delayed M time, reported time of their movement. If they do, then it is not the case that an aspect of sense of agency has been impacted by a lack of sleep as much as that more general attentional capacities are impaired with a lack of sleep. In any case, insofar as action control involves initiation of action and preparation for action, there are good grounds for thinking that sleep deprivation will reduce action control, at both the subpersonal and personal levels.

To answer the question of how sleep deprivation impacts sense of agency specifically, we can turn to how agents' performance on the intentional binding task might be affected by insufficient sleep. Previous research found that when participants were subject to twenty-four hours of forced wakefulness, participants gave enhanced explicit sense of agency judgments for others' actions (Hon & Po, 2016). But given the self-serving bias moderating explicit sense of agency judgments, it would be useful to know how sleep deprivation impacts implicit sense of agency. This issue can be considered with respect to both the action-binding and effect-binding elements of intentional binding under conditions of sleep deprivation (or sufficient sleep). *Action-binding* refers to the perceived shift forward of the time of action toward the outcome in voluntary conditions. *Effect-binding* refers to the perceived shift backward of the time of the outcome toward the action (cause) in voluntary conditions. We predicted above that agents who are acutely sleep deprived or chronically sleep restricted should on average have more positive RPs (with smaller amplitude in a negative direction) than non-sleep-deprived agents. Interestingly, more positive RPs are associated with attenuated intentional binding, particularly for effect binding (Jo et al.,

2014). Hence, we expect that insufficient sleep should lead to attenuated intentional binding, especially for effect binding. If so, agents who act under sleep deprivation would evidence weakened (implicit) sense of agency.

This weakened sense of agency at the level of perception of event timing may well be significant for morally weighty exercises of agency, as the phenomenology of agency—the experience of agency—is taken to ground the capacity for choice (cf. the debate on deliberative compatibilism; e.g., Nelkin, 2004) and the capacity for free decisions and actions. For instance, the seeming that one is deciding in the face of genuinely open alternatives and is in control of actions and outcomes drives the motivation and development of accounts of free will (McKenna & Coates, 2016). Changes in perception of one's agency, then, may impact the kinds of decisions and actions that agents attribute to themselves and hence whether observers attribute responsibility for those decisions and actions to those agents. Indeed, given that certain subpopulations are subject to insufficient sleep non-culpably as part of the conditions of their employment, we ought to pay attention to how sleep deprivation can impact agency and responsibility judgments in moral contexts and also what mediates these relationships. The latter question of mediation is beyond the scope of this review but may be relevant for how to fit together changes in agency and sense of agency with changes in moral judgment under sleep deprivation. In fact, here, there is a growing neuroscience literature on sleep deprivation and moral reasoning and judgment to which we now turn to conclude our review.

3.3.3   Sleep Deprivation and Moral Reasoning: What We Know and Future Directions

Though limited, a handful of studies to date have observed impairment in moral reasoning under sleep deprivation in healthy civilians (Killgore, Killgore, et al., 2007; Tempesta et al., 2012) and military officers (Olsen, 2013; Olsen, Palleson, & Eid, 2010). These studies did not explore the neural basis of their behavioral findings. However, similarities between the brain areas that are affected by sleep deprivation and those that are used for moral judgments suggest that sleep deprivation could have profound effects on moral judgment.

In the most recent decade, substantial salience has been placed on the study of PFC processes and their role in impaired performance under sleep deprivation. A simple keyword search for "sleep deprivation AND prefrontal cortex AND humans" in the National Library of Medicine database (PubMed) revealed 169 articles. The negative consequences of sleep deprivation on PFC processes that have been discovered in the last year are multidimensional and include decreased oxygenation (Borragán et al., 2019), altered small-molecule metabolites (Bourdon et al., 2018), and decreased functional connectivity (Yang et al., 2018). Among subcortical regions, the nucleus accumbens (NAc) is also highly recruited and activated under conditions of sleep deprivation, making an evidence-based argument for sleep deprivation as a reward-seeking state (Volkow et al., 2012).

Similar processes and areas have been found to be important in moral judgment.[17] Regions of the frontal cortex that show altered activity while considering moral dilemmas include the ventromedial PFC and the superomedial prefrontal cortex (from de Oliveira-Souza, Zahn, & Moll, 2015). Among subcortical regions, the NAc shows altered functional connectivity in response to moral dilemmas (de Oliveira-Souza et al., 2015). Lesion and psychopathology studies have identified the NAc as a key modulator of reward-seeking behavior (de Oliveira-Souza et al., 2015). Convergence of these separate studies thus lends credence to the hypothesis that decreased moral reasoning under sleep-deprived states is neurobiologically manifest from hyperactivation of rapid reward-seeking states.

Advanced neuroimaging techniques have also shown direct linkage of prefrontal and anterior subcortical regions with the subcortical limbic system, including the amygdala and hippocampus, and the surrounding temporal cortex, including the superior temporal sulcus (de Oliveira-Souza et al., 2015). These medial regions are altered by moral reasoning (from de Oliveira-Souza et al., 2015) and also sleep deprivation, as shown in independent studies (Goldstein et al., 2013).

Another brain region worthy of attention is the septum. Lesion studies in animal models have found a phenomenon known as "septal rage" (Caplan, 1973) as well as dysregulation of sleep processes (Gerashchenko, Salin-Pascual, & Shiromani, 2001), but parallel studies in humans are limited. Nevertheless, the identification of the septum as a modulator of moral behavior (Gorman & Cummings, 1992) suggests that emotional states related to the septum are involved in moral judgment and decision making. This connection lends further credence to the hypothesis that sleep deprivation can unravel moral judgment and decision making along with related emotions.

Additional evidence comes from neurochemical studies. Sleep–wake transitions are largely mediated by acetylcholinergic, monoaminergic, and GABA-ergic pathways (reviewed in Torterolo, Monti, & Vanini, 2019). Experiments have also demonstrated the impact of altered neurochemical tone on moral reasoning processes. The greatest emphasis has been placed on alteration of neurochemical and neuroendocrine tone through psychotropic medications that can dually serve as sleep- or wake-promoting medications, including: (1) serotonergic tone through citalopram, (2) norepinephrine tone through propranolol, (3) dopaminergic tone through L-DOPA and modafinil, (4) intranasal delivery of oxytocin, and (5) oral delivery of testosterone (Crockett, Clark, Hauser, et al., 2010; Crockett, Clark, Lieberman, et al., 2010; Terbeck et al., 2013; Levy et al., 2014; De Dreu et al., 2011; Montoya et al., 2013). The indirect consequences of these existing dual psychotropic and sleep–wake medications are a foundation for future research efforts.

Several of these neurotransmitters and brain areas are related to affect or emotion. So, that might be one key to understanding the relation between sleep and moral judgment. Studies of acute sleep deprivation of more than forty-eight hours in healthy individuals have found modest yet subclinical deviations in atypical emotional or affective states (Kahn-Greene, Killgore, et al., 2007) linked to moral reasoning, including: (1) reductions in perceived emotional intelligence (Killgore, Lipizzi, et al., 2008) and reactivity to moral dilemmas (Zohar et al., 2005); (2) elevations in interpersonal frustration (Kahn-Greene, Lipizzi, et al., 2006); and (3) poor moral judgment (Killgore, Killgore, et al., 2007). In all of these studies of acute sleep deprivation, these affective deviations in healthy individuals subside with adequate sleep during the recovery period, emphasizing that sleep alone is necessary and sufficient to stabilize affective states linked to moral reasoning.

Much more remains to be discovered. However, these studies at least provide tantalizing hints of new directions for research on how sleep deprivation can alter neurochemicals, neuronal activity, and brain networks that are essential for moral judgments, decisions, and emotions.

3.4 Concluding Remarks

In this chapter, we have argued that action initiation and taking oneself to be the author of events in the world are two hallmarks of what it is to be an agent exercising control over one's actions and making one eligible for morally reactive attitudes. We then reviewed two well-established neuro-scientific frameworks for the investigation of action initiation and sense of agency: the intentional binding paradigm and the Libet paradigm. We next outlined how the subpopulation of agents who act under sleep deprivation might perform in those paradigms via an assessment of the current state of the literature on cognitive performance and the functional neuroanatomy of sleep deprivation. Finally, we reviewed the extant work on sleep depriva-tion and moral reasoning, with an eye toward how otherwise neurotypical agents who suffer from insufficient sleep might alter their moral reasoning and moral behavior in extended wakefulness.

We believe that such a review fills a lacuna in research and that future work connecting the neuroscience of agency paradigms with sleep paradigms is imperative. Several subpopulations suffer from insufficient sleep, including military personnel, shift workers, college students, doctors, and others in high-stress occupations. Where insufficient sleep is a necessary aspect of the job, these individuals are not culpable for the cognitive changes induced by sleep deprivation. Suppose, as is hypothesized, that weakened initiation of agency and reduced sense of agency impacts moral judgment and morally significant action. In the philosophical literature on moral responsibility, other stressors are taken as affecting control of practical decision making and actions and so are excusing or exempting conditions for moral and legal responsibility (e.g., extreme emotion, certain psychiatric conditions; see P. F. Strawson, 1974, for the inspiration for work on excusing and exempting conditions). Hence, insofar as one is not culpable for one's sleep deprivation, ought we to mitigate blame for morally bad actions under sleep deprivation? This question is largely unexplored theoretically. The synthesized research here promises to spark a new strand of critical debate, both in philosophy and in applied contexts such as criminal law, as to the control and responsi-bility of agents for their actions under sleep deprivation. It could also moti-vate the development of more effective interventions for these populations who must endure sleep deprivation.

Notes

1. Nonhuman animals engage in goal-directed behaviors in some sense, but here the focus is a subset of goal-directed behaviors, human intentional actions.

2. So, too, this kind of control is distinctive of human nature insofar as we do not hold other species to be morally responsible for what they do.

3. One prominent example to illustrate the difference is Locke's (1975) locked room case. I am sitting in a room chatting with a friend. I prefer to stay chatting longer. Unbeknown to me, the door to leave is locked. While I can freely decide to stay (and can freely decide to leave), I am not free to leave.

4. Here, we may worry that, given these results, to some degree, intentional binding as a marker of sense of agency also falls prey to a self-serving bias effect.

5. Libet and colleagues (1983) took this item of interest, discussed here as the conscious intention to flex, to be the broader disjunction of intention, urge, or desire to flex.

6. The Libet clock is same time perception measure clock adopted by the intentional binding paradigm, as referenced in II.1.

7. Libet did posit a more restricted role for conscious intentions in the action production process. In his view, although conscious intentions do not initiate neural preparation for movement, an agent may consciously veto that neural preparation prior to movement. That is, the agent can consciously decide not to move and so not move (Libet, 1999, 2005).

8. For a recent wide-ranging review of the neuroscience of free will based on Libet's studies, see Waller (2019). For a book anthology of empirical and philosophical work on the Libet paradigm, see Sinnott-Armstrong and Nadel (2011). For a book-length treatment of science and free will, see Mele (2009); in popular writing format, Mele (2014a); and in dialogue form, Mele (2014b).

9. We will not discuss here what one of the present authors (and others) have labeled elsewhere the dualist threat or the epiphenomenalism threat (e.g., Mele, 2014c; Nahmias, 2010; Waller, forthcoming). These are interesting and important issues but do not directly relate to the subpopulation of agents that are the focal point of this chapter in section 3.3, sleep-deprived agents.

10. This claim is independent from the claim that conscious intentions do not causally contribute to action production. It is consistent with the unconscious initiation threat that conscious intentions are causally efficacious downstream in either producing or vetoing bodily movement post unconscious preparation for movement. To deny this further causal role of conscious intentions in action production on the basis of the Libet results is to embrace the epiphenomenalism threat (see, e.g., Waller, 2019).

11. To be more precise, for those actions to be free in the basic sense. For an action to be a basically free one, it does not derive its free status from other earlier basically free actions. Hence, it is open to those who endorse this metaphysical claim to also

hold the following: if conscious mental states must be among the causes of a basically free decision, then other unconsciously produced decisions and actions can be free in virtue of their relation to those basically free decisions.

12. As Mele (2009) highlights, there is empirical work by Gollwitzer and colleagues supporting the causal efficacy of conscious distal intentions. Distal intentions are intentions to do something later—as opposed to intentions to do something now. Participants who form intentions on how and when to accomplish a goal later are significantly more likely to follow through on those plans (Gollwitzer, 1996; Gollwitzer & Brandstätter, 1997; Gollwitzer, 1999).

13. For a recent discussion of whether W, an average of reported times, accurately reflects the actual onset of a conscious intention to act now, see Mele (2010a, 2010b, 2010c, 2012). Herdova (2016) examines whether such experimental setups require the participant to have any intentions to act now at all.

14. Committed meditators—the experimental group in Jo, Hinterberger, and colleagues (2015)—were those who had at least three years of experience practicing mindfulness meditation.

15. See also Schultze-Kraft and colleagues (2020) for another innovative cueing approach that supports a relationship between the RP and presence of a conscious intention to move.

16. To take the RP as indicative of noncommittal preparation or bias toward voluntary action still conflicts with those who argue that the RP represents passive, not active (i.e., not true preparatory), activity. In particular, Schurger, Sitt, and Dehaene (2012) have pushed this skepticism about the active nature of the RP further, proposing the integration-to-bound model of RP. This model holds that, roughly, random walk type fluctuation leads to neural activity crossing the threshold for motor response (i.e., movement). The jury is still out on whether the integration-to-bound model accurately deflates the active, voluntary nature of the RP. For instance, Khalighinejad and colleagues (2018) found that the integration-to-bound model fits the recorded EEG signature of neural activity during endogenously generated movements *only if* an additional parameter corresponding to attention to task demands and cognitive effort—active neural noise—is added to the model.

17. See chapter 1 in this volume.

References

Abhang, P. A., Gawali, B. W., & Mehrotra, S. C. (2016). *Introduction to EEG-and speech-based emotion recognition.* Cambridge, MA: Academic Press.

Alexander, P., Schlegel, A., Sinnott-Armstrong, W., Roskies, A., Tse, P. U., & Wheatley, T. (2015). Dissecting the readiness potential: An investigation of the relationship between readiness potentials, conscious willing, and action. In A. Mele (Ed.), *Surrounding free will: Philosophy, psychology, neuroscience* (pp. 203–233). New York: Oxford University Press.

Alexander, P., Schlegel, A., Sinnott-Armstrong, W., Roskies, A., Wheatley, T., & Tse, P. (2016). Readiness potentials driven by non-motoric processes. *Consciousness and Cognition, 39*, 38–47.

Bandura, A. (1982). Self-efficacy mechanism in human agency. *American Psychologist, 37*(2), 122.

Bandura, A. (2001). Social cognitive theory: An agentic perspective. *Annual Review of Psychology, 52*(1), 1–26.

Barlas, Z., Hockley, W. E., & Obhi, S. S. (2018). Effects of free choice and outcome valence on the sense of agency: Evidence from measures of intentional binding and feelings of control. *Experimental Brain Research, 236*(1), 129–139.

Barlas, Z., & Obhi, S. (2013). Freedom, choice, and the sense of agency. *Frontiers in Human Neuroscience, 7*, 514.

Beyer, F., Sidarus, N., Bonicalzi, S., & Haggard, P. (2017). Beyond self-serving bias: Diffusion of responsibility reduces sense of agency and outcome monitoring. *Social Cognitive and Affective Neuroscience, 12*(1), 138–145.

Bonicalzi, S., & Haggard, P. (2019). From freedom from to freedom to: New perspectives on intentional action. *Frontiers in Psychology, 10*, 1193.

Borhani, K., Beck, B., & Haggard, P. (2017). Choosing, doing, and controlling: Implicit sense of agency over somatosensory events. *Psychological Science, 28*(7), 882–893.

Borragán, G., Guerrero-Mosquera, C., Guillaume, C., Slama, H., & Peigneux, P. (2019). Decreased prefrontal connectivity parallels cognitive fatigue-related performance decline after sleep deprivation. An optical imaging study. *Biological Psychology, 144*, 115–124.

Bourdon, A. K., Spano, G. M., Marshall, W., Bellesi, M., Tononi, G., Serra, P. A., . . . & Cirelli, C. (2018). Metabolomic analysis of mouse prefrontal cortex reveals upregulated analytes during wakefulness compared to sleep. *Scientific Reports, 8*(1), 1–17.

Bradley, G. W. (1978). Self-serving biases in the attribution process: A re-examination of the fact or fiction question. *Journal of Personality and Social Psychology, 36*(1), 56–71.

Buehner, M. J. (2012). Understanding the past, predicting the future: Causation, not intentional action, is the root of temporal binding. *Psychological Science, 23*(12), 1490–1497.

Buehner, M. J. (2015). Awareness of voluntary and involuntary causal actions and their outcomes. *Psychology of Consciousness: Theory, Research, and Practice, 2*(3), 237.

Buehner, M. J., & Humphreys, G. R. (2009). Causal binding of actions to their effects. *Psychological Science, 20*(10), 1221–1228.

Caplan, M. (1973). An analysis of the effects of septal lesions on negatively reinforced behavior. *Behavioral Biology, 9*(2), 129–167.

Caspar, E. A., Christensen, J. F., Cleeremans, A., & Haggard, P. (2016). Coercion changes the sense of agency in the human brain. *Current Biology, 26*(5), 585–592.

Caspar, E. A., Cleeremans, A., & Haggard, P. (2015). The relationship between human agency and embodiment. *Consciousness and Cognition, 33*, 226–236.

Chambon, V., Moore, J. W., & Haggard, P. (2015). TMS stimulation over the inferior parietal cortex disrupts prospective sense of agency. *Brain Structure and Function, 220*(6), 3627–3639.

Christensen, J. F., Di Costa, S., Beck, B., & Haggard, P. (2019). I just lost it! Fear and anger reduce the sense of agency: A study using intentional binding. *Experimental Brain Research, 237*(5), 1205–1212.

Crockett, M. J., Clark, L., Hauser, M. D., & Robbins, T. W. (2010). Serotonin selectively influences moral judgment and behavior through effects on harm aversion. *Proceedings of the National Academy of Sciences of the United States of America, 107*(40), 17433–17438.

Crockett, M. J., Clark, L., Lieberman, M. D., Tabibnia, G., & Robbins, T. W. (2010). Impulsive choice and altruistic punishment are correlated and increase in tandem with serotonin depletion. *Emotion, 10*(6), 855.

De Dreu, C. K., Greer, L. L., Van Kleef, G. A., Shalvi, S., & Handgraaf, M. J. (2011). Oxytocin promotes human ethnocentrism. *Proceedings of the National Academy of Sciences of the United States of America, 108*(4), 1262–1266.

de Oliveira-Souza, R., Zahn, R., & Moll, J. (2015). 11 neural correlates of human morality: An overview. In J. Decety & T. Wheatley (Eds.), *The moral brain: A multidisciplinary perspective* (pp. 183–196). Cambridge, MA: MIT Press.

De Pirro, S., Lush, P., Parkinson, J., Duka, T., Critchley, H. D., & Badiani, A. (2019). Effect of alcohol on the sense of agency in healthy humans. *Addiction Biology, 25*(4), e12796.

Dominik, T., Dostál, D., Zielina, M., Šmahaj, J., Sedláčková, Z., & Procházka, R. (2017). Libet's experiment: Questioning the validity of measuring the urge to move. *Consciousness and Cognition, 49*, 255–263.

Ehlen, J. C., Brager, A. J., Baggs, J., Pinckney, L., Gray, C. L., DeBruyne, J. P., . . . & Paul, K. N. (2017). Bmal1 function in skeletal muscle regulates sleep. *eLife, 6*, e26557.

Ehlen, J. C., Jefferson, F., Brager, A. J., Benveniste, M., & Paul, K. N. (2013). Period-amplitude analysis reveals wake-dependent changes in the electroencephalogram during sleep deprivation. *Sleep, 36*(11), 1723–1735.

Fischer, J. M., & Ravizza, M. (1998). *Responsibility and control: A theory of moral responsibility*. Cambridge: Cambridge University Press.

Fischer, J. M., & Tognazzini, N. (2011). The physiognomy of responsibility. *Philosophy and Phenomenological Research, 82(2)*, 381–417.

Frenda, S. J., Berkowitz, S. R., Loftus, E. F., & Fenn, K. M. (2016). Sleep deprivation and false confessions. *Proceedings of the National Academy of Sciences of the United States of America, 113*(8), 2047–2050.

Fried, I., Katz, A., McCarthy, G., Sass, K. J., Williamson, P., Spencer, S. S., & Spencer, D. D. (1991). Functional organization of human supplementary motor cortex studied by electrical stimulation. *Journal of Neuroscience, 11*(11), 3656–3666.

Fried, I., Mukamel, R., & Kreiman, G. (2011). Internally generated preactivation of single neurons in human medial frontal cortex predicts volition. *Neuron, 69*(3), 548–562.

Gerashchenko, D., Salin-Pascual, R., & Shiromani, P. J. (2001). Effects of hypocretin–saporin injections into the medial septum on sleep and hippocampal theta. *Brain Research, 913*(1), 106–115.

Goel, N., Abe, T., Braun, M. E., & Dinges, D. F. (2014). Cognitive workload and sleep restriction interact to influence sleep homeostatic responses. *Sleep, 37*(11), 1745–1756.

Goel, N., Basner, M., Rao, H., & Dinges, D. F. (2013). Circadian rhythms, sleep deprivation, and human performance. *Progress in Molecular Biology and Translational Science, 119*, 155–190.

Goldstein, A. N., Greer, S. M., Saletin, J. M., Harvey, A. G., Nitschke, J. B., & Walker, M. P. (2013). Tired and apprehensive: Anxiety amplifies the impact of sleep loss on aversive brain anticipation. *Journal of Neuroscience, 33*(26), 10607–10615.

Gollwitzer, P. M. (1996). The volitional benefits of planning. In P. M. Gollwitzer & J. A. Bargh (Eds.), *The psychology of action: Linking cognition and motivation to behavior* (pp. 287–312). New York: Guilford Press.

Gollwitzer, P. M. (1999). Implementation intentions: Strong effects of simple plans. *American Psychologist, 54*, 493–503.

Gollwitzer, P. M., & Brandstätter, V. (1997). Implementation intentions and effective goal pursuit. *Journal of Personality and Social Psychology, 73*, 186–199.

Gorman, D. G., & Cummings, J. L. (1992). Hypersexuality following septal injury. *Archives of Neurology, 49*(3), 308–310.

Greenberg, J., Pyszczynski, T., & Solomon, S. (1982). The self-serving attributional bias: Beyond self-presentation. *Journal of Experimental Social Psychology, 18*(1), 56–67.

Haggard, P. (2017). Sense of agency in the human brain. *Nature Reviews Neuroscience, 18*, 197–208.

Haggard, P. (2019). The neurocognitive bases of human volition. *Annual Review of Psychology, 70,* 9–28.

Haggard, P., & Clark, S. (2003). Intentional action: Conscious experience and neural prediction. *Consciousness and Cognition, 12,* 695–707.

Haggard, P., Clark, S., & Kalogeras, J. (2002). Voluntary action and conscious awareness. *Nature Neuroscience, 5,* 282–385.

Haggard, P., & Eimer, M. (1999). On the relation between brain potentials and the awareness of voluntary movements. *Experimental Brain Research, 126*(1), 128–133.

Herdova, M. (2016). Are intentions in tension with timing experiments? *Philosophical Studies, 173*(3), 573–587.

Hon, N., & Poh, J. H. (2016). Sleep deprivation produces feelings of vicarious agency. *Consciousness and Cognition, 40,* 86–92.

Honn, K., Hinson, J. M., Whitney, P., & Van Dongen, H. (2018). Cognitive flexibility: A distinct element of performance impairment due to sleep deprivation. *Accident Analysis and Prevention, 18,* 30070–30078.

Jahanshahi, M., Jenkins, I. H., Brown, R. G., Marsden, C. D., Passingham, R. E., & Brooks, D. J. (1995). Self-initiated versus externally triggered movements: I. An investigation using measurement of regional cerebral blood flow with PET and movement-related potentials in normal and Parkinson's disease subjects. *Brain, 118*(4), 913–933.

Jo, H.-G., Hinterberger, T., Wittmann, M., & Schmidt, S. (2015). Do meditators have higher awareness of their intentions to act? *Cortex, 65,* 149–158.

Jo, H.-G., Wittman, M., Hinterberger, T., & Schmidt, S. (2014). The readiness potential reflects intentional binding. *Frontiers in Human Neuroscience, 8,* 421.

Kahn-Greene, E. T., Killgore, D. B., Kamimori, G. H., Balkin, T. J., & Killgore, W. D. (2007). The effects of sleep deprivation on symptoms of psychopathology in healthy adults. *Sleep Medicine, 8*(3), 215–221.

Kahn-Greene, E. T., Lipizzi, E. L., Conrad, A. K., Kamimori, G. H., & Killgore, W. D. (2006). Sleep deprivation adversely affects interpersonal responses to frustration. *Personality and Individual Differences, 41*(8), 1433–1443.

Khalighinejad, N., Schurger, A., Desantis, A., Zmigrod, L., & Haggard, P. (2018). Precursor processes of human self-initiated action. *NeuroImage, 165,* 35–47.

Killgore, W. D., Killgore, D. B., Day, L. M., Li, C., Kamimori, G. H., & Balkin, T. J. (2007). The effects of 53 hours of sleep deprivation on moral judgment. *Sleep, 30*(3), 345–352.

Killgore, W. D., Lipizzi, E. L., Killgore, D. B., Rupp, T. L., Kamimori, G. H., & Balkin, T. J. (2008). Emotional intelligence predicts declines in emotion-based decision-making following sleep deprivation. *Sleep, 31,* A134.

Kirsch, W., Kunde, W., & Herbort, O. (2019). Intentional binding is unrelated to action intention. *Journal of Experimental Psychology: Human Perception and Performance, 45*(3), 378.

Kornhuber, H. H., & Deecke, L. (1965). Hirnpotentialänderungen bei Willkürbewegungen und passiven Bewegungen des Menschen: Bereitschaftspotential und reafferente Potentiale. *Pflüger's Archiv für die gesamte Physiologie des Menschen und der Tiere, 284*(1), 1–17.

Latane, B., & Darley, J. M. (1968). Group inhibition of bystander intervention in emergencies. *Journal of Personality and Social Psychology, 10(3)*, 215–221.

Levy, N., Douglas, T., Kahane, G., Terbeck, S., Cowen, P. J., Hewstone, M., & Savulescu, J. (2014). Are you morally modified?: The moral effects of widely used pharmaceuticals. *Philosophy, Psychiatry, and Psychology, 21*(2), 111–125.

Libet, B. (1985). Unconscious cerebral initiative and the role of conscious will in voluntary action. *Behavioral and Brain Sciences, 8*, 529–566.

Libet, B. (1999). Do we have free will? *Journal of Consciousness Studies, 6*(8–9), 47–57.

Libet, B. (2005). Do we have free will? In R. Kane (Ed.), *The Oxford handbook of free will* (pp. 551–564). Oxford: Oxford University Press.

Libet, B., Gleason, C. A., Wright, E. W., & Pearl, D. K. (1983). Time of conscious intention to act in relation to onset of cerebral activity (readiness-potential). *Brain, 106*(3), 623–642.

Libet, B., Wright, E. W., & Gleason, C. A. (1982). Readiness potentials preceding unrestricted spontaneous pre-planned voluntary acts. *Electroencephalography and Clinical Neurophysiology, 54*, 322–335.

Lo, J. C., Groeger, J. A., Santhi, N., Arbon, E. L., Lazar, A. S., Hasan, S., . . . Dijk, D.-J. (2012). Effects of partial and acute total sleep deprivation on performance across cognitive domains, individuals and circadian phase. *PLoS One, 7*(9), e45987.

Locke, J. (1975). *An essay concerning human understanding.* (P. Nidditch, Ed.) Oxford: Oxford University Press. (Original work published 1689)

McCauley, P., Kalachev, L. V., Mollicone, D. J., Banks, S., Dinges, D. F., & Van Dongen, H. P. (2013). Dynamic circadian modulation in a biomathematical model for the effects of sleep and sleep loss on waking neurobehavioral performance. *Sleep, 36*, 1987–1997.

McKenna, M., & Coates, J. D. (2016). Compatibilism. In E. N. Zalta (Ed.), *The Stanford encyclopedia of philosophy* (Winter 2016 ed.). Retrieved from https://plato.stanford.edu/archives/win2016/entries/compatibilism/

Mele, A. R. (2009). *Effective intentions: The power of conscious will.* Oxford: Oxford University Press.

Mele, A. R. (2010a). Conscious deciding and the science of free will. In R. F. Baumeister, A. Mele, & K. Vohs (Eds.), *Free will and consciousness: How might they work* (pp. 43–65). Oxford: Oxford University Press.

Mele, A. R. (2010b). Libet on free will: Readiness potentials, decisions, and awareness. In W. Sinnott-Armstrong & L. Nadel (Eds.), *Conscious will and responsibility: A tribute to Benjamin Libet* (pp. 23–33). Oxford: Oxford University Press.

Mele, A. R. (2010c). Conscious intentions. In J. K. Campbell, M. O'Rourke, & H. Silverstein (Eds.), *Action, ethics, and responsibility* (pp. 85–108). Cambridge, MA: MIT Press.

Mele, A. R, (2012). Free will and neuroscience: Revisiting Libet's studies. In A. Suarez & P. Adams (Eds.), *Is science compatible with free will?*. New York: Springer.

Mele, A. R. (2014a). *Free: Why science hasn't disproved free will.* New York: Oxford University Press.

Mele, A. R. (2014b). *A dialogue on free will and science.* New York: Oxford University Press.

Mele, A. R. (2014c). Free will and substance dualism: The real scientific threat to free will? In W. Sinnot-Armstrong (Ed.), *Moral psychology: Free will and responsibility* (Vol. 4, pp. 195–207). Cambridge, MA: MIT Press.

Montoya, E. R., Terburg, D., Bos, P. A., Will, G. J., Buskens, V., Raub, W., & van Honk, J. (2013). Testosterone administration modulates moral judgments depending on second-to-fourth digit ratio. *Psychoneuroendocrinology, 38*(8), 1362–1369.

Moore, J. W., Ruge, D., Wenke, D., Rothwell, J., & Haggard, P. (2010). Disrupting the experience of control in the human brain: Pre-supplementary motor area contributes to the sense of agency. *Proceedings of the Royal Society B: Biological Sciences, 277*(1693), 2503–2509.

Moreton, J., Callan, M. J., & Hughes, G. (2017). How much does emotional valence of action outcomes affect temporal binding? *Consciousness and Cognition, 49*, 25–34.

Nahmias, E. (2010). Scientific challenges to free will. In C. Sandis & T. O'Connor (Eds.), *A companion to the philosophy of action* (pp. 345–356). Chichester, UK: Wiley-Blackwell.

Nelkin, D. K. (2004). Deliberative alternatives. *Philosophical Topics, 32*(1/2), 215–240.

Olsen, O. K. (2013). The impact of partial sleep deprivation on military naval officers' ability to anticipate moral and tactical problems in a simulated maritime combat operation. *International Maritime Health, 64*(2), 61–65.

Olsen, O. K., Palleson, S., & Eid, J. (2010). Impact of partial sleep deprivation on moral reasoning in military officers. *Sleep, 33*, 1086–1090.

Parés-Pujolràs, E., Kim, Y. W., Im, C. H., & Haggard, P. (2019). Latent awareness: Early conscious access to motor preparation processes is linked to the readiness potential. *NeuroImage, 202*, 116140.

Raś, M., Nowik, A. M., Klawiter, A., & Króliczak, G. (2019). When is the brain ready for mental actions? Readiness potential for mental calculations. *Acta Neurobiologia Experimentalis, 79*, 386–398.

Sanford, P., Lawson, A. L., King, A. N., & Major, M. (2020). Libet's intention reports are invalid: A replication of Dominik et al. (2017). *Consciousness and Cognition, 77*, 102836.

Schlegel, A., Alexander, P., Sinnott-Armstrong, W., Roskies, A. Tse, P., & Wheatley, T. (2013). Barking up the wrong free: Readiness potentials reflect processes independent of conscious will. *Experimental Brain Research, 229*(3), 329–335.

Schlegel, A., Alexander, P., Sinnott-Armstrong, W. Roskies, A. Tse, P., & Wheatley T. (2015). Hypnotizing Libet: Readiness potentials with non-conscious volition. *Consciousness and Cognition, 33*, 196–203.

Schultze-Kraft, M., Parés-Pujolràs, E., Matić, K., Haggard, P., & Haynes, J. D. (2020). Preparation and execution of voluntary action both contribute to awareness of intention. *Proceedings of the Royal Society B: Biological Sciences, 287*(1923), 20192928.

Schurger, A., Sitt, J. D., & Dehaene, S. (2012). An accumulator model for spontaneous neural activity prior to self-initiated movement. *Proceedings of the National Academy of Sciences of the United States of America, 109*(42), E2904–E2913.

Shepherd, J. (2015). Scientific challenges to free will and moral responsibility. *Philosophy Compass, 10*(3), 197–207.

Simon, E. B., Lahav, N., Shamir, R., Hendler, T., & Maron-Katz, A. (2017). Tired and misconnected: A breakdown of brain modularity following sleep deprivation. *Biological Psychiatry, 81*(10), S322.

Sinnott-Armstrong, W., & Nadel, L. (Eds.). (2011). *Conscious will and responsibility: A tribute to Benjamin Libet*. Oxford: Oxford University Press.

Soon, C. S., Brass, M., Heinze, H.-J., & Haynes, J.-D. (2008). Unconscious determinants of free decisions in the human brain. *Nature Neuroscience, 11*, 543–545.

Strawson, P. F. (1974). *Freedom and resentment, and other essays*. London: Egmont Books.

Suzuki, K., Lush, P., Seth, A. K., & Roseboom, W. (2019). Intentional binding without intentional action. *Psychological Science, 30*(6), 842–853.

Tempesta, D., Couyoumdjian, A., Moroni, F., Marzano, C., De Gennaro, L., & Ferrara, M. (2012). The impact of one night of sleep deprivation on moral judgments. *Social Neuroscience, 7*(3), 292–300.

Terbeck, S., Kahane, G., McTavish, S., Savulescu, J., Levy, N., Hewstone, M., & Cowen, P. J. (2013). Beta adrenergic blockade reduces utilitarian judgement. *Biological Psychology, 92*(2), 323–328.

Torterolo, P., Monti, J. M., & Vanini, G. (2019). Neurochemistry and pharmacology of sleep. In E. Murillo-Rodríguez (Ed.), *The behavioral, molecular, pharmacological, and clinical basis of the sleep–wake cycle* (pp. 45–83). Cambridge, MA: Academic Press.

Travers, E., Khalighinejad, N., Schurger, A., & Haggard, P. (2020). Do readiness potentials happen all the time? *NeuroImage, 206*, 116286.

Volkow, N. D., Tomasi, D., Wang, G. J., Telang, F., Fowler, J. S., Logan, J., . . . & Ferré, S. (2012). Evidence that sleep deprivation downregulates dopamine D2R in ventral striatum in the human brain. *Journal of Neuroscience, 32*(19), 6711–6717.

Vyazovskiy, V. V., Olcese, U., Hanlon, E. C., Nir, Y., Cirelli, C., & Tononi, G. (2011). Local sleep in awake rats. *Nature, 472*(7344), 443–447.

Waller, R. R. (2019). *Recent work on agency, freedom, and responsibility: A review*. White Paper. Conshohocken, PA: John Templeton Foundation.

Waller, R. R. (forthcoming). Science of free will: Neuroscience. In J. Campbell, K. Mickelson, & V. A. White (Eds.), *Blackwell guide to free will*. New York: Wiley-Blackwell.

Watson, G. (1996). Two faces of responsibility. *Philosophical Topics, 24*, 227–248.

Wegner, D. M. (2002). *The illusion of conscious will*. Cambridge, MA: MIT Press.

Wegner, D. M., & Wheatley, T. (1999). Apparent mental causation: Sources of the experience of will. *American Psychologist, 54*(7), 480–492.

Whitney, P., Hinson, J. M., Satterfield, B. C., Grant, D. A., Honn, K. A., & Van Dongen, H. P. (2017). Sleep deprivation diminishes attentional control effectiveness and impairs flexible adaptation to changing conditions. *Scientific Reports, 7*(1), 1–9.

Yang, L., Lei, Y., Wang, L., Chen, P., Cheng, S., Chen, S., . . . & Yang, Z. (2018). Abnormal functional connectivity density in sleep-deprived subjects. *Brain Imaging and Behavior, 12*(6), 1650–1657.

Yoshie, M., & Haggard, P. (2013). Negative emotional outcomes attenuate sense of agency over voluntary actions. *Current Biology, 23*(20), 2028–2032.

Yoshie, M., & Haggard, P. (2017). Effects of emotional valence on sense of agency require a predictive model. *Scientific Reports, 7*(1), 1–8.

Zhao, R., Zhang, X., Fei, N., Zhu, Y., Sun, J., Liu, P., . . . & Qin, W. (2019). Decreased cortical and subcortical response to inhibition control after sleep deprivation. *Brain Imaging and Behavior, 13*(3), 638–650.

Zohar, D., Tzischinsky, O., Epstein, R., & Lavie, P. (2005). The effects of sleep loss on medical residents' emotional reactions to work events: A cognitive-energy model. *Sleep, 28*(1), 47–54.

4 Me, My (Moral) Self, and I

Jim A. C. Everett, Joshua August Skorburg, and Jordan L. Livingston

4.1 Introduction

What makes you the same person you were five years ago? Will you be the same person that will exist in five years' time? Would you be the same person if, by some fantastical freak of nature, you woke up in someone else's body? These questions have fascinated humans for thousands of years, and with the specter of new technological advances such as human–machine cyborgs looming, they will continue to do so. They will continue to do so because they get at a fundamental human question: What makes you *you*?

In this chapter, we outline the interdisciplinary contributions that philosophy, psychology, and neuroscience have provided in the understanding of the self and identity, focusing on one specific line of burgeoning research: the importance of morality to perceptions of self and identity.[1] Of course, this rather limited focus will exclude much of what psychologists and neuroscientists take to be important to the study of self and identity (that plethora of self-hyphenated terms seen in psychology and neuroscience: self-regulation, self-esteem, self-knowledge, self-concept, self-perception, and more; Katzko, 2003; Klein, 2012). We will likewise not engage with many canonical philosophical treatments of self and identity. But we will lay out a body of research that brings together classic themes in philosophy, psychology, and neuroscience to raise empirically tractable philosophical questions, and philosophically rigorous empirical questions about self and identity.

More specifically, in section 4.2, we will review some recent research that has treated traditional philosophical questions about self and identity as empirical questions. Within this body of work, we will be primarily concerned with the finding that morality (more so than memory) is perceived to be at the core of self and identity. Then, in section 4.3, we raise and respond to a variety of questions and criticisms: first, about the operationalization of identity concepts in the empirical literature; second, about the generalizability of the moral self effect; third, about the direction of change; fourth, about connections with recent work in neuroscience; and fifth, about the target of evaluation. Finally, in section 4.4, we consider a variety of implications and applications of this work on the moral self. Throughout, we aim to highlight connections between classical themes in philosophy, psychology, and neuroscience, while also suggesting new directions for interdisciplinary collaboration.

4.2   From the Armchair to the Lab

Given our limited focus, we follow Shoemaker (2019) in dividing contemporary philosophical approaches to morality and personal identity into four broad categories: psychological views, biological views, narrative views, and anthropological views. For present purposes, much of the research we describe below is situated within the domain of psychological views. This family of views about personal identity generally holds that person X at Time 1 is the same person as Y at Time 2 if and only if X is in some sense psychologically continuous with Y (henceforth, psychological continuity views). Of course, there is much debate about what such psychological continuity consists of, but following Locke (1975), memory is traditionally thought to be central in this regard. So, X at Time 1 is the same person as Y at Time 2 if X can, at one time, remember an experience Y had at another time. This vague formulation is subject to a number of objections (see Olson, 2019, for an overview), but our concern here is not to vindicate or impugn any particular version of a psychological theory of personal identity. Instead, we want to focus on the ways in which such philosophical theorizing might be informed by work in psychology and neuroscience—and also how work in psychology and neuroscience can shape philosophical theorizing.

As Shoemaker and Tobia (forthcoming) point out, if a psychological continuity view depends on a certain view of memory or consciousness, then experimental methods might eventually be able to clarify the precise mechanisms that support the relevant kinds of memory or consciousness. But by and large, this is not the preferred strategy in the literature. Instead, much of the relevant interdisciplinary work has focused on judgments of identity change and the psychological processes that underlie such judgments, and our work is a contribution to this trend.

In this vein, some of the early work in experimental philosophy took psychological continuity views as testable hypotheses. For example, Blok, Newman, and Rips (2005) asked participants to imagine that a team of doctors removed a patient's brain (call him "Jim") and destroyed his body. Then, in one version of the vignette, participants were told that the doctors successfully transplanted Jim's brain into a new body, such that the memories in Jim's brain are preserved in the new body. In another version of the vignette, participants were told that the memories were not preserved during the transplant.

In these kinds of scenarios, psychological continuity theories predict that in cases where *memories* were preserved, participants should judge that the brain recipient is "still Jim," but in cases where the memories are not preserved, participants should disagree that the brain recipient is "still Jim." And indeed, this is precisely what Blok and colleagues (2005) found, and it was also replicated by Nichols and Bruno (2010). There are, as we will see below, concerns about the adequacy of this methodology. Still, these studies helped to spark a line of research in which armchair philosophical speculation about self and identity formed the basis of empirical inquiry. In turn, this methodology opened a range of new questions about specific components of psychological continuity views. Is there something special or unique about memory? Or might other psychological features be important for understanding judgments about self and identity?

4.2.1   The Unbearably Moral Nature of the Self

Even in the earliest accounts of psychological continuity, morality has played an important role. After all, for Locke, personal identity could also be understood as a "forensic" concept: it is tied to memories, but it is also crucial for moral concepts such as responsibility, praise, and blame.

In recent years, an influential line of work has suggested that morals, more so than memories, are actually perceived to be at the heart of self and identity. Sparking this line of research, Prinz and Nichols (2017), Strohminger and Nichols (2014), and others presented participants with a wide variety of traits and asked them to imagine how much a change to a specific trait would influence whether someone is perceived to be the same person (e.g., "Jim can no longer remember anything that happened before the accident. Aside from this, he thinks and acts the same way as before the accident. Is Jim still the same person as before the accident?").

To test the importance of different kinds of psychological traits, Strohminger and Nichols (2014) used a variety of items that detailed different changes that a person could go through. Some were moral changes (e.g., now Jim is a psychopath or pedophile); others were personality changes (e.g., now Jim is shy, or absentminded); others were a loss of memories (e.g., now Jim cannot remember time spent with parents, or cannot remember how to ride a bike); others were changes to desires and preferences (e.g., now Jim desires to eat healthily or wants to quit smoking); and yet others were perceptual (e.g., now Jim has lost his ability to feel pain or see color).

According to classic psychological continuity views that prioritize memories, one should expect changes to memories to be more impactful on perceived identity persistence than other kinds of changes. Instead, across the various studies and specific items, results have consistently supported a moral self effect: when someone changes in terms of moral traits such as honesty, empathy, or virtuousness, they are rated as more of a different person than when they change in terms of memories, preferences, or desires. Thus, "moral traits are considered more important to personal identity than any other part of the mind" (Strohminger & Nichols, 2014, p. 168).

This finding is not a one-off. The moral self effect holds across a wide variety of scenarios (Strohminger, Knobe, & Newman, 2017; Prinz & Nichols, 2017) and has been replicated across contexts. For example, the effect has been replicated across age groups, such that eight- to ten-year-olds rate moral changes as most core to identity (Heiphetz, Strohminger, Gelman, et al., 2018), as well as across cultures, such that Buddhist monks in India also rate moral changes as most core to identity (Garfield et al., 2015). The moral self effect has even been replicated in real-world contexts such that family members of patients with neurodegenerative diseases tend to rate changes to moral faculties as more disruptive to identity than changes to memories (as in Alzheimer's disease) or changes to physical motor functions (as in amyotrophic lateral sclerosis; Strohminger & Nichols, 2015). In fact, we know of only one group of individuals in whom the moral self effect does *not* appear to replicate—in psychopaths (Strohminger, 2018).

4.3   Some Questions and Criticisms from Philosophy, Psychology, and Neuroscience

Having reviewed some of the relevant work on the moral self effect, we now turn to in-depth considerations of some questions and criticisms of this research program. First, we will consider prominent criticisms about the operationalization of identity concepts in the moral self literature, and we will detail some responses based on our recent work. Next, we raise questions about the applicability and generalizability of the moral self effect. Then, we will consider questions about the direction of change and connections with related research on the "true self." Finally, we draw novel connections with recent work in neuroscience. Throughout, given the interdisciplinary nature of the topic, we argue that philosophy, psychology, and neuroscience can each contribute to addressing these questions, albeit to varying degrees. In the section that follows, we review the ways in which each discipline has contributed to addressing these questions, noting the strength and weaknesses of each approach, and incorporating a review of our own interdisciplinary work along the way.

4.3.1   Is Identity Quantitative, Qualitative, or Both?

A prominent question that has been raised in response to the findings canvassed above is whether, when faced with these thought experiments, participants are really reporting that someone has become a *different person* or if instead they are just reporting *dissimilarity*. (Berniūnas & Dranseika, 2016; Dranseika, 2017; Starmans & Bloom, 2018). In the philosophical literature, something like this idea has been variously conceptualized as the difference between quantitative identity, on the one hand, and qualitative identity, on the other (Parfit, 1984; Schechtman, 1996).

When faced with a question such as "Is *X* the same as before?" Schechtman (1996) notes that there are two different approaches. The first concerns re-identification: What are the necessary and sufficient conditions under which *X* at Time 1 is identical to *Y* at Time 2? These kinds of questions are about the logical relation of identity, and are often discussed in terms of quantitative or numerical identity. The second concerns characterization: What makes *X* the person that they are? These questions are about which actions, beliefs, values, desires, and traits are properly attributable to a person, and are often discussed under the heading of qualitative identity.

Similarly, Parfit (1984) highlights the importance of distinguishing qualitative versus numerical identity:

> There are two kinds of sameness, or identity. I and my Replica are qualitatively identical, or exactly alike. But we may not be numerically identical, or one and the same person. Similarly, two white billiard balls are not numerically but may be qualitatively identical. If I paint one of these balls red, it will cease to be qualitatively identical with itself as it was. But the red ball that I later see and the white ball that I painted red are [numerically] identical. They are one and the same ball. (p. 201)

When it comes to people, twins, for example, can be qualitatively identical but still numerically distinct people. You might not be able to tell them apart, but they still need two passports to travel abroad. Variants of this distinction have been used to criticize the early studies on the importance of memories to identity persistence (e.g., Blok et al., 2005; Blok, Newman, & Rips, 2007; Nichols & Bruno, 2010).

As Berniūnas and Dranseika (2016) argue, these experimental designs potentially conflate qualitative and numerical concepts of identity. When Nichols and Bruno (2010) ask participants, "What is required for some person in the future to be the same person as you?" it is possible that participants are interpreting this question in a qualitative not numerical sense. To illustrate this, Berniūnas and Dranseika draw on a convenient pair of Lithuanian phrases that disambiguate the two: *tas pats* and *toks pats*.[2] When explicitly disambiguating these two senses to their Lithuanian participants, they found that participants were significantly more likely to agree that someone was the "same person" after losing their memories, suggesting that "retention of memory may not be so critical to the preservation of individual numerical identity" (Berniūnas & Dranseika, 2016, p. 115).

In the context of work on the moral self effect, Starmans and Bloom (2018) similarly leverage this distinction between quantitative and qualitative identity. They claim that while Strohminger and colleagues have sought to make claims about quantitative identity ("After changing morally, can I identify X as the same person?"), they are actually obtaining participants' intuitions about qualitative identity ("After changing morally, is X dissimilar to how they were before?"). While it makes sense, they argue, that if Jim lost his moral conscience after an accident, he would seem like a qualitatively different person, it wouldn't make sense to suggest that post-accident Jim is numerically distinct from pre-accident Jim, such that pre-accident Jim's debts are now forgiven or that post-accident Jim must now get a new passport. Jim is still the same person; he's just dissimilar from before. The worry is that the measures typically used in the moral self effect studies cannot clearly differentiate between these different senses of identity. Starmans and Bloom (2018) thus suggest that "we cannot tell whether these data reflect people's intuitions about similarity or about numerical identity . . . but we think that the most natural reading of these questions leads participants to answer in terms of similarity, not personal identity. In the real world, nobody sees moral changes as influencing identity" (p. 567).

Here, we want to consider a few potential responses to these criticisms.[3] First, we are not sure that it is entirely possible to separate qualitative from quantitative identity in the way Starmans and Bloom's criticism seems to require. Second, we are not convinced that morality is unimportant to identity in either folk psychology or philosophy.

More recent psychological continuity accounts have tended to focus on the *degree* of psychological connectedness. According to Parfit (1984), such psychological connectedness does include memories but also psychological dispositions, attitudes, personality, preferences, and so on. This has elements of qualitative identity (a greater degree of psychological connectedness means that someone is more similar) but also numerical identity because it is the degree of psychological connectedness that allows us to identify a person at different times as the same. Both of these points suggest that a strict division between qualitative and quantitative identity may be untenable.

There is also an empirical response to the criticisms raised above. One might think that judgments of identity change should have concomitant practical consequences. Presumably, if you judge someone to be a different person now than they were before, you would also judge that they are likely to behave differently than before. In the context of the moral self effect, a person's loss of morals should then lead participants to expect more, or worse, practical consequences for that person than with equivalent losses of memories, preferences, desires, and so on. Indeed, in some of our work, we have shown that compared to memories, moral changes not only affected perceptions of identity persistence (as in previous studies) but, crucially, such changes also led participants to infer a range of practical consequences subsequently, including changes in behavior, evaluations by third parties, and reductions in relationship quality (Everett, Skorburg, Livingston, et al., unpublished manuscript).

To address this question of whether moral changes are affecting judgments of (in Starmans & Bloom's terms) quantitative identity or mere similarity, we drew on the idea of special obligations: the duties we have to someone simply because of who they are. One might think, for example, that someone has obligations to their mother that they don't have to a stranger, and these special obligations toward their mother do not change, even if she were to suffer a severe debilitating illness that changed her personality. Someone's obligations and duties to her are the same because she herself *is* the same person, however dissimilar she is now to how she was in her prime. If participants judge that their own special moral obligations toward a loved one are more affected when their loved one loses their morality (compared to losing their memories), this might suggest that participants are not thinking *solely* in terms of similarity. In fact, our results do suggest that something like this is the case (Everett, Skorburg, & Savulescu, 2020).

In our studies, we presented participants with a classic "body switch" thought experiment in which a loved one undergoes a brain transplant with a stranger and, as a consequence, experiences no psychological change (the control condition), loses all their memories, or completely loses their moral conscience. After assessing perceptions of identity persistence, we presented a moral dilemma, asking participants to imagine that one of the patients must die (Study 1) or be left alone in a care home for the rest of their life (Study 2). In our studies, participants were made to decide who they would save or care for: the patient with their partner's brain and the stranger's body, or the patient with the stranger's brain and the patient's body.

This enabled us to test two things. First, it enabled us to replicate and extend previous empirical studies looking at whether people see psychological continuity as more important than physical continuity. Indeed, in line with previous work, in our control condition, we found that participants were much more likely both to judge that the person with their partner's brain and the stranger's body was the "real" partner, and to think that their moral duties toward this person were stronger than to the person with the stranger's brain and their partner's body. More importantly, though, by also including two conditions, one where, after the patient either lost all their memories or completely lost their moral conscience, we could also see whether changes to morals would be more disruptive than changes to memories for participants' perceived moral duties toward the patient.

If Starmans and Bloom's criticism (i.e., that previous studies on the moral self effect are *only* about similarity, not identity) is on the right track, then we should see no change to perceptions of moral duties, depending on whether someone lost their memories or morals. The partners described in the vignettes might be perceived as more dissimilar if they lost their morals (i.e., qualitatively different), but they would still be judged as the same person (i.e., numerically identifiable as the same person), and so presumably the special obligations would remain intact.

Indeed, we found some evidence that participants thought their moral duties toward the partner were, in fact, decreased when their partner experienced changes to their morality compared to when they experienced changes to memories or experienced no psychological (but only physical) change. These results suggest that participants, to some extent, do perceive a person's identity to be disrupted by their loss of memories or morality, and that previous results are not only about perceived similarity (Everett, Skorburg, & Savulescu, 2020).

Taken together, the conceptual and empirical responses to challenges about the operationalization of identity concepts in the moral self literature suggests that folk intuitions about self and identity likely involve (again, using Starmans & Bloom's terms) both identity and similarity, and these are likely flexibly activated and focused on both qualitative and quantitative identity in different contexts, depending on the task at hand. We address the philosophical implications of this suggestion in section 4.3 below.

4.3.2   Is the Moral Self Effect Generalizable? Is It Only Applicable to Fantastical Thought Experiments?

For all the work that philosophers and psychologists have done on the experimental philosophy of self and identity, one might still wonder whether this is meaningful. Much of the literature (our contributions included) tends to use vignette-based studies employing far-fetched, science-fiction-like examples involving, for example, brain transplants, body switches, magic pills, time machines, and reincarnation. While these scenarios might be helpful to clarify intuitions about philosophical thought experiments, it is also important to explore whether and to what extent the moral self effect holds in more common everyday cases. Here, again, our work has focused on the practical consequences of judgments about self and identity.

One way of testing the generalizability of the moral self effect outside the realm of thought experiments is to look at real-life cases characterized by changes to morals and memories. In this vein, Strohminger and Nichols (2015) looked at ratings of identity persistence by family members of patients with different neurodegenerative diseases. By including patients with different kinds of diseases and different symptoms, they could look at how family members judged that someone was the same person after they experienced changes to moral faculties (as in some cases of frontotemporal dementia) compared to changes to memories (as in Alzheimer's disease) or changes to physical motor functions (as in amyotrophic lateral sclerosis). Mirroring the findings from thought experiments, their results show that family members of patients with frontotemporal dementia, which was associated with moral changes, rated the patient as more of a different person than did family members of patients with other kinds of diseases. Of course, while such studies provide evidence of how the moral self effect emerges in real-world contexts, their high ecological validity does come with less control: these patients will all have different presentations of symptoms, will differ in severity, and so on.

In a recent paper, we extended this line of work by taking a blended approach using an experimental philosophy method with tightly controlled vignettes but focusing on real-life cases of addiction (Earp et al., 2019). Why addiction? A common refrain from family members and friends of addicts is that the person they knew before is not the same as the addict now. As one mother put it, "Six years have passed since I discovered that my son was using drugs. I [was] devastated, not to mention how worried I was about his well-being. My son *was not the same person anymore*" (Urzia, 2014, emphasis added).

As Tobia (2017) notes, such stories are all too common: "Many have witnessed someone they loved change so profoundly that the person remaining seems an entirely different one." Moreover, addiction is highly moralized in a way that, say, dementia is not. As a result, we hypothesized that the processes at play in the moral self effect might also arise in the context of addiction. That is, if an agent's becoming addicted to drugs leads to the perception that they are a "different person," this may be due to a presumed diminishment in moral character that such addiction stereotypically brings about. Across six studies, we found that participants judged an agent who became addicted to drugs as being closer to "a completely different person" than "completely the same person" and that these judgments of identity change are indeed driven by perceived negative changes in the moral character of drug users (Earp et al., 2019).

We take these results (along with others discussed below) as evidence that the moral self effect does indeed generalize to various contexts beyond the tightly controlled and sometimes far-fetched realm of philosophical thought experiments. We discuss some implications and potential applications in section 4.4.

4.3.3 Does the Direction of Moral Change Matter?

An important question has been looming in the background so far. The moral self effect assumes that morality is at the heart of self and identity, such that changes to morals are more disruptive than other kinds of changes. This conclusion has almost exclusively been drawn from studies focusing on perceived identity persistence following either a *loss* a morals or a *loss* of memories. But what if certain features are *gained* instead?

Work on the "true self" suggests that the direction of moral change could matter. A number of studies have suggested that people typically regard others' true selves as being fundamentally good (Newman, Bloom, & Knobe, 2014; De Freitas et al., 2018; Newman, De Freitas, & Knobe, 2015; Bench et al., 2015). As people become more moral, they are perceived to become closer to their true self or their "essence," whereas when they become less moral, they are perceived to move further away from their true self (Bench et al., 2015; Tobia, 2016, 2017).

Tobia (2015) draws on the well-worn (if potentially apocryphal) case study of Phineas Gage, a railroad worker who experienced brain damage in a horrific accident, after which he was reported to have become cruel and impulsive—so much so that "he was no longer Gage." In his work, Tobia gave participants one of two versions of this story. In one condition, participants saw the "standard" case of Phineas Gage, where he was kind before the accident but cruel afterwards—that is, where Gage morally *deteriorated*. In another condition, holding the magnitude of the change constant, participants saw a vignette where Gage was described as cruel before the accident but kind afterward—that is, where Gage morally *improved*. In both conditions, Tobia asked participants to judge whether Phineas Gage was the same person as before the accident. He found that Gage was less likely to be judged as identical to his pre-accident self when the change was in a "bad" direction (deteriorating from kind to cruel) than when the change was in a "good" direction (improving from cruel to kind), even when the magnitude of the change was held constant.

These findings were further substantiated by a study demonstrating that moral enhancement is less disruptive to perceptions of identity than moral degradation and that moral degradation is especially disruptive to perceptions of identity when people expect moral enhancement (Molouki & Bartels, 2017). That said, Prinz and Nichols (2017) also report findings suggesting that whether moral changes were in a positive or negative direction did not matter—that "moral changes are regarded as threats to identity, regardless of whether those changes are good or bad" (p. 454). While somewhat mixed, these findings at least raise the interesting suggestion that judgments of identity change are not solely a function of the magnitude of the change but could be importantly related to the direction of the change. When people are perceived as deteriorating (and especially when they are perceived to deteriorate morally), they might be judged to be more of a different person than when they improve or change in a positive direction.

While this work suggests that the direction of moral change could play an important role in the moral self effect (in line with what would be predicted based on the true self literature), more work is necessary on how direction and the type of change interact—and how both of these interact depending on the target of the judgment. Perhaps, for example, gaining new memories is more disruptive than both losing memories *and* losing morals, or perhaps all of this depends on whether participants are thinking about themselves, a friend, a stranger, or an enemy. In the same study mentioned above, Prinz and Nichols (2017) focus on judgments of the self and others, and found that the pattern for others replicated when thinking of the self: that it mattered more when the changes were moral, but it didn't matter which direction the change were in. This, of course, goes against the suggestion in other work (e.g., Tobia, 2015; Molouki & Bartels, 2017). Perhaps different results would be obtained with a within-subjects "one change" paradigm used by Strohminger and Nichols (2014), and perhaps it matters who specifically the target is (see next section).

In recent work, we sought to shed more light on how the direction of change and target might interact. We asked participants to imagine someone changing in a variety of ways (morality, memories, warmth, and competence), where some participants read that someone increased the trait (i.e., became more moral or gained new memories), while others, as in other studies, read that someone deteriorated (i.e., becomes less moral or lost memories). Our results showed that changes to morality were most disruptive to perceived identity but that the direction of change mattered too: a friend became more of a different person when they became *less* moral, but a foe became more of a different person when they became *more* moral.

Together, all these results suggest that the direction of change does matter—that even if morals tend to be more disruptive for identity than memories, losing morals can be more impactful than gaining morals. Intriguingly, though, this also seems sensitive to the target of the thought experiment: who are we thinking of?

4.3.4   Does the Target Matter?

From a philosophical perspective, one might think that the primacy of morality or memories (or whatever else) for identity should be insensitive to who we're thinking about. If memories are at the core of psychological continuity, then this finding should hold not only for judgments about the self, but also for strangers, friends, or even enemies. Indeed, much of the work we discussed in the previous sections has found that the moral self effect does not tend to depend on the target of evaluation (e.g., first- vs. third-person judgments). Yet, these findings are somewhat surprising in light of traditional findings in social psychology that show pervasive effects of target, such that perceptions of self are often biased in certain ways in comparison to perceptions of another (and vice versa). For example, people typically think of themselves differently from others with a strong actor–observer bias (e.g., Ross, Amabile, & Steinmetz, 1977) and think of those close to them differently from others (e.g., Alves, Koch, & Unkelbach, 2016; Simon, 1992), and tend to rate morality as being more important for the self and people close to them than for strangers or people they dislike (e.g., Brown, 1986; Epley & Dunning, 2000; Leach, Ellemers, & Barreto, 2007).

Given that these first- and third-person asymmetries are so pervasive in the field of psychology, similar findings would be expected within the subfield of empirical approaches to self and identity.[4] However, to date, first- and third-person asymmetries have not been observed for studies investigating identity change. One of the first studies to consider the role of target showed that a body-switching paradigm yielded similar results regardless of target (Nichols & Bruno, 2010), and some of the earliest work on the moral self effect demonstrated that moral change was more important than memories whether it was presented in a first- or third-person perspective (Prinz & Nichols, 2017). More recently, no difference was observed in a series of studies directly comparing the moral self effect for self and a hypothetical other ("Chris"), albeit showing stronger effects of changes in certain moral traits on other compared to self (Heiphetz, Strohminger,

& Young, 2017). Moreover, our work examining the effect of target across many different categories showed that the moral self effect holds across self, known friend, and unknown stranger, but not for a known foe (although the self and friend condition were not compared directly), again showing a small to negligible effect of target for self and (most) others (Everett, Skorburg, Livingston, et al., unpublished manuscript).

It is possible that the lack of asymmetry between self and other reflects the implicit positive nature of the moral self across most targets (Strohminger et al., 2017). Indeed, the "true self" literature has demonstrated that although our own true selves are deemed to be inherently good, so are the true selves of others (Bench et al., 2015). This lack of the actor–observer bias for moral traits may be one reason that the effect is not observed in the studies reviewed above. However, there are other plausible avenues of explanation. Although there are well-known asymmetries for perceiving self and other, there are also asymmetries and biases for perceiving the self in the past, present, and the future (Ersner-Hershfield, Wimmer, & Knutson, 2008; Quoidbach, Gilbert, & Wilson, 2013; Bartels & Urminsky, 2011), such that thinking about the self in a hypothetical thought experiment may not be the same as thinking about the self in the here and now. Moreover, many of the moral self studies to date have not used specified targets, and using a more concrete, known target might influence the results (see Everett, Skorburg, & Savulescu, 2020; Everett, Skorburg, Livingston, et al., unpublished manuscript). Regardless of whether the judgments change in different circumstances, the mechanisms driving the effects, or lack thereof, remain to be explored in the future.

4.3.5 What Are the Neural Mechanisms of the Target of the Moral Self Effect?

Neuroimaging can potentially help to clarify some of the questions posed above. For example, neuroimaging studies investigating self-referential processing (in the traditional psychological sense) have found that when individuals are asked to indicate whether a series of trait words describe themselves or another individual, this processing tends to recruit the cortical midline structures of the brain (precuneus and medial prefrontal cortex [mPFC]), with self-referential activity recruiting activity in more ventral regions of the mPFC and other-referential processing recruiting activity in more dorsal regions of the mPFC (Denny et al., 2012; Wagner, Haxby, & Heatherton, 2012).

It is not entirely clear why this pattern of activity exists, and it is unlikely that the differential activity is entirely a result of the target of investigation (self or other). For example, information about the self tends to be inherently more positive than information about other individuals, and this difference is also reflected at the neural level, with self-referential neural activity sharing highly overlapping patterns with both positively valenced information (Chavez, Heatherton, & Wagner, 2016) and value-based processing (Berkman, Livingston, & Kahn, 2017; Yarkoni et al., 2011). Information about the self, by its very nature, is also more familiar than information about other individuals, and brain regions tracking self-relevant information may be tied not only to value but also to familiarity (Lin, Horner, & Burgess, 2016).

Regardless of what is being tracked in these regions, the constellation of activity can be informative, particularly along the ventral-to-dorsal gradient. Indeed, studies have shown that the closeness of a target to the self can be tracked such that targets closer to the self activate more ventral regions of the mPFC, whereas targets less close to the self activate more dorsal regions of the mPFC (Mitchell, Banaji, & Macrae, 2006), with very close targets demonstrating overlapping activity (Zhu et al., 2007).

Moreover, the activity within these regions varies, to some extent, with elements of hypothetical thought. Counterfactual thinking for self and for other has demonstrated a similar ventral-to-dorsal gradient in the brain (De Brigard et al., 2015). Thinking about the self in the future, too, tends to recruit regions of the cortical midline structures as part of the default mode network (Buckner & Carroll, 2007; Buckner, Andrews-Hanna, & Schacter, 2008). Whereas some studies have found that thinking about the self in the future activates more dorsal (compared to ventral) regions of the mPFC (Packer & Cunningham, 2009), others have found that thinking about the self in the future compared to the present simply recruits the ventromedial prefrontal cortex (vmPFC) to a lesser degree (Ersner-Hershfield et al., 2008; Tamir & Mitchell, 2011; D'Argembeau et al., 2010).

This body of research offers a number of potential resources for generating hypotheses about the neural mechanisms underlying the moral self effect. At a general level, given that thinking about identity is very person centered, activation in cortical midline structures of the brain is likely still expected. More specifically, if self–other asymmetries are absent from the moral self literature because thinking about a hypothetical self is akin to thinking about another person, overlapping activity in dorsal regions of the mPFC might be expected.

Alternatively, if self–other asymmetries are absent from the empirical literature on self and identity because thinking about another individual changing on a moral dimension is valuable to the self, overlapping activity in ventral regions of the mPFC might be expected. In any case, identifying the neural processes underlying the moral self effect could help to clarify why the primacy of morality in identity does not seem to depend on the target.

4.3.6   What Are the Neural Mechanisms of Values and Traits?

Identifying the neural mechanisms underlying the moral self effect could also help to clarify questions concerning why morality, above and beyond other personality traits, appears to be so essential to identity.

Among other things, morality can be understood as sets of norms within a society that are distinguished because they are somehow more important or more valued (see, e.g., Copp, 2001), and the social nature of morality helps to highlight its value. For example, a person is likely to care about a change to her friend because she may no longer receive the benefits of being socially tied to her friend. However, the same person is also likely to care about her own morality, given that it impacts her own social reputation. In both scenarios, the social features of morality ensure that changes to friends or changes to self are quite important, or valuable, to the self, and recent work supports the idea that perceived importance to self mediates judgments of identity change (Heiphetz, Strohminger, & Young, 2017).

Incorporating evidence from neuroscience can help to assess whether the moral self effect is indeed driven by value-based processing (see May et al., this volume, for an overview of the value-based mechanisms posited by much recent work in the moral judgment and moral learning literature). Broadly, evidence increasingly suggests that moral cognition is best understood not as a unique and modular set of capacities, but rather in terms of more domain-general processes of reasoning, value integration, reward processing, and decision making.

For example, Shenhav and Greene (2010, 2014) have argued that value-based processing is crucial for understanding moral judgment such that when participants are asked to imagine classic sacrificial dilemmas in which they can either do nothing and risk the death of a large group of people or do something at the expense of a single individual, value-based subregions of the brain are largely involved in making these types of calculations. A related set of studies have implicated value-based computation in the vmPFC during charitable giving (Hare et al., 2010; Hutcherson, Bushong, & Rangel, 2015), and yet another found that moral transgressions disrupt neural representations of value in this same region (Crockett et al., 2017). Based on this body of evidence, it seems likely that some sort of value-based processing also underpins the moral self effect such that moral traits might elicit activity in value-based regions of the brain, either selectively or more strongly than other nonmoral (e.g., personality traits).

Although a value-based mechanism for the moral self effect would not be entirely surprising, it could help to push forward the field in two important ways. First, a mechanistic understanding of the moral self effect could help to elucidate the broader relationship between morality and identity. One recent commentary called into question whether morality is as important to identity as identity is to morality (Strohminger, 2018). We propose that the relationship between the two may be best clarified by assessing their common mechanisms.

Identity, like morality, relies on value-based processing. Recent studies note a strong overlap between neural activity associated with thinking about identity and value (Kim & Johnson, 2015; Northoff & Hayes, 2011), and a meta-analysis of neuroimaging studies on self and value reveals a large cluster of overlapping activation in the vmPFC (Berkman et al., 2017). Morality and identity, then, are likely intimately related via their value-based processing, and evaluating morality and identity in terms of their value-based processing may provide another currency with which to assess the nature of their relationship.

In addition to clarifying the nature of the relationship between morality and identity, a value-based approach to understanding the moral self effect could help to push forward the field of neuroscience itself. Within the vast set of neuroscience studies that have taken up issues of self and identity, most have utilized a task that prompts participants to rate the extent to which different personality trait words describe themselves (Denny et al., 2012; Wagner et al., 2012)—an approach that remains the gold standard.

However, despite the task's popularity, very few studies[5] have reported any neural differences in assessing the effect of trait type. One of the reasons why differences across traits are not reported is because most effects of traits may not become evident when using traditional univariate analysis.

In contrast, more nuanced multivariate techniques are revealing neural differences in, for example, the types of information we use to organize our representations of other people (Tamir et al., 2016). Insights from the moral self effect highlighting the privileged status of moral (vs. nonmoral) trait words and their distinct neural mechanisms (e.g., value-based processing) could motivate future neuroscience studies to continue in this tradition, using more advanced techniques to pay attention to differences between trait words.

In this sense, not only can neuroscience help to clarify mechanisms underlying the moral self effect, but insights from the moral self literature that stem from philosophy and psychology can make important contributions to the ongoing work on the neuroscience of self and identity.

4.4   Practical and Philosophical Implications

In this final section, we will consider a few implications and applications for the moral self research program. In particular, we attempt to contextualize how the nuanced questions and concerns outlined in the previous section might have bearing on a broader range of practical issues.

4.4.1   Behavior Change and Self-Regulation

One way that clarifying the psychological and neural processes involved in judgments about self and identity is important is that this research might have the potential to provide insights into translational work in domains such as self-regulation.[6]

One example comes from the intriguing possibility that identity judgments could be leveraged as a tool for behavior change. For instance, the identity-value model of self-regulation holds that identity serves as a salient value input for facilitating successful self-regulation and that stable value-laden sources of identity are strongest (Berkman et al., 2017). If morality is, as the foregoing results suggest, core to identity and is driven by value-based processing, it may be a candidate target for interventions seeking to promote behavior change. Of course, given that morality is so essential to identity, it may also be tougher to manipulate than other aspects of identity. To this effect, one recent paper found that people do not desire to be more moral, in part because they already perceive themselves to be quite moral (Sun & Goodwin, 2019). However, just because people do not desire to be *more* moral does not mean that moral identity cannot be used as an effective motivator.

Indeed, identity has often been used as a source of moral motivation (e.g., Hardy & Carlo, 2005, 2011), and appealing to moral reasoning has been shown to motivate compliance on certain behaviors such as paying taxes (e.g., Blumenthal et al., 2001; Ariel, 2012) and environmental conservation (Bolderdijk et al., 2013; Hopper & Nielson, 1991). Given that many self-regulatory failures are often moralized (Rozin & Singh, 1999; Frank & Nagel, 2017), appealing to moral identities and values may be an effective strategy for motivating successful self-regulation as well.

Moreover, counterfactual thought experiments, such as those traditionally used by philosophers, might also play a key role in motivating self-regulation. Many effective self-regulation techniques already draw upon hypothetical and imaginative cognitive techniques encouraging individuals to think about themselves in new and alternative ways (Kross et al., 2014; White et al., 2016). Encouraging participants to imagine the degree to which they would become a new person if they were to achieve a goal (e.g., "become a whole new you!") could provide an avenue for examining ways in which identity and value facilitate self-regulation. Exploring the mechanisms underlying the traditional moral self effect, although not directly related to translational applications, may be able to help motivate work in this direction.

4.4.2 Punishment and Responsibility

It is evident that, philosophically, identity has long been thought to be connected to moral concepts of blame, punishment, and responsibility. Practically, the research discussed in this chapter highlights how judgments about blame and responsibility are affected by perceptions of identity continuity—or disruption. We have already discussed work showing how someone's becoming addicted to drugs leads to the perception that they are a "different person," seemingly caused by perceived negative changes in the moral character of drug users (Earp et al., 2019). As another example, Gomez-Lavin and Prinz (2019) have examined the moral self effect in the context of parole decisions, finding that participants were significantly more likely to grant parole to offenders who underwent changes in their moral values compared to mere behavioral changes. It will be interesting for future work to consider in more depth the way that appeals to identity are used in both legal and forensic settings to justify responsibility and punishment.

4.4.3 Bioethics and New Technologies

Another example which has garnered much attention as of late involves patients undergoing neurostimulation, such as deep brain stimulation (DBS) treatments for Parkinson's disease. In one case report, after eighteen months of DBS, a patient's motor symptoms were improved, but it was reported that "she was no longer able to work, had a loss of inspiration and a taste for her work and for life in general," and she said, "Now I feel like a machine, I've lost my passion. I don't recognize myself anymore" (Schüpbach et al., 2006, p. 1812). In light of this story (and many others like it), Skorburg and Sinnott-Armstrong (forthcoming) have suggested that because people tend to see moral traits as especially identity conferring, measuring changes to moral functioning pre and post DBS should be a priority for neuroethicists concerned with identity changes brought about by DBS and other forms of neurostimulation.

4.4.4 Philosophical Implications

So far, most of our discussion has centered on experimental approaches to questions about self and identity. Here, we want to consider the extent to which such empirical work has a bearing on more traditional philosophical theorizing about personal identity. Consider the following argument, adapted from Berniūnas and Dranseika (2016) and Nichols and Bruno (2010):

1. If it is appropriate to consider folk intuitions in assessing theories of personal identity, then, ceteris paribus, folk intuitions that are more robust ought to be given more weight in assessing theories of personal identity.

2. Folk intuitions favoring psychological continuity accounts of personal identity are especially robust.

3. Therefore, if it is appropriate to rely on folk intuitions in assessing theories of personal identity, then, ceteris paribus, folk intuitions favoring psychological continuity accounts of personal identity ought to be given special weight in assessing theories of personal identity

Of course, one important question here is whether it is indeed appropriate to rely on folk intuitions about personal identity. After all, one might reasonably hold that what laypersons think is irrelevant to the fundamental metaphysical questions about personal identity. Perhaps the fact that ordinary people perceive morality as important to personal identity simply misses the point. What *really* matters is some metaphysical account of continuity. But this can only hold if we assume a "deep" metaphysical notion of identity, which need not be the only game in town (Prinz & Nichols, 2017). There is surely a sense in which some philosophical problems of identity are deeply metaphysical in this way. But as Prinz and Nichols (2017) point out, other questions about personal identity are not deep in the sense that they don't "depend on some hidden fact about the structure of reality." Instead, they argue, "It depends on us." We classify, label, and apply various notions of identity:

> If we are right that questions of personal identity are settled by how we do, in fact, classify, then this is a case where experimental philosophy can actually contribute to metaphysical debates. Surveys, in this case, do not just tell us what ordinary people think; they reveal the actual correlates of identity, because ordinary practices of classification determine conditions of identity. (Prinz & Nichols, 2017, p. 450)

If all of this is on the right track and it is indeed appropriate to rely on the robust folk intuitions for philosophical debates, then we need to be clear on *which* folk intuitions are robust. In this chapter, we have considered numerous challenges and criticisms of the moral self effect, along with a wide range of replications and extensions. We think there is ample evidence to support the claim that the moral self effect is robust. As a result, we contend that within psychological continuity views of personal identity, theories that emphasize the importance of morality ought to be given pride of place. Similarly, insofar as the results we have described are robust, then this could also entail that the prominent position afforded to memories within psychological continuity views may need to be revised and reconsidered. In any case, by drawing on the philosophical, psychological, and neuroscientific work on the moral self, we hope to have shown that a topic as complex and important as the moral self will surely require collaborations among philosophers, psychologists, and neuroscientists, among others.

Notes

1. In this chapter, we use "self and identity" as a catchall for these various approaches. However, when quoting other research or engaging with specific criticisms, terms such as "personal identity," "identity persistence," etc. will inevitably crop up as well.

But unless otherwise noted, we take ourselves to be primarily engaging with the (relatively narrow) literature described in more detail below, which examines folk judgments of identity change.

2. When contrasted with *toks pats*, *tas pats* means "the same" in the sense of numerical identity, while *toks pats* means "the same" in the sense of qualitative identity. As they reminded their participants in the studies, "If we have two white billiard balls, we can say that they are *toks pats*, but they are not *tas pats* billiard ball. If we paint one of the balls red, it is not *toks pats* billiard ball as before, but it is still *tas pats*, only painted" (Berniūnas & Dranseika, 2016, p. 114).

3. This section draws from and expands upon Everett, Skorburg, and Savulescu (2020).

4. This section (4.3.4) and the section on neural mechanisms of target (4.3.5) are drawn from and expand upon Jordan Livingston's doctoral dissertation.

5. Aside from studies that successfully investigate differences in positive and negative trait information (e.g., Glisky, & Marquine, 2009; Fossati et al., 2004).

6. Sections 4.4.1 and 4.3.6 are drawn from Jordan Livingston's doctoral dissertation.

References

Alves, H., Koch, A., & Unkelbach, C. (2016). My friends are all alike—The relation between liking and perceived similarity in person perception. *Journal of Experimental Social Psychology, 62*, 103–117.

Ariel, B. (2012). Deterrence and moral persuasion effects on corporate tax compliance: Findings from a randomized controlled trial. *Criminology, 50*(1), 27–69.

Bartels, D. M., & Urminsky, O. (2011). On intertemporal selfishness: How the perceived instability of identity underlies impatient consumption. *Journal of Consumer Research, 38*(1), 182–198.

Bench, S. W., Schlegel, R. J., Davis, W. E., & Vess, M. (2015). Thinking about change in the self and others: The role of self-discovery metaphors and the true self. *Social Cognition, 33*(3), 169–185.

Berkman, E. T., Livingston, J. L., & Kahn, L. E. (2017). Finding the "self" in self-regulation: The identity-value model. *Psychological Inquiry, 28*(2–3), 77–98.

Berniūnas, R., & Dranseika, V. (2016). Folk concepts of person and identity: A response to Nichols and Bruno. *Philosophical Psychology, 29*(1), 96–122.

Blok, S. V., Newman, G. E., & Rips, L. J. (2005). Individuals and their concepts. In W.-k. Ahn, R. L. Goldstone, B. C. Love, A. B. Markman, & P. Wolff (Eds.), *APA decade of behavior series. Categorization inside and outside the laboratory: Essays in honor of Douglas L. Medin* (pp. 127–149). Washington, DC: American Psychological Association.

Blok, S. V., Newman, G. E., & Rips, L. J. (2007). Postscript: Sorting out object persistence. *Psychological Review, 114*(4), 1103–1104.

Blumenthal, M., Christian, C., Slemrod, J., & Smith, M. G. (2001). Do normative appeals affect tax compliance? Evidence from a controlled experiment in Minnesota. *National Tax Journal, 54*, 125–138.

Bolderdijk, J. W., Steg, L., Geller, E. S., Lehman, P. K., & Postmes, T. (2013). Comparing the effectiveness of monetary versus moral motives in environmental campaigning. *Nature Climate Change, 3*(4), 413.

Brown, J. D. (1986). Evaluations of self and others: Self-enhancement biases in social judgments. *Social Cognition, 4*(4), 353–376.

Buckner, R. L., Andrews-Hanna, J. R., & Schacter, D. L. (2008). The brain's default network. *Annals of the New York Academy of Sciences, 1124*(1), 1–38.

Buckner, R. L., & Carroll, D. C. (2007). Self-projection and the brain. *Trends in Cognitive Sciences, 11*(2), 49–57.

Chavez, R. S., Heatherton, T. F., & Wagner, D. D. (2016). Neural population decoding reveals the intrinsic positivity of the self. *Cerebral Cortex, 27*(11), 5222–5229.

Copp, D. (2001). *Morality, normativity, and society.* New York: Oxford University Press.

Crockett, M. J., Siegel, J. Z., Kurth-Nelson, Z., Dayan, P., & Dolan, R. J. (2017). Moral transgressions corrupt neural representations of value. *Nature Neuroscience, 20*(6), 879.

D'Argembeau, A., Stawarczyk, D., Majerus, S., Collette, F., Van der Linden, M., & Salmon, E. (2010). Modulation of medial prefrontal and inferior parietal cortices when thinking about past, present, and future selves. *Social Neuroscience, 5*(2), 187–200.

De Brigard, F., Spreng, R. N., Mitchell, J. P., & Schacter, D. L. (2015). Neural activity associated with self, other, and object-based counterfactual thinking. *NeuroImage, 109*, 12–26.

De Freitas, J., Sarkissian, H., Newman, G. E., Grossmann, I., De Brigard, F., Luco, A., & Knobe, J. (2018). Consistent belief in a good true self in misanthropes and three interdependent cultures. *Cognitive Science, 42*, 134–160.

Denny, B. T., Kober, H., Wager, T. D., & Ochsner, K. N. (2012). A meta-analysis of functional neuroimaging studies of self-and other judgments reveals a spatial gradient for mentalizing in medial prefrontal cortex. *Journal of Cognitive Neuroscience, 24*(8), 1742–1752.

Dranseika, V. (2017). On the ambiguity of "the same person." *AJOB Neuroscience, 8*(3), 184–186.

Earp, B. D., Skorburg, J. A., Everett, J. A., & Savulescu, J. (2019). Addiction, identity, morality. *AJOB Empirical Bioethics, 10*(2), 136–153.

Epley, N., & Dunning, D. (2000). Feeling "holier than thou": Are self-serving assessments produced by errors in self-or social prediction? *Journal of Personality and Social Psychology, 79*(6), 861.

Ersner-Hershfield, H., Wimmer, G. E., & Knutson, B. (2008). Saving for the future self: Neural measures of future self-continuity predict temporal discounting. *Social Cognitive and Affective Neuroscience, 4*(1), 85–92.

Everett, J. A. C., Skorburg, J. A., Livingston, J. L., Chituc, V., & Crockett, M. J. (Unpublished manuscript). Morality dominates perceived identity persistence for the self, a stranger, a friend, and a foe.

Everett, J. A. C., Skorburg, J. A., & Savulescu, J. (2020). The moral self and moral duties. *Philosophical Psychology, 33*(7), 924–945.

Fossati, P., Hevenor, S. J., Lepage, M., Graham, S. J., Grady, C., Keightley, M. L., Craik, F., & Mayberg, H. (2004). Distributed self in episodic memory: Neural correlates of successful retrieval of self-encoded positive and negative personality traits. *Neuroimage, 22*(4), 1596-1604.

Frank, L. E., & Nagel, S. K. (2017). Addiction and moralization: The role of the underlying model of addiction. *Neuroethics, 10*(1), 129–139.

Garfield, J. L., Nichols, S., Rai, A. K., & Strohminger, N. (2015). Ego, egoism and the impact of religion on ethical experience: What a paradoxical consequence of Buddhist culture tells us about moral psychology. *The Journal of Ethics, 19*(3–4), 293–304.

Gomez-Lavin, J., & Prinz, J. (2019). Parole and the moral self: Moral change mitigates responsibility. *Journal of Moral Education, 48*(1), 65–83.

Glisky, E. L., & Marquine, M. J. (2009). Semantic and self-referential processing of positive and negative trait adjectives in older adults. *Memory, 17*(2), 144-157.

Hardy, S. A., & Carlo, G. (2005). Identity as a source of moral motivation. *Human Development, 48*(4), 232–256.

Hardy, S. A., & Carlo, G. (2011). Moral identity: What is it, how does it develop, and is it linked to moral action? *Child Development Perspectives, 5*(3), 212–218.

Hare, T. A., Camerer, C. F., Knoepfle, D. T., O'Doherty, J. P., & Rangel, A. (2010). Value computations in ventral medial prefrontal cortex during charitable decision making incorporate input from regions involved in social cognition. *Journal of Neuroscience, 30*(2), 583–590.

Heiphetz, L., Strohminger, N., Gelman, S. A., & Young, L. L. (2018). Who am I? The role of moral beliefs in children's and adults' understanding of identity. *Journal of Experimental Social Psychology, 78*, 210–219.

Heiphetz, L., Strohminger, N., & Young, L. L. (2017). The role of moral beliefs, memories, and preferences in representations of identity. *Cognitive Science, 41*(3), 744–767.

Hopper, J. R., & Nielsen, J. M. (1991). Recycling as altruistic behavior: Normative and behavioral strategies to expand participation in a community recycling program. *Environment and Behavior, 23*(2), 195–220.

Hutcherson, C. A., Bushong, B., & Rangel, A. (2015). A neurocomputational model of altruistic choice and its implications. *Neuron, 87*(2), 451–462.

Katzko, M. W. (2003). Unity versus multiplicity: A conceptual analysis of the term "self" and its use in personality theories. *Journal of Personality, 71*(1), 83–114.

Kim, K., & Johnson, M. K. (2015). Activity in ventromedial prefrontal cortex during self-related processing: Positive subjective value or personal significance? *Social Cognitive and Affective Neuroscience, 10*(4), 494–500.

Klein, S. B. (2012). The self and its brain. *Social Cognition, 30*(4), 474–518.

Kross, E., Bruehlman-Senecal, E., Park, J., Burson, A., Dougherty, A., Shablack, H., . . . Ayduk, O. (2014). Self-talk as a regulatory mechanism: How you do it matters. *Journal of Personality and Social Psychology, 106*(2), 304–324.

Leach, C. W., Ellemers, N., & Barreto, M. (2007). Group virtue: The importance of morality (vs. competence and sociability) in the positive evaluation of in-groups. *Journal of Personality and Social Psychology, 93*(2), 234.

Lin, W. J., Horner, A. J., & Burgess, N. (2016). Ventromedial prefrontal cortex, adding value to autobiographical memories. *Scientific Reports, 6*, 28630.

Locke, J. (1975). Of identity and diversity. In essay concerning human understanding. In J. Perry (Ed.), *Personal identity* (pp. 33–52). Berkeley, CA: University of California Press. (Original work published 1694).

Mitchell, J. P., Macrae, C. N., & Banaji, M. R. (2006). Dissociable medial prefrontal contributions to judgments of similar and dissimilar others. *Neuron, 50*(4), 655–663.

Molouki, S., & Bartels, D. M. (2017). Personal change and the continuity of the self. *Cognitive Psychology, 93*, 1–17.

Newman, G. E., Bloom, P., & Knobe, J. (2014). Value judgments and the true self. *Personality and Social Psychology Bulletin, 40*(2), 203–216.

Newman, G. E., De Freitas, J., & Knobe, J. (2015). Beliefs about the true self explain asymmetries based on moral judgment. *Cognitive Science, 39*(1), 96–125.

Nichols, S., & Bruno, M. (2010). Intuitions about personal identity: An empirical study. *Philosophical Psychology, 23*(3), 293–312.

Northoff, G., & Hayes, D. J. (2011). Is our self nothing but reward? *Biological Psychiatry, 69*(11), 1019-1025.

Olson, E. T. (2019). Personal identity. In E. N. Zalta (Ed.), *The Stanford encyclopedia of philosophy* (Fall 2019 ed.). Retrieved from https://plato.stanford.edu/archives/fall2019/entries/identity-personal/

Packer, D. J., & Cunningham, W. A. (2009). Neural correlates of reflection on goal states: The role of regulatory focus and temporal distance. *Social Neuroscience, 4*(5), 412–425.

Parfit, D. (1984). *Reasons and persons.* Oxford: Oxford University Press.

Prinz, J. J., & Nichols, S. (2017). Diachronic identity and the moral self. In J. Kiverstein (Ed.), *The Routledge handbook of philosophy of the social mind* (pp. 465–480). New York: Routledge.

Quoidbach, J., Gilbert, D. T., & Wilson, T. D. (2013). The end of history illusion. *Science, 339*(6115), 96–98.

Ross, L. D., Amabile, T. M., & Steinmetz, J. L. (1977). Social roles, social control, and biases in social-perception processes. *Journal of Personality and Social Psychology, 35*(7), 485.

Rozin, P., & Singh, L. (1999). The moralization of cigarette smoking in the United States. *Journal of Consumer Psychology, 8*(3), 321–337.

Schechtman, M. (1996). *The constitution of selves.* Ithaca, NY: Cornell University Press.

Schüpbach, M., Gargiulo, M., Welter, M., Mallet, L., Béhar, C., Houeto, J. L., . . . & Agid, Y. (2006). Neurosurgery in Parkinson disease: A distressed mind in a repaired body? *Neurology, 66*(12), 1811–1816.

Shenhav, A., & Greene, J. D. (2010). Moral judgments recruit domain-general valuation mechanisms to integrate representations of probability and magnitude. *Neuron, 67*(4), 667–677.

Shenhav, A., & Greene, J. D. (2014). Integrative moral judgment: Dissociating the roles of the amygdala and ventromedial prefrontal cortex. *Journal of Neuroscience, 34*(13), 4741–4749.

Shoemaker, D. (2019). Personal identity and ethics. In E. N. Zalta (Ed.), *The Stanford encyclopedia of philosophy* (Winter 2019 ed.). Retrieved from https://plato.stanford .edu/archives/win2019/entries/identity-ethics/

Shoemaker, D., & Tobia, K. (forthcoming). Personal identity. In J. M. Doris & M. Vargas (Eds.), *Oxford handbook of moral psychology.* Oxford: Oxford University Press.

Simon, B. (1992). The perception of ingroup and outgroup homogeneity: Reintroducing the intergroup context. *European Review of Social Psychology, 3*(1), 1–30.

Skorburg, J. A., & Sinnott-Armstrong, W. (2020). Some ethics of deep brain stimulation. In D. Stein & I. Singh (Eds.), *Global mental health and neuroethics* (pp. 117–132). New York: Elsevier.

Starmans, C., & Bloom, P. (2018). Nothing personal: What psychologists get wrong about identity. *Trends in Cognitive Sciences, 22*(7), 566–568.

Strohminger, N. (2018). Identity is essentially moral. In K. Gray & J. Graham (Eds.), *Atlas of moral psychology* (pp. 141–148). New York: Guilford Press.

Strohminger, N., Knobe, J., & Newman, G. (2017). The true self: A psychological concept distinct from the self. *Perspectives on Psychological Science, 12*(4), 551–560.

Strohminger, N., & Nichols, S. (2014). The essential moral self. *Cognition, 131*(1), 159–171.

Strohminger, N., & Nichols, S. (2015). Neurodegeneration and identity. *Psychological Science, 26*(9), 1469–1479.

Sun, J., & Goodwin, G. (2019). Do people want to be more moral? *Psychological Science, 31*(3), 243–257.

Tamir, D. I., & Mitchell, J. P. (2011). The default network distinguishes construals of proximal versus distal events. *Journal of Cognitive Neuroscience, 23*(10), 2945–2955.

Tamir, D. I., Thornton, M. A., Contreras, J. M., & Mitchell, J. P. (2016). Neural evidence that three dimensions organize mental state representation: Rationality, social impact, and valence. *Proceedings of the National Academy of Sciences, 113*(1), 194–199.

Tobia, K. (2015). Personal identity and the Phineas Gage effect. *Analysis, 75*(3), 396–405.

Tobia, K. (2016). Personal identity, direction of change, and neuroethics. *Neuroethics, 9*(1), 37–43.

Tobia, K. (2017, September 19). Change becomes you. Retrieved from https://aeon.co/essays/to-be-true-to-ones-self-means-changing-to-become-that-self

Urzia, V. (2014). *Anthony and me*. Bloomington, IN: Xlibris.

Wagner, D. D., Haxby, J. V., & Heatherton, T. F. (2012). The representation of self and person knowledge in the medial prefrontal cortex. *Wiley Interdisciplinary Reviews: Cognitive Science, 3*(4), 451–470.

White, R. E., Prager, E. O., Schaefer, C., Kross, E., Duckworth, A. L., & Carlson, S. M. (2016). The "batman effect": Improving perseverance in young children. *Child Development, 88*(5), 1563–1571.

Yarkoni, T., Poldrack, R. A., Nichols, T. E., Van Essen, D. C., & Wager, T. D. (2011). Large-scale automated synthesis of human functional neuroimaging data. *Nature Methods, 8*(8), 665.

Zhu, Y., Zhang, L., Fan, J., & Han, S. (2007). Neural basis of cultural influence on self-representation. *NeuroImage, 34*(3), 1310–1316.

5 Neuroscience and Mental Illness

Natalia Washington, Cristina Leone, and Laura Niemi

The fast-developing field of neuroscience has given philosophy, as well as other disciplines and the public broadly, many new tools and perspectives for investigating one of our most pressing challenges: addressing the health and well-being of our mental lives. In some cases, neuroscientific innovation has led to clearer understanding of the mechanisms of mental illness and precise new modes of treatment. In other cases, features of neuroscience itself, such as the enticing nature of the data it produces compared to previous behavioral methods, together with its costliness and "coldness" have complicated understanding of mental illness and decision making about mental illness. Taking neuroscientific information into account can leave practitioners in psychiatry and law with difficult questions, stemming not only from the complexity of these fields, but also from our rapidly evolving understanding of and facility with neuroscience. In this chapter, we will review several examples of the insights and dilemmas that have unfolded as mental illness has been examined through the lens of neuroscience, covering diagnoses such as obsessive-compulsive disorder (OCD), schizophrenia, addiction, and severe mood disorders.

In the first section, we will consider issues surrounding the introduction of the tools and methodologies of neuroscience to clinical research and treatment. We illustrate first with the case of treatment-resistant OCD, where being able to move from behavioral to biological explanations has led to clinical breakthroughs. Our second illustration considers efforts to identify a unifying explanation for the symptoms of schizophrenia. In this case, the contributions of neuroscience have led to little agreement, leading some to worry whether a single cohesive disorder explains the various phenomena.

In our second section, we explore how this kind of uneven progress in the identification of biological explanations for mental illness has affected our theoretical understanding both of what mental illnesses are and the appropriate ways of investigating them. We describe change at the level of national mental health policy and institutional guidelines stemming from the influence of neuroscience. In particular, we examine the National Institute of Mental Health's (NIMH) recent challenge of the ascendancy of the *Diagnostic and Statistical Manual of Mental Disorders* (DSM) as a guiding framework for psychiatric research with their release of the Research Domain Criteria (RDoC) framework, aimed at understanding mental illness as dysfunction in general psychological and neural systems (Kraemer, 2015). This move highlights a deeper issue about cognitive ontology or the frameworks we use to delineate the mind and brain into relevant component parts, using terms, for example, such as "functional system," "neural mechanism," or "brain area." These frameworks can shape what qualifies as causal information and what does not.

Finally, our third section looks at how neuroscientific information about mental illness has affected both the clinic and the public more generally. For one, consumption of neuroscientific evidence in the courtroom has become a hot button issue, since it appears that these kinds of explanations can increase stigmatizing judgments of those suffering with mental illness. At the same time, there are areas in which neuroscience itself is viewed with strong skepticism—neurobiological and neuropharmacological interventions being characterized as too expensive, too inhuman, or even outright harmful. Like genomics, artificial intelligence, and other recently developed scientific technologies, neuroscientific findings meant to contribute to decision making about mental illness face significant public debate and controversy.

We summarize with an overview of how neuroscience has informed the philosophical understanding of mental illness, despite and because of the complexity it has brought with it. We describe our expectations for the future direction of neuroscience on the topic of mental illness across disciplines and with the public.

5.1   Mental Illness and Neuroscientific Methodology

Neuroscientific research in a medical context evokes images of complicated and expensive machines and apparatuses from the late twentieth and early twenty-first century. You are probably thinking of lying in the center of a large white drum that can scan your head by using magnets (as in functional magnetic resonance imaging [fMRI]), emitting x-rays (commonly known as computed tomography), or detecting radioactive "tracer" substances (as in positron emission tomography). Or maybe you are thinking of wearing a cap of electrodes that can record electrical activity, as in electroencephalography (EEG). Each of these tools is a method of neuroimaging that can tell us something different about the brain and about mental illness. But before diving into *what* they can tell us, it is useful to think about *why* clinical scientists have come to rely on them, rather than on earlier tools involving behavioral methods.

Briefly, the 1970s ushered in a paradigm shift in thinking about mental illness. From a theoretical standpoint, this meant a change in focus from behavioral to biological explanations, as viewing the brain as a biological organ composed of cells and organized in a particular structure allowed mental illnesses to be understood as biologically grounded. Institutionally, this meant a restructuring of the standard document used to describe and diagnose mental disorders, the DSM. By the time the DSM-III was published in 1980, it had been reorganized based on descriptive definitions of mental disorders, introducing what are known as "operational diagnostic criteria"— lists of symptoms that could be independently and reliably identified by clinicians. Diagnosis of a mental disorder could now be accomplished through assessment of presenting symptoms such as depressed mood, fatigue, or insomnia—symptoms with biological or "biomedical" explanations.

Whether the DSM or the broader discipline of psychiatry has landed on the "right" concept of what mental disorders are is fiercely debated (Aragona, 2015). For now, we only need the uncomplicated point that both mental disorders and the symptoms that underlie them can be usefully targeted with the tools and methods of biomedical science—more specifically, neuroscience, the science of the brain. It is widely understood that mental illness involves other more complicated factors such as genetic and socio-environmental influences.[1] Nevertheless, when trying to understand what has gone wrong with the mind, looking at its major organ is an appropriate place to start. In the next section, we look at an example where this starting orientation has provided us with major breakthroughs: tools for intervention in treatment-resistant OCD.

5.1.1 Neurostimulation and OCD

OCD is a mental disorder commonly characterized by repetitive intrusive thoughts or obsessions that the individual tries to suppress and/or repetitive behavioral or mental rituals performed in an effort to neutralize distress. It is thought to affect approximately 2–3 percent of the population or about one in fifty people (Milad & Rauch, 2012). Like many familiar DSM diagnoses, OCD is highly comorbid with other mental disorders, in particular anxiety and depression, and is closely related to a few idiosyncratic behavior patterns that are understood as their own diagnoses (including hoarding, trichotillomania or hair pulling, and some forms of body dysmorphia). Perhaps unsurprisingly, studies of healthy populations have found that as many as 50–75 percent of people have experienced intrusive thoughts from time to time (Clark & Rhyno, 2005). The key difference between clinical and nonclinical presentations of such thoughts lies not in the quantity of intrusive thoughts alone, but rather in the appraisals of those thoughts and the cognitive effects of the appraisals. That is, while most individuals do not consider intrusive thoughts relevant to the self, do not feel much guilt about them, and are less concerned about suppressing them, individuals with OCD find the thoughts salient and unacceptable, experience intense guilt, and spend a good deal of time and concentration attempting to suppress or resist them[2] (Clark & Rhyno, 2005; Collardeau, Corbyn, & Abramowitz, 2019).

The neuroscientific study of OCD has provided a wide range of information about the mechanisms of the disorder. For our purposes, here are three key insights:

1. *Evidence of cortical–subcortical circuit disruption*—Malfunctioning cortical–striatal–thalamic–cortical (CSTC) circuits are known to be key in OCD, as is corticostriatal–limbic activation (Milad & Rauch, 2012; Stein, Goodman, & Rauch, 2000). For instance, Stein and colleagues (2000) theorize that striatal lesions lead to certain evolutionarily supported behaviors, such as hoarding supplies or washing one's hands, being performed in inappropriate excess. Moreover, it has long been known that the orbitofrontal cortex (OFC), anterior cingulate cortex (ACC), and the caudate nucleus have been shown to be overactive among patients with OCD in resting state as well as being particularly distinctive after a patient has been presented with a stimulus related to one of their obsessions or fears (Rauch et al., 1998).

2. *High activity in language areas*—Using fMRI, Kuhn and colleagues (2013) found that language circuits in the brain, and Broca's area in particular, were associated with intrusive thoughts, indicating that they may be represented linguistically. Furthermore, in participants more prone to experiencing recurrent intrusive thoughts, these areas showed more activation during resting state (Kuhn et al., 2013).

3. *Memory and imagination in OCD*—Studies of the neural correlates of counterfactual imagination also offer insight into the mechanisms of OCD in the brain. In particular, the brain encodes mental imagery similarly to when one is actually seeing the scene. Thus, when a patient with OCD replays a certain scene in order to check the veracity of a memory, the brain treats what is being pictured as interchangeable with the actual memory. As such, increased checking paradoxically decreases certainty and only serves to exacerbate the memory distrust that is symptomatic of OCD (Radomsky, Rachman, & Hammond, 2001; Radomsky, Gilchrist, & Dussault, 2006; Moritz & Jaeger, 2018; Tyron & McKay, 2009).

Fortunately, when treated early, youth with OCD have up to a 40 percent chance of significantly recovering as they reach adulthood, while adults with OCD have a 20 percent chance of recovery over forty years. In order to maximize chances of recovery, the most common treatment plan includes exposure and response prevention, a type of cognitive behavioral therapy (CBT), alongside pharmaceuticals such as selective serotonin reuptake inhibitors (SSRIs), clomipramine (a tricyclic antidepressant), or antipsychotics. Exposure and response prevention targets the anxiety caused by increased checking and other mental rituals. During the intervention, patients are hierarchically exposed to aversive situations and stimuli while the clinician encourages them to resist performing compulsions and to instead wait until the anxiety passes. This allows the brain an opportunity to habituate to the stimuli and to a sense of calm.

Exposure and response prevention programs are effective for less severe cases of OCD. This has been demonstrated with neural findings, including decreased overactivity in the OFC after treatment (Shin et al., 2013). In addition, for a broad range of severity levels, patients' OFC, ACC, and caudate nucleus activation more closely resembles that of healthy controls after combined pharmacotherapy and exposure and response prevention (Rauch et al., 1998). Hauner and colleagues (2012) likewise found distinct changes in the ACC, insula, amygdala, and visual cortex when patients were confronted with previously aversive stimuli after exposure and response prevention treatment.

What about severe, treatment-refractory cases? If patients show no improvement after thirty or more hours of CBT intervention and various courses of SSRIs and other pharmaceuticals, more serious intervention may be warranted (Veale et al., 2014). Because we can use neuroimaging techniques to observe areas of high activity directly in treatment-refractory cases, it is possible to target these areas directly. Indeed, transcranial magnetic stimulation, an intervention that painlessly transmits an electromagnetic pulse through the scalp to superficial brain targets, including the prefrontal cortex, supplementary motor area, and OFC, has shown promising results in treating OCD, as measured by both behavioral and neural outcome variables in an array of studies, including blinded randomized controlled trials (Cocchi et al., 2018; Goyal, Roy, & Ram, 2019).

Another method of directly targeting overactive areas of the brain comes in the form of deep brain stimulation (DBS). This technique, in which microelectrodes are surgically implanted in the patient's brain, has gradually become preferred over more invasive methods used in the past such as lesions of the OFC (Alonso et al., 2015). Recently, DBS devices have been used experimentally with success in areas outside the OFC, including the nucleus accumbens (NAc) and some areas of the striatum (Figee et al., 2013; Huys et al., 2019). One DBS intervention study saw at least a 50 percent reduction in symptoms in 85 percent of the otherwise treatment-refractory patients (Barcia et al., 2019).

Clearly, OCD is an excellent example of an illness where clinical insight has been directly gained from neural findings associated with individuals' symptoms, and where innovative, minimally invasive therapies have improved even the most severe cases. Unfortunately, there are other less encouraging examples where the neural mechanisms of psychiatric illness are less clear. In the next section, we will take a look at recent efforts to understand schizophrenia from a neuroscientific perspective.

5.1.2   The Heterogeneity of Schizophrenia

In the DSM-5, schizophrenia is a disorder characterized by psychotic states (periods of acute hallucinations, delusions, paranoia, and disorganized speech and behavior), as well as, in some cases, lack of affect, avolition, and catatonia. Outside of any psychotic episodes, schizophrenia is characterized by deficits in a number of cognitive abilities such as working, episodic, and declarative memory, semantic processing, attention, and verbal learning.[3] As might be imagined, a disorder with such heterogeneous symptoms makes for a difficult research target. While human research populations in psychiatry are commonly drawn from those diagnosed using DSM diagnostic criteria and rated according to a standard scale, schizophrenia researchers are often interested in particular symptoms in isolation—one lab may be focused on disordered speech, and another on hearing voices. Indeed, schizophrenia is not just phenotypically but also genetically heterogeneous (Liang & Greenwood, 2015), as genome-wide association studies over the past decades have failed to find anything like the "genes for schizophrenia" (Adriaens, 2008). That is, rather than finding any particular alleles that account for the diagnosis, schizophrenia appears to be merely weakly related to hundreds of different mutations that shape individual neurocognitive variation. This overall picture has led some theorists to argue that what we know as one disorder is really a category that subsumes many related though distinct disorders (cf. Adriaens, 2008; Gruzelier, 2011; Liang & Greenwood, 2015).[4]

Nevertheless, individuals diagnosed with schizophrenia often receive effective neuropharmacological intervention in the form of antipsychotics. This strategy may be more like the proverbial hammer than the scalpel, but its efficacy is strong evidence that the dopaminergic system is one important factor in schizophrenia (because both first- and second-generation antipsychotics work to inhibit dopamine activity in the brain; Davis et al., 1991; Grace & Gomes, 2018; McCutcheon et al., 2019). However, the exact role that dopamine plays in neural systems in schizophrenia is unclear. This may be because of the inherent limitations on research on human subjects.

As Kesby and colleagues (2018) point out, basic research is often conducted using rodents as animal models (Jones, Watson, & Fone, 2011; Marcotte, Pearson, & Srivastava, 2001; Tomasella et al., 2018). In these studies, positive symptoms are induced in the animal subjects using psychostimulants. Unfortunately, there are key differences between these models and findings in clinical research with humans. While dopaminergic activity in subcortical regions is associated with positive symptoms both in humans and in nonhuman animal models, the precise location of the activity has been found to differ: dopaminergic activity in the striatum is an important correlate of positive symptoms in humans, whereas animals' symptoms are associated with dopaminergic activity in the limbic system (Kesby et al., 2018). Kesby and colleagues propose using more appropriate animal models in future studies so that the molecular-level causes can be better understood.[5]

Another approach to the neural understanding of schizophrenia focuses on electrical activity in the brain and what is referred to as the oscillatory connectome—the back and forth "harmonic" patterns of electrical activity in the brain during its resting and working states. Using EEG to measure event-related potentials, the electrical response of the brain to a particular stimulus or "event," Light and Swerdlow (2015) report that a reliable bio-marker of schizophrenia severity is low mismatch negativity. Mismatch nega-tivity occurs in response to an unexpected stimulus (e.g., a red square amid a series of blue circles, a consonant amid a series of vowels, or a "no-go" signal amid a series of "go" signals). Briefly, response to an expected stimulus gener-ally involves a rise in potential to around 1–2 mV as recorded by EEG and a return to zero within the first 400 milliseconds. An unexpected stimulus, on the other hand, involves a rise above or around 1–2 mV, generally followed by a drop *below* zero. This difference in amplitude constitutes the mismatch negativity. In short, the brain reacts more strongly when it is surprised.

In individuals with a diagnosis of schizophrenia, difference in response amplitude between expected and unexpected stimuli is much smaller than in controls. More recent studies have also associated this low mismatch nega-tivity with particular brain regions. Among individuals with schizophrenia, expected stimuli prompted lessened thalamocortical activation compared to controls, and unexpected stimuli produced reduced cortical connectivity in comparison to controls (Lee et al., 2019). Synthesizing related research, Nagai and colleagues (2013) report that "mismatch negativity amplitude reduction reflects sensory network dysfunction in schizophrenia" (p. 2).

What are we to make of this conclusion? Is sensory network dysfunction an additional neurological "symptom" of the schizophrenia diagnosis to be added to the list of cognitive behavioral deficits that these individuals experi-ence? Does it underlie one or more of those symptoms, or play a causal role in the development of psychosis? What can we learn from the more general pattern of decreased functional connectivity in individuals with schizophre-nia? With luck, greater specificity in network-level analyses in these special populations will improve the future of computer-based cognitive therapy programs, and help patients better understand and respond to a range of sensory and affective stimuli. This line of inquiry remains ongoing.

5.2   Mental Illness and Neurocognitive Ontology

Our analysis of current research on OCD and schizophrenia in the last sec-tion highlights an important limitation of neuroscientific methodology. Tools such as fMRI and EEG can reveal areas of high and low activity in the brain, but it is left to the scientific theorist to infer from this evidence whether something is working well or poorly, and up to the clinician to determine the usefulness of this information.

In the case of OCD, we discussed overactivity in several areas but were able to pass by the question of what exact neural mechanism or mechanisms that overactivity represents. Safe to say, the brain likely has no dedicated "intrusive thought" or "ritual behavior" circuitry that can be identified across individuals. At most, we can say that the OFC, ACC, and likely the CSTC loop play a role in the production of these symptoms. This gap in our understanding is eased by the fact that we can leverage the tools at our disposal—microelectrodes for instance—to mitigate behavioral symptoms without necessarily knowing whether we have a case of "neural mechanism gone wrong" or how those neural mechanisms support associated cognitive functions. Unfortunately, we have no such elegant solutions for schizophrenia at this time. Thus, we are left asking *how* exactly the dopaminergic system, on the one hand, is causally related to, say, working memory deficits or auditory hallucinations on the other (and then, hopefully, how we can leverage that understanding in a therapeutic setting).

This inferential gap is a problem about cognitive ontology—the search for the structure or, more accurately, the structural components of the mind. An ontology of the mind would provide an answer to the question "What are the mind's parts?" Notice that this is a separate question from "What are the parts or systems of the brain?" A major goal of cognitive neuroscience, then, is to orient these frameworks with respect to each other or, as psychologists Russ Poldrack and Tal Yarkoni (2016) put it, to "delineate how brain systems give rise to mental function[s]." The introduction of psychiatric categories—such as "symptoms," "dysfunctions," "diagnoses," and "disorders"—introduces additional layers of complexity to the situation. In the remainder of this section, we will discuss the extent to which disciplinary interests in psychiatry and neuroscience overlap and whether any degree of theoretical reductionism is warranted. That is, we will take stock of whether the mechanisms of interest to neuroscientists are the same as the ones relevant to mental illness in the clinic and consider what happens if not.

5.2.1 Neurocognitive Ontology and Diagnosis

One influential answer to questions about neurocognitive ontology has recently been defended by Ken Kendler, Peter Zachar, and Carl Craver in their paper "What Kinds of Things are Psychiatric Disorders?" (2011). Drawing an analogy with Richard Boyd's work on property clusters as definitive of biological species, they argue that "the complex and multi-level causal mechanisms that produce, underlie and sustain psychiatric syndromes" are researchers' most appropriate targets (p. 1146). In making this proposal, Kendler and colleagues are distinctly aware that "no one level is likely to capture the full complexity of the mechanisms sustaining or underlying . . . our best-codified diagnostic categories"—that is, that mental disorders will not be fully understood by excluding the causal contributions of our physical and social environments (p. 1148). Nonetheless, they wish to emphasize that "there are more or less general modes of functioning in the human mind/brain and [neurocognitive] mechanisms that sustain those different modes of functioning" (p. 1147). Thus, Kendler and colleagues, and others who adopt this view, are making a kind of methodological bet—namely, that continued research in cognitive neuroscience is the best route toward providing satisfactory explanations of mental illness. That is, when trying to understand the underpinnings of the diagnostic categories we are most familiar with (those in the DSM and so on), they insist neural mechanisms are the best place to shine the light.

This assumption has recently been challenged by several authors. Washington (2016) and Murphy (2017) note that the high variability and plasticity of human minds and brains suggest that the causal contributions of our physical and social environments may be more important than previously assumed. Tekin (2016) laments that much of current psychiatric research fails to incorporate first-person accounts of those with mental disorders and information about the role of the self in particular. But most pressing for our purposes here might be a kind of methodological circularity or what philosopher Kathryn Tabb (2015) calls the assumption of diagnostic discrimination—in her words, the assumption that "our diagnostic tests group patients together in ways that allow for relevant facts about mental disorder to be discovered" (p. 1049).

To understand the worry here, imagine that you have a large pile of shiny rocks that you presume are gold and wish to use in experiments. Some are gold, but some are pyrite or "fool's gold," and so they have different underlying chemical structures. On the assumption that everything in your pile is of the same kind, you might run your experiments assuming common properties in your specimens. Your research might consistently achieve only partial success and, in particular, never be very illuminating about what makes gold *gold*. In fact, what distinguishes it from different rocks might always seem mysterious. In the same way, neuroscientific research in psychiatry for the most part proceeds by focusing on populations that have already been grouped by diagnostic category. Yet, as we have discussed, current diagnostic standards neither easily accommodate comorbidity nor involve precise tracking of the heterogeneity involved in mental disorders such as schizophrenia. It is an open question then, for example, whether examining the role of dopamine in working memory in individuals diagnosed with schizophrenia, or otherwise conducting diagnosis-based research, is the right way to proceed. Results might say nothing about how to help any particular patient, or why our population has been grouped together in the first place. One alternative would be to focus research on the connection between neurotransmitter function and observable behavioral dysfunction. In other words, Tabb (2015) and others see a future where current psychiatric diagnoses do not neatly reduce to dysfunction in neurocognitive mechanisms—where our current nosology either is reimagined to fit more neatly with neuroscientific findings or is validated by emphasizing causal pathways that crosscut neuroscientific ontological distinctions (pathways in our social, cultural, and environmental ecologies for instance).

How we best proceed from here remains open. Some argue that psychiatry should stay away from so-called mechanisms entirely (Hartner & Theuerer, 2018). On the other hand, the NIMH has doubled down on the search for mechanistic dysfunction in psychological systems. In the wake of the publication of the DSM-5, NIMH released the RDoC—not itself a diagnostic manual but rather a framework for structuring psychiatric research that stays away from diagnostic categories. Those seeking funding from the NIMH may now propose to investigate one or more domains or constructs (e.g., "reward learning" or "visual perception") from one or more levels of analysis (e.g., "cells" or "circuits"). If successful, these efforts may reshape psychiatric categories as we know them.

5.2.2 Neurocognitive Ontology and the Roots of Addiction and Substance Abuse

To cap off this discussion, let us examine a concrete case. Addiction and substance abuse are serious contemporary mental health concerns. To what extent has neuroscientific evidence advanced our understanding of these conditions or given us new tools for intervention? To begin with, we know that substance use causes immediate chemical changes in the brain. Dopamine, for example, which is associated with changes in the NAc and amygdala, plays a significant role in the reinforcement of addiction and may drive drug reward (Koob, Sanna, & Bloom, 1998). Nutt and McLellan (2013) also note that dopamine receptor density plays a role in determining the extent to which an individual will enjoy stimulants, which might explain why certain individuals develop dependence. However, they also emphasize that because the specifics of different drugs' actions in the brain are different—opioids mimic endogenous opioid transmitters, while alcohol blocks glutamate and enhances GABA—the notion that a single neurotransmitter mechanism explains drug dependence is no longer viable. We also know that chronic drug use can adapt brains in ways that are enduring and complex. Using network analysis, McHugh and colleagues (2013) found that individuals addicted to cocaine had reduced connectivity between the bilateral putamen and posterior insula and right postcentral gyrus and scored higher in impulsivity than control subjects. Drug abuse has also been linked to impaired glutamate homeostasis, which impairs prefrontal regulation of striatal circuitry, potentially explaining why drug users are unable to control their drug seeking (Kalivas, 2009).

In sum, it seems unlikely that any one neural mechanism can explain all of the symptoms of addiction. The neurochemical drivers of cravings, relapse, and withdrawal and long-term damage are different among both different substances and different individuals. Indeed, approximately 40–60 percent of variation in levels of addiction can be attributed to either genetic factors alone or gene–environment interactions (Volkow & Ting-Kai, 2005). Thus, while effective treatment may be available in a narrow set of situations—the immediate use of an opioid overdose reversal intervention such as naloxone, for example—treatment for long-term care and recovery from substance abuse is a difficult prospect. Standard care, for those who have access, generally takes the form of psychosocial support such as family care, residential treatment ("rehab clinics"), expert counseling, and support group activity.

Perhaps this is not surprising. Addiction, like other mental disorders, is grounded in the brain, but it is also uniquely influenced by features of the social environment. Indeed, the current opioid crisis is the prime case study for the link between substance abuse and social capital. Deaths by prescription opioid overdose have tripled between 2001 and 2016 in people aged fifteen to thirty-four years (Samet & Kertesz, 2018), and there has been a pronounced increase in the prevalence of overdose and other "diseases of despair"—for instance, alcohol abuse and suicide—among middle-aged white people without a college degree (Dasgupta, Beletsky, & Ciccarone,

2018). Furthermore, Heilig and colleagues (2016) found there is a strong link between addiction and social integration, with social exclusion and addiction linked to activity in the insula.

In effect, then, the diagnostic category "addiction" groups individuals together who share important social features but who vary widely with respect the neurocognitive mechanisms that sustain their behavior. One important ontology—that which identifies neurocognitive dysfunctions in the brain—appears to crosscut another—that which groups clinically relevant populations. Is addiction, then, best understood through the lens of neurocognitive dysfunction? More pressingly, where can we best exert leverage in order to help those struggling with substance abuse? These questions are left open by current neuroscientific theory and methodology.

5.3   Neuroscience, Science Communication, and Impacts on Treatment

In our final section, we would like to return to applied issues. Mental health is a concern for all. Thus, it is worth asking how neuroscientific information is communicated to the public more generally and how it is received. To begin, we will look at a domain where neuroscientific information about mental illness is thought to be relevant to our practices of blame and punishment.

5.3.1   The Perceived Credibility of Neuroscience in the Courtroom

In 2005, Grady Nelson brutally murdered his wife and her eleven-year-old daughter in Miami-Dade County, Florida.[6] His sentence of life in prison rather than the death penalty stirred controversy in 2010, when several jurors indicated that the presentation of EEG "scans" (i.e., images of recorded waveform amplitude readings) influenced their decision making. As one juror put it, "The technology really swayed me . . . After seing [sic] the brain scans, I was convinced this guy had some sort of brain problem" (Ovalle, 2010, p. 2). As this case highlights, neural evidence can be of weighty importance in judgments of responsibility. Briefly, this is because our folk notions of responsibility typically require that actors have some degree of control or agency over their behavior, and neural evidence of a mental disorder suggests threats to a person's agency (for more, see King & May, 2018; Murphy & Washington, forthcoming). If, as Grady's defense attorney argued, "the moment this crime occurred, Grady had a broken brain," this can sometimes be seen as a reason to mitigate punishment (Ovalle, 2010, p. 1).

One may reasonably wonder what EEG readings recorded much later than the events of 2005 can tell us. Even if they are evidence of neuroatypicality in Grady, they might not be evidence of any particular diagnosis, supposing we understood how to draw such inferences.[7] They might not be evidence of whether Grady suffered from mental illness at the earlier time when he committed his crimes. And even if this is known, there is ongoing debate about whether a diagnosis itself is sufficient to mitigate blame in these kinds of cases or whether it must be demonstrated that mental illness was causally relevant to a subject's degree of control in the particular behavior (Sifferd, Hirstein, & Fagan, 2016). For example, it may be known that I suffer from compulsions, but unknown whether compulsive behavior was implicated in a particular act of theft (e.g., maybe I intended to get back at a rival). In Grady's case, prosecuting attorneys insisted that testimony from neuroscientists "was a lot of hocus pocus and bells and whistles" and that "When you look[ed] at the facts of the case, there was nothing impulsive about this murder" (Ovalle, 2010, p. 2).

The big worry here is that these questions reveal subtleties that can be overlooked by jurors and other non-experts. This is becoming a pressing problem, as the number of court cases that involve the presentation of neuroscientific evidence is growing rapidly (Farahany, 2015). Worse, recent analyses from psychologists Nick Haslam and Erlend Kvaale (2015) reveal a link between "biogenetic" forms of explanation (e.g., recourse to neural dysfunction to explain mental disorder) and increased stigma surrounding mental illness. That is, at the same time that a neurally based assessment can diminish attributions of blame for an individual, it increases perceived dangerousness and desire for social distance as well as imputing a kind of prognostic pessimism (Haslam & Kvaale, 2015). With such high stakes, there is increasing pressure on neuroscientists concerning the misinterpretation of their research.

5.3.2   The Perceived Efficacy of ECT

When aimed at ameliorating mental illness, neuroscience and neuroscientific tools themselves also face their share of criticism. In most stories in movies and television, things have taken a rotten turn for the protagonist when doctors wheel out the electrodes and restraints. It may therefore surprise you to hear that since the 1940s, electroconvulsive therapy[8] (ECT) has been considered a safe and effective treatment for severe mood disorders, with widespread clinical support. During an ECT treatment, a subject first undergoes general anesthesia. Then, a small electric current is used to induce a brief seizure, triggering chemical changes that reverse severe symptoms. Slade and colleagues (2017) found that among patients with major depressive disorder, bipolar disorder, or schizoaffective disorder, ECT is associated with a 6.6 percent thirty-day readmission risk compared to 12.3 percent for individuals who did not receive ECT. Moreover, Bjølseth and colleagues (2015) found that ECT can be effective for elderly patients with major depression, and that both bifrontal and unilateral ECT are associated with statistically significant decreases in symptom severity. ECT can also be used to treat agitation in dementia patients. Tang and colleagues (2014) found that dementia patients who received ECT experienced a significant reduction in agitation based on Pittsburgh Agitation Scale scores (Glass, Forester, & Hermida, 2017).

In movies and on TV, unfortunately, ECT is continually portrayed as violent and frightening. In a 2016 study of TV programs and movies, Sienaert found that the patient had not given consent in 57.3 percent of scenes portraying ECT, the patient was not given an anesthetic in 72 percent, and the apparatus was used for torture in 13.4 percent. ECT has also been portrayed as a way to erase memories. This frightening picture is many people's first introduction to ECT. Ithman and colleagues (2018) found that 94 percent of medical students prior to training had learned about ECT from either a film or word of mouth, and 24.05 percent reported they were frightened by the procedure. ECT's historically low use may therefore be driven by patient and practitioner stigma or by beliefs of its negative cognitive effects[9] (Sackeim, 2017).

Better information access is associated with a reduction in negative assessments of ECT. In the study by Ithman and colleagues (2018), only 2.53 percent of medical students surveyed reported that they continued to fear ECT after clerkship. Exposure to ECT, for both the individual and their family, has been shown to be effective in reducing patient fears. For example, Elias, Ang, Schneider, and George (2019) similarly found that when family members watched ECT procedures, a majority (76 percent) of family members reported that the experience was reassuring and rewarding, and 71 percent of families reported that it lessened their fears and improved their knowledge of ECT.

5.3.3   Conclusions and Future Directions

Each development in clinical neuroscience affects the lives of people in mental distress, whether through our public institutions or in private life as people try to make sense of the latest findings conveyed by the news, entertainment media, and word of mouth. Scientists, philosophers, and the public can neither keep up with the pace of novel findings in clinical neuroscience nor curate the findings consistently. Some diagnoses—or behavioral, cognitive, or emotional disruptions—will tend to receive more focus than others and will be better investigated with more valid results. We have reviewed several areas where neuroscientific findings have contributed directly to human welfare, including the treatment of severe OCD, ECT for severe depression, and in the understanding of addictive processes. We have described ways that neuroscience at multiple levels of analyses has complicated the understanding of mental disorders, including schizophrenia and addiction, and how this might also lead to a fundamental restructuring of how we conceptualize mental illnesses and create guidelines for diagnosis, treatment, and research.

The tools of neuroscience are enticing, and this is not a sin that needs to be punished. These tools have, in some cases, shown us new paths toward recovery and symptom alleviation when we thought none were available. Managing this powerful feature while conducting ethical research with those in mental distress is the responsibility of clinical scientists. We note that accuracy in news reporting and the presentation of findings to the public in an unbiased manner remains a problematic issue if baseline neuroscience knowledge in the public is low, as it is. Justice in the courtroom is questionable when fates are determined by scientific information that is understandable for a minority of a jury, for example.

There is currently broad support for fuller integration of neuroscience in research on mental health, and clinical interventions are rapidly developing. Ultimately, neuroscience has the potential to reveal better ways of understanding the causes of mental distress and to transform how we categorize mental illness and health.

Acknowledgments

Many thanks to Natalie Boychuk for research assistance.

Notes

1. One influential model of mental illness, known as the biopsychosocial model, understands psychiatric symptoms as the products of biological, psychological, and social causes.

2. An important note is that in the case of OCD, such thoughts are ego-dystonic, meaning they align neither with who the individual is nor with their intentions/ wishes. In other disorders wherein these thoughts might be ego-syntonic, a different approach to treatment is warranted.

3. Though these are not reflected in the DSM criteria.

4. Many thanks to Niklas Andersson (Washington University, St. Louis, MO) for discussions on this point.

5. Kanyuch & Anderson (2017), for instance, propose using the marmoset monkey, given that marmosets' prefrontal cortices more closely resemble those of humans, as well as the fact that marmosets have already been used as animal models in studies of working memory, anhedonia, fear generalization, and cognitive flexibility.

6. State v. Nelson, No. F05–846 (11th Fla. Cir. Ct. Dec. 2, 2010), archived at https:// perma.cc/7XA5-2JXG?type=pdf; Judge Okays QEEG Evidence, supra note 8; Miller, supra note 6. For more information, see generally Transcript of Opening Statement, Nelson, No. F05–846, archived at https://perma.cc/6TZZ-NZHA?type=pdf

7. One reason Grady's case stands out to us is the availability of testimony from those directly involved in the hearings. A drawback of relying on it here is that we are limited to focusing on the influence of the EEG evidence rather than on the influence of any particular psychiatric diagnosis. However, other recent high-profile cases do highlight the diagnosis itself (see, e.g., Sifferd et al.'s, 2018, treatment of mass murderer Anders Brevik).

8. Commonly referred to by the misnomer, "electroshock" therapy.

9. The most common side effects of ECT include disorientation and impairments in learning and anterograde and retrograde memory. This can be mitigated by using brief pulse waveform over sine wave simulation and by conducting unilateral ECT rather than bilateral. The adverse cognitive effects are also attributable to an individual's seizure threshold—as there is no way to predict accurately how much electricity an individual can take, some patients (especially young, small women) may receive excess current (Prudic, 2008). This said, it is necessary to weigh these effects against the continuation of severe symptoms requiring ongoing hospitalization.

References

Adriaens, P. R. (2008). Debunking evolutionary psychiatry's schizophrenia paradox. *Medical Hypotheses, 70*, 1215–1222.

Alonso, P., Cuadras, D., Gabriëls, L., Denys, D., Goodman, W., Greenberg, B. D., . . . Menchon, J. M. (2015). Deep brain stimulation for obsessive-compulsive disorder: A meta-analysis of treatment outcome and predictors of response. *PLoS One, 24*, 10(7). doi: 10.1371/journal.pone.0133591

Aragona, M. (2015). Rethinking received views on the history of psychiatric nosology: Minor shifts, major continuities. In P. Zachar, D. Stoyanov, M. Aragona, & A. Jablensky (Eds.), *Alternative perspectives on psychiatric validation: DSM, ICD, RDoC, and beyond* (pp. 27–46). Oxford: Oxford University Press.

Barcia, J. A., Avecillas-Chasin, J. M., Nombela, C., . . . Strange, B. A. (2019). Personalized striatal targets for deep brain stimulation in obsessive-compulsive disorder. *Brain Stimulation, 12*(3), 724–734.

Bjølseth, T. M., Engedal, K., Benth, J. S., Dybedal, G. S., Gaarden, T. L., & Tanum, L. (2015). Clinical efficacy of formula-based bifrontal versus right unilateral electroconvulsive therapy (ECT) in the treatment of major depression among elderly patients: A pragmatic, randomized, assessor-blinded, controlled trial. *Journal of Affective Disorders, 175*, 8–17.

Clark, D. A., & Rhyno, S. (2005). Unwanted intrusive thoughts in nonclinical individuals: Implications for clinical disorders. In D. A. Clark (Ed.), *Intrusive thoughts in clinical disorders: Theory, research, and treatment* (pp. 1–29). New York: Guilford Press.

Cocchi, L., Zalesky, A., Nott, Z., Whybird, G., Fitzgerald, P. B., & Breakspear, M. (2018). Transcranial magnetic stimulation in obsessive-compulsive disorder: A focus on network mechanisms and state dependence. *NeuroImage: Clinical, 19*, 661–674. https://doi.org/10.1016/j.nicl.2018.05.029

Collardeau, F., Corbyn, B., & Abramowitz, J. (2019). Maternal unwanted and intrusive thoughts of infant-related harm, obsessive-compulsive disorder and depression in the perinatal period: study protocol. *BMC Psychiatry, 19*(94). https://doi.org/10.1186/s12888-019-2067-x

Dasgupta, N., Beletsky, L., & Ciccarone, D. (2018). Opioid crisis: No easy fix to its social and economic determinants. *American Journal of Public Health, 108*(2), 182–186.

Davis, K. L., Kahn, R. S., Ko, G., & Davidson, M. (1991). Dopamine in schizophrenia: A review and reconceptualization. *The American Journal of Psychiatry, 148*(11), 1474–1486. https://doi.org/10.1176/ajp.148.11.1474.

Elias, A., Ang, A., Schneider, A., George, K. (2019). Family presence during electroconvulsive therapy. *Journal of ECT, 35*(2), 91–94. doi: 10.1097/YCT.0000000000000559.

Farahany, N. A. (2015). Neuroscience and behavioral genetics in US criminal law: An empirical analysis. *Journal of Law and the Biosciences, 2*(3), 485–509.

Figee, M., Luigjes, J., Smolders, R., Valencia-Alfonso, C. E., van Wingen, G., de Kwaasteniet . . . & Denys, D. (2013). Deep brain stimulation restores frontostriatal network activity in obsessive-compulsive disorder. *Nature Neuroscience, 16*(4), 386–387. doi: 10.1038/nn.3344.

Glass, O. M., Forester, B. P., & Hermida, A. P. (2017). Electroconvulsive therapy (ECT) for treating agitation in dementia (major neurocognitive disorder)—A promising option. *International Psychogeriatrics, 29*(5), 717–726.

Goyal, N., Roy, C., & Ram, D. (2019). FMRI correlates of neuromodulation of the dorsolateral prefrontal cortex using transcranial magnetic stimulation in patients with resistant obsessive compulsive disorder. *Brain Stimulation, 12*(2), P587.

Grace, A. A., & Gomes, F. V. (2018). The circuitry of dopamine system regulation and its disruption in schizophrenia: Insights into treatment and prevention. *Schizophrenia Bulletin, 45*(1), 148-157. doi:10.1093/schbul/sbx199

Gruzelier, J. (2011). The heterogeneity of schizophrenia: The inconvenient truth. *Neuroscience Letters, 500*, e16.

Hartner, D. F., & Theuerer, K. L. (2018). Psychiatry should not seek mechanism of disorder. *Journal of Theoretical and Philosophical Psychology, 38*(4), 189–204.

Haslam, N., & Kvaale, E. P. (2015). Biogenetic explanations of mental disorder: The mixed-blessings model. *Current Directions in Psychological Science, 24*(5), 399–404.

Hauner, K. K., Mineka, S., Voss, J. L., & Paller, K. A. (2012). Exposure therapy triggers lasting reorganization of neural fear processing. *Proceedings of the National Academy of Sciences of the United States of America, 109*(23), 9203–9208.

Heilig, M., Epstein, D. H., Nader, M. A., & Shaham, Y. (2016). Time to connect: Bringing social context into addiction neuroscience. *Nature Reviews Neuroscience, 17*(9), 592.

Huys, D., Kohl, S., Baldermann, J. C., Timmermann, L., Sturm, V., Visser-Vandewalle, V., & Kuhn, J. (2019). Open-label trial of anterior limb of internal capsule–nucleus accumbens deep brain stimulation for obsessive-compulsive disorder: Insights gained. *Journal of Neurology, Neurosurgery, and Psychiatry, 90*(7), 805–812. doi: 10.1136/jnnp-2018-318996.

Ithman, M., O'Connell, C., Ogunleye, A., Lee, S., Chamberlain, B., & Ramalingam, A. (2018). Pre- and post-clerkship knowledge, perceptions, and acceptability of electroconvulsive therapy (ECT) in 3rd year medical students. *Psychiatry Quarterly, 89*(4), 869–880.

Jones, C. A., Watson, D. J. & Fone, K. C. (2011). Animal models of schizophrenia. British Journal of Pharmacology, 164(4), 1162–1194.

Kalivas, P. W. (2009). The glutamate homeostasis hypothesis of addiction. *Nature Reviews Neuroscience, 10*(8), 561.

Kanyuch, N., & Anderson, S. (2017). Animal models of developmental neuropathology in schizophrenia. *Schizophrenia Bulletin, 43*(6), 1172–1175. doi: 10.1093/schbul/sbx116.

Kendler, K. S., Zachar, P., & Craver, C. (2011). What kinds of things are psychiatric disorders?. *Psychological medicine, 41*(6), 1143–1150. https://doi.org/10.1017/s0033291710001844.

Kesby, J. P., Eyles, D. W., McGrath, J. J., & Scott, J. G. (2018). Dopamine, psychosis and schizophrenia: The widening gap between basic and clinical neuroscience. *Translational Psychiatry, 8*(1), 30.

King, M., & May, J. (2018). Moral responsibility and mental illness: A call for nuance. *Neuroethics, 11*(1), 11–22.

Koob, G. F., Sanna, P. P., & Bloom, F. E. (1998). Neuroscience of addiction. *Neuron, 21*(3), 467–476.

Kraemer, H.C. (2015). Research domain criteria (RDoC) and the *DSM*—two methodological approaches to mental health diagnosis. JAMA Psychiatry, 72(12), 1163–1164. doi:10.1001/jamapsychiatry.2015.2134

Kühn, S., Schmiedek, F., Brose, M., Schott, B. H., Lindenberger, U., & Lövden, M. (2013). The neural representation of intrusive thoughts. *SCAN, 8*, 688–693.

Lee, W. H., Moser, D. A., Ing, A., Doucet, G. E., & Frangou, S. (2019). Behavioral and Health Correlates of Resting-State Metastability in the Human Connectome Project (2019). *Brain Topography, 32*(1), 80-86. doi: 10.1007/s10548-018-0672-5.

Liang, G. S., & Greenwood, T. A. (2015). The impact of clinical heterogeneity in schizophrenia on genomic analyses. *Schizophrenia Research, 161*(2–3), 490–495.

Light, G. A., & Swerdlow, N. R. (2015). Neurophysiological biomarkers informing the clinical neuroscience of schizophrenia: Mismatch negativity and prepulse inhibition of startle. *Current topics in behavioral neurosciences, 21*, 293–314. https://doi.org/10.1007/7854_2014_316.

Marcotte, E. R., Pearson, D. M., & Srivastava, L. K. (2001). Animal models of schizophrenia: a critical review. *Journal of psychiatry & neuroscience: Journal of Psychiatry and Neuroscience, 26*(5), 395–410.

McCutcheon, R. A., Abi-Dargham, A., & Howes, O. D. (2019). Schizophrenia, dopamine and the striatum: From biology to symptoms. *Trends in Neurosciences, 42*(3), 205–220. https://doi.org/10.1016/j.tins.2018.12.004.

McHugh, M. J., Demers, C. H., Braud, J., Briggs, R., Adinoff, B., & Stein, E. A. (2013). Striatal-insula circuits in cocaine addiction: Implications for impulsivity and relapse risk. *The American Journal of Drug and Alcohol Abuse, 39*(6), 424–432.

Milad, M. R., & Rauch, S. L. (2012). Obsessive-compulsive disorder: Beyond segregated cortico-striatal pathways. *Trends in cognitive sciences, 16*(1), 43–51. https://doi.org/10.1016/j.tics.2011.11.003.

Moritz, S., & Jaeger, A. (2018). Decreased memory confidence in obsessive-compulsive disorder for scenarios high and low in responsibility: Is low still too high? *European Archives of Psychiatry and Clinical Neuroscience, 268*(3), 291–299.

Murphy, D. (2017). Are all mental disorders culture-bound syndromes? In K. Kendler and J. Parmas (Eds.), *Philosophical issues in psychiatry IV: Psychiatric nosology in the light of DSM-5* (pp. 152–165). Baltimore, MD: Johns Hopkins University Press.

Murphy, D., & Washington, N. (forthcoming). Agency in mental illness and disability. In J. M. Doris & M. Vargas (Eds.), *Oxford handbook of moral psychology*. Oxford: Oxford University Press.

Nagai, T., Tada, M., Kirihara, K., Araki, T., Jinde, S., & Kasai, K. (2013). Mismatch negativity as a "translatable" brain marker toward early intervention for psychosis: A review. *Frontiers in Psychiatry, 4*, 115. https://doi.org/10.3389/fpsyt.2013.00115.

Nutt, D., & McLellan, A. T. (2013). Can neuroscience improve addiction treatment and policies? *Public Health Reviews, 35*, 5.

Ovalle, D. (2010, December 11). Novel defense helps spare perpetrator of grisly murder. *The Miami Herald*, pp. 1–2.

Poldrack, R. A., & Yarkoni, T. (2016). From brain maps to cognitive ontologies: Informatics and the search for mental structure. *Annual Review of Psychology, 67*(1), 587–612.

Prudic, J. (2008). Strategies to minimize cognitive side effects with ECT: Aspects of ECT technique. *The Journal of ECT, 24*(1), 46–51.

Radomsky, A. S., Gilchrist, P. T., & Dussault, D. (2006). Repeated checking really does cause memory distrust. *Behaviour Research and Therapy, 44*(2), 305–316. https://doi.org/10.1016/j.brat.2005.02.005.

Radomsky, A. S., Rachman, S., & Hammond, D. (2001). Memory bias, confidence and responsibility in compulsive checking. *Behaviour Research and Therapy, 39*(7), 813–822. https://doi.org/10.1016/s0005-7967(00)00079-6.

Rauch, S. L, Whalen, P. J, Dougherty, D., Jenike, M. A. (1998). Neurobiological models of obsessive–compulsive disorder. In M. A. Jenike, L. Baer, & W. E. Minichiello (Eds.), *Obsessive–compulsive disorders practical management* (pp 222–253). St. Louis, MO: Mosby.

Sackeim, H. A. (2017). Modern electroconvulsive therapy: Vastly improved yet greatly underused. *JAMA Psychiatry, 74*(8), 779–780.

Samet, J. H., & Kertesz, S. G. (2018). Suggested paths to fixing the opioid crisis: Directions and misdirections. *JAMA Network Open, 1*(2), e180218.

Schilman, E. A., Klavir, O., Winter, C., Sohr, R., & Joel, D. (2010). The role of the striatum in compulsive behavior in intact and orbitofrontal-cortex-lesioned rats: Possible involvement of the serotonergic system. *Neuropsychopharmacology, 35*(4), 1026–1039. https://doi.org/10.1038/npp.2009.208.

Shin, L. M., Davis, F. C., VanElzakker, M. B, Dahlgren, M. K., & Dubois, S. J. (2013). Neuroimaging predictors of treatment response in anxiety disorders. *Biology of Mood and Anxiety Disorders, 3*(15). https://doi.org/10.1186/2045-5380-3-15.

Sienaert, P. (2016). Based on a true story? The portrayal of ECT in international movies and television programs. *Brain Stimulation, 9*, 882–891.

Sifferd, K., Hirstein, W., & Fagan, T. (2016). Legal insanity and executive function. In M. White (Ed.), *The insanity defense: Multidisciplinary views on its history, trends, and controversies* (pp. 215–242). Westport, CT: Praeger.

Sifferd, K. L. (2018). Non-eliminative reductionism: Not the theory of mind some responsibility theorists want, but the one they need. In B. D. Lazarov (Ed.), *Neurolaw and responsibility for action: concepts, crimes, and courts* (pp. 71–103). Cambridge: Cambridge University Press.

Slade, E. P., Jahn, D. R., Regenold, W. T., & Case, B. G. (2017). Association of electroconvulsive therapy with psychiatric readmissions in U.S. Hospitals. *JAMA Psychiatry, 74*(8), 798–804.

Stein, D. J., Goodman, W. K., & Rauch, S. L. (2000). The cognitive-affective neuroscience of obsessive-compulsive disorder. *Current Psychiatry Reports, 2*, 341–346. doi:10.1007/s11920-000-0079-2

Tabb, K. (2015). Psychiatric progress and the assumption of diagnostic discrimination. *Philosophy of Science, 82*(5), 1047–1058.

Tang, Y. L., Hermida, A. P., Ha, K., Laddha, S. R., & McDonald, W. M. (2014). Efficacy and safety of ECT for behavioral and psychological symptoms in dementia (BPSD): A retrospective chart review. *American Journal of Geriatric Psychiatry, 22*, S114–S115.

Tekin, Ş. (2016). Are mental disorders natural kinds?: A plea for a new approach to intervention in psychiatry. *Philosophy, Psychiatry, & Psychology, 23*, 147–163.

Tomasella, E., Bechelli, L., Ogando, M. B., Mininni, C., Di Guilmi, M. N., De Fino, F., . . . Gelman, D. M. (2018). Deletion of dopamine D2 receptors from parvalbumin interneurons in mouse causes schizophrenia-like phenotypes. *Proceedings of the National Academy of Sciences of the United States of America, 115*(13), 3476–3481. https://doi.org/10.1073/pnas.1719897115.

Tyron, W. W., & McKay, D. (2009). Memory modification as an outcome variable in anxiety disorder treatment. *Journal of Anxiety Disorders, 23*, 546–556.

Veale, D., Miles, S., Smallcombe, N., Ghezai, H., Goldacre, B., & Hodsoll, J. (2014). Atypical antipsychotic augmentation in SSRI treatment refractory obsessive-compulsive disorder: A systematic review and meta-analysis. *BMC Psychiatry, 14*, 317. https://doi.org/10.1186/s12888-014-0317-5.

Volkow, N. D., & Li, T. (2005). The neuroscience of addiction. *Nature Neuroscience, 8*(11), 1429.

Washington, N. (2016). Culturally unbound: Cross-cultural cognitive diversity and the science of psychopathology. *Philosophy, Psychiatry, & Psychology, 23*(2), 165–179.

6 Ethical Implications of Neurobiologically Informed Risk Assessment for Criminal Justice Decisions: A Case for Pragmatism

Eyal Aharoni, Sara Abdulla, Corey H. Allen, and Thomas Nadelhoffer

6.1 Introduction

The criminal justice system has a problem: it is tasked with protecting society from dangerous offenders, but it cannot reliably predict who will and who will not reoffend. Given that its predictions are imperfect, efforts to deter or incapacitate dangerous offenders will sometimes encroach on the rights of non-dangerous individuals, and conversely, efforts to protect the non-dangerous will sometimes harbor some "bad apples." To the extent that predictive errors can be reduced, our society could become safer and more just.

One potential way to reduce predictive error is to develop better risk assessment tools. In the criminal justice context, risk assessment refers to any technique that generates probabilistic predictions about the ostensible offender's behavior—such as the probability of reoffending, relapsing, or responding to treatment—by querying information about their attributes or circumstances. Risk assessment is employed throughout the criminal justice pipeline, from intake and sentencing to release, mainly to inform decisions about supervision (such as security level assignments or eligibility for parole) and treatment. It is also used in civil commitment settings such as involuntary hospitalization for individuals found to be dangerous but not guilty by reason of insanity. Traditionally, offender risk estimates were determined using unstructured clinical judgment by trained forensic psychologists—a subjective technique that has since been shown to perform little better than chance (Monahan & Silver, 2003). Recently, structured actuarial risk assessment techniques have improved the accuracy of offender placement decisions by quantifying the offender's degree of fit with a known validation sample. As a result, more than twenty states now require their courts to use these statistical tools in service of offender sentencing and placement decisions (Starr, 2014).

Despite the increasing popularity of actuarial risk assessment, its use in legal settings has been the subject of much scholarly debate. When is it justified to use statistical models based on group data to determine legal outcomes for individual offenders? Risk assessment research suggests that a variety of biological indicators of risk—such as genes and brain function—may carry some degree of statistical weight in predictions of risk. Are some statistical indicators of risk more ethically problematic than others? Given the rapid advances in risk assessment research, it is imperative to anticipate and clarify when, if ever, statistical indicators should inform justice-related risk assessment.

Many ethicists suggest that all actuarial risk assessment is too problematic for use in justice settings, cautioning about violations of beneficence (e.g., unjustified harm to offenders), justice (e.g., unfair distribution of sanctions), and respect for persons (e.g., unjustified restrictions of the offender's freedom or exposure of his private mental life), among other problems. In this analysis, we examine some of the main ethical concerns, with an eye toward so-called neuroprediction technologies—that is, the use of neurobiological markers in offender risk assessment. We attempt to glean insight into some of the normative intuitions surrounding such technologies by evaluating their strengths and weaknesses relative to other alternative strategies. In this way, our analysis is fundamentally contrastive.

We conclude that while some uses of actuarial risk assessment might potentially violate individual rights to beneficence, justice, and respect for persons, these problems arise not just for evidence-based tools but for any decision procedure that society adopts to protect the public safety and civil rights of its members by trying to identify potentially dangerous individuals. We therefore attempt to shift the debate from *whether* actuarial risk assessment is justified to *when* (see also Specker et al., 2018). We argue that appeals to individual rights alone are not sufficient to distinguish between the ethically appropriate and inappropriate applications of actuarial risk assessment. Any principled attempt to evaluate the appropriateness of risk tools must, *for each application*, evaluate its unique costs *relative* to its benefits and *relative* to traditional clinical approaches (i.e., the status quo). When applied to various uses by the law, we find that actuarial risk assessment often fares better on key ethical criteria than traditional clinical methods.

Broadly, we find that appreciation of these relational contexts in which risk assessments are solicited carries the potential to clarify the underlying ethical concerns about risk assessment.

6.2   Legal Applications of Risk Assessment

Before we discuss the internal characteristics of common risk assessment techniques, it is important to understand the variety of ways in which these techniques can be, and often are, employed in legal settings.

- After a person is arrested, risk information might be useful to prosecutors making charging decisions. Prosecutors have wide discretion in such decisions. So, if a defendant is perceived to be particularly dangerous, a prosecutor might seek to file more charges or more severe charges.

- Bail decisions may also be informed by risk information. A defendant assessed to be a low flight risk may be offered bail or probation instead of jail time (the default). Likewise, bail fines can be increased or decreased depending on whether a defendant has a high or low risk score.

- Risk assessments can also inform whether the individual is placed on probation or instead sentenced to prison. Risk scores are not typically used to influence the *length* of a prison sentence, but may be used in pre-sentencing reports and could be considered relevant to actual sentencing decisions— for example, in Texas, capital sentencing hinges in part on a determination of future dangerousness (Tex. Code Crim. Proc., Art. 37.071).

- Within a facility, risk information can determine security level assignments: Authorities may assign low, medium, or high supervision based on the offender's recent behavior in custody, not as punishment but as a protective measure.

- A prison inmate could be released to parole prior to his sentence end date on grounds of a low risk assessment score. Similar uses are invoked when deciding to incarcerate a defendant via criminal versus civil commitment.

- Treatment interventions can be based on risk assessment. In some states, such as Florida, judges can mandate treatments such as anti-androgen therapy (i.e., "chemical castration") to certain sex offenders in addition to their original sentence based on their risk of reoffending (Fla. § 794.0235). Risk assessment can also inform decisions to administer treatment (typically noninvasive treatment) as a sentencing diversion.

In all of these ways, risk assessment is integral to many types of justice decisions either in service of or in opposition to the offender's private interests (see also Baum & Savulescu, 2013; Coppola, 2018; Tonry, 2014).

6.3 Risk Assessment Methods

6.3.1 Traditional Risk Assessment

Two broad types of risk assessment are employed within the legal system. Traditional risk assessment techniques rely heavily on unstructured clinical judgment to assess risk of antisocial behavior. Here, the driving factors in a clinician's determination of risk are the clinicians' (1) specific training, (2) experience with similar cases in the past, and (3) subjective assessment of the offender based on the case file, the crime, and (sometimes) an interview with the offender. The poor predictive accuracy of traditional clinical techniques has been demonstrated repeatedly. As Monahan (1981/1995) pointed out more than thirty years ago:

> It would be fair to conclude that the "best" clinical research currently in existence indicates that *psychiatrists and psychologists are accurate in no more than one out of three predictions of violent behavior over a several-year period among institutionalized populations that had both committed violence in the past (and thus had high base rates for it) and who were diagnosed as mentally ill.* (pp. 48–49; emphasis in original)

In an effort to explain why clinical risk assessment is so inaccurate and unreliable, Monahan (1981/1995, p. 32) identified what he takes to be "the four most common blind spots" of the clinical method: (1) a lack of specificity in defining the criteria being used, (2) a failure to take statistical base rates into consideration, (3) a reliance on bogus correlations, and (4) a failure to account for situational and environmental factors.

Given that clinical assessment is inherently subjective, it is unsurprising that the resulting predictions of future dangerousness are so unreliable (Lidz, Mulvey, & Gardner, 1993; Monahan, Brodsky, & Shan, 1981). In light of these types of problems, some commentators have even gone so far as to suggest that relying on clinical risk assessment for the purposes of the law is tantamount to "flipping coins in the courtroom" (Ennis & Litwack, 1974). Worse still, the unguided and intuitive nature of the process also makes it possible for the biases and prejudices of clinicians to influence their assessments. As Monahan (1981/1995) points out, "It is important to distinguish between the factors clinicians believe they are using—correctly or incorrectly—to predict violent behavior and the factors that actually appear to influence their decisions" (p. 31). In summarizing the primary weaknesses of clinical risk assessment, Krauss and Sales (2001) make the following observation:

> In addition to relying on cognitive biases and heuristics that affect the judgments of ordinary people under conditions of uncertainty . . . mental health practitioners have been found to poorly combine information, use irrelevant information, and inappropriately vary the information they use in formulating predictions for an individual. Worse, their propensity for gathering excessive and irrelevant information also likely leads mental health practitioners to have greater confidence in their conclusions than is warranted. (p. 279; references omitted)

As a result, clinical risk assessments, perhaps unsurprisingly, tend not to be consistent from one mental health professional to the next.

Indeed, the American Psychiatric Association (APA) filed an amicus brief in the landmark case *Barefoot v. Estelle* (463 U.S. 883 [1983]). On their view, "the large body of research in this area indicates that, even under the best of conditions, psychiatric predictions of long-term future dangerousness are wrong in at least two out of every three cases." Because the ability of psychiatrists to reliably predict future dangerousness was "unproved," the APA claimed that "psychiatric testimony on future dangerousness impermissibly distorts the fact-finding process in capital cases." The court nevertheless dismissed the concerns of the APA, with the following reasoning:

> Neither petitioner nor the Association suggests that psychiatrists are always wrong with respect to future dangerousness, only most of the time. Yet the submission is that this category of testimony should be excised entirely from all trials. We are unconvinced, however, at least as of now, that the adversary process cannot be trusted to sort out the reliable from the unreliable evidence and opinion about future dangerousness, particularly when the convicted felon has the opportunity to present his own side of the case.[1]

In short, the court concluded that the adversarial nature of our legal system was sufficient to address the concerns raised by the petitioner with respect to the general unreliability of predictions of future dangerousness. Thus, the decision in *Barefoot* established that violence risk assessments are admissible in capital sentencing. The court has subsequently permitted these assessments in lower stake settings as well. The current state of the law is such that clinical risk assessments, despite their serious problems, are admissible and widely used in the criminal law. Given that risk assessment is here to stay for the foreseeable future, the key question is: Can we improve upon the kinds of unstructured assessments that have traditionally been used?

6.3.2 Actuarial Risk Assessment

Whereas clinical risk assessment employs intuitive and subjective methods, actuarial risk assessment employs explicit, statistical, algorithmic methods. These methods work by matching features of the individual with an existing data set with known risk levels. These features can include clinical assessment of major mental illness, antisocial behavior, social support, and substance abuse history, to name a few. However, developers of actuarial risk assessment go beyond the clinical assessments by quantifying and validating exactly which types of data are to be considered in the prediction (Monahan, 1981/1995, p. 64). These quantitative approaches have demonstrated marked improvement in predictive accuracy (Dawes, Faust, & Meehl, 1989; Steadman et al., 2000; Banks et al., 2004), and meta-analyses have confirmed the superiority of evidence-based methods to clinical ones (Grove & Meehl, 1996).

Consider, for instance, the Violence Risk Appraisal Guide (VRAG), a twelve-item actuarial tool for predicting future dangerousness (Quinsey et al., 1998). VRAG uses the following static and historical predictor variables that have each been shown to confer risk for violence:

1. whether the offender lived with both biological parents as a child
2. whether the offender exhibited maladjustment in elementary school
3. whether the offender has history of alcohol problems
4. the offender's marital status
5. the offender's nonviolent criminal history prior to offense
6. whether the offender has failed on prior conditional release
7. the offender's age at the time of offense
8. the nature of the victim's injuries (e.g., death, hospitalization, treated and released, or slight)
9. whether the victim was a female
10. whether the offender meets the DSM criteria for a personality disorder
11. whether the offender meets the DSM criteria for schizophrenia
12. the offender's score on the Psychopathy Checklist Revised (PCL-R).

Not all predictor variables carry the same weight, as some variables (e.g., PCL-R score) are more predictive than others (e.g., alcohol problems). Offenders are categorized into the following risk levels: low, medium, and high risk. Each level consists of three more fine-grained "bins." Quinsey and colleagues (1998) found that "VRAG score was positively related to the probability of at least one violent reoffense, to the severity of the reoffenses that occurred, and to the speed with which violent reoffenses occurred" (p. 163). Since its development, VRAG has been tested within diverse populations and across a wide variety of settings.

One common way of testing the validity of a predictive model uses a receiver operating characteristic (ROC) analysis. The main reason for implementing a ROC analysis is that this method of representing predictive accuracy "is independent of the base rate of violence in the study sample" (Monahan et al., 2001, p. 95). The statistic used to summarize a ROC analysis is called the area under the curve (AUC), that is, "the probability that a randomly selected violent patient will have been assessed . . . as being in a higher risk group than a randomly selected nonviolent patient" (Monahan et al., 2001, p. 95). When one performs a ROC analysis, the AUC varies from 0.5—that is, accuracy is no better than chance—to 1.00—that is, perfect accuracy. Across multiple studies, Quinsey and colleagues (1998) found that the AUC for VRAG ranges from 0.73 to 0.77.[2] While this isn't ideal, it outperforms the clinical models we discussed earlier by a wide margin.

Given the increased predictive accuracy of models such as VRAG, some jurisdictions have begun to utilize actuarial risk assessment in decisions regarding criminal sentencing, supervised release, and the treatment and support programs required during supervision. Assessing individual offenders' risk levels, matching them to risk-appropriate programs, and targeting their unique criminal risk needs increases success, reduces antisocial behavior, and enhances public safety (Andrews, 2006; Aos, Miller, & Drake, 2006; MacKenzie, 2006; Taxman, 2002). Indeed, the traditional "one-size-fits-all" group treatment programs that ignore individual risk level can have damaging effects. One traditional group treatment program, for instance, reduced recidivism for high-risk offenders by more than 25 percent but led to an increase in reincarceration of low-risk offenders (Latessa, Lovins, & Smith, 2010). Because of the behavioral benefits of risk-appropriate treatment, some states have even begun to require the use of actuarial risk assessment in legal decisions (e.g., Arkansas SB750; Public Safety Improvement Act 2011, Pub. L. No. SB750).

Though actuarial predictions represent an improvement over clinical judgment, they are by no means perfect. However, every additional percentage increase in risk assessment accuracy could reduce victimizations by targeting high-risk offenders while at the same time reducing our heavy reliance on mass incarceration by diverting low-risk offenders to treatment programs outside of prisons. For these reasons, scientists are beginning to examine whether the inclusion of neurobiological and/or genetic markers can improve predictive models based solely on non-biological evidence.

6.3.3 Neurobiologically Informed Risk Assessment

Neuroprediction refers to the use of neuroscientific measures to characterize biological markers of human behavior that increase the ability to classify particular behavioral outcomes accurately (e.g., Aharoni, Vincent, et al., 2013; Berns & Moore, 2012; Camchong, Stenger, & Fein, 2012; Delfin et al., 2019; Janes et al., 2010; Just et al., 2017; Pardini et al., 2014; Paulus, Tapert, & Schuckit, 2005; Sinha & Li, 2007; Steele, Fink, et al., 2014; Steele, Claus, et al., 2015; Steele, Maurer, et al., 2017). Currently, their use has been largely limited to research settings rather than legal settings.

One reason why courts have not used neuroprediction might be that neuroprediction and other biopredictive techniques have an unwholesome history. In the nineteenth century, it was believed that the morphology of a person's skull revealed information about their personality and behavior—a now-obsolete practice known as phrenology. Phrenology also inspired theories about criminal behavior, including one theory by Italian physician Cesare Lombroso that criminal offenders can be identified from a unique physiological signature, and that their criminal dispositions are biologically inherited, inferior, and unchangeable (Ellwood, 1912; Lombroso, 2006; Verplaetse, 2009). On this essentialist and deterministic view, many offenders are "born criminal"—a view that had a problematic relationship with eugenics and scientific racism. In the wake of Nazism, the explanatory pendulum swung toward situational models for human behavior that didn't appeal to biomarkers. These theories assumed that people are born into this world as blank slates and that any differences between them are the result of environmental and social factors such as poor parenting or poverty (e.g., Heider, 1958; for a critical analysis, see Pinker, 2016).

It is now widely held that human behavior is the result of a complex interplay of biological, psychological, and social factors aptly named the biopsychosocial model (Raine, 2013). For example, early lead exposure has been shown to increase the risk of later violent and aggressive behavior by impairing the development of brain areas such as the anterior cingulate cortex (ACC; Cecil et al., 2008; Liu et al., 2011). Several biomarkers have been developed to estimate the presence of lead poisoning (Sanders et al., 2009). However, while brain dysfunction caused by lead exposure represents a biomarker, it is also shaped by environmental factors as well, since people who live in homes or neighborhoods that have high levels of lead tend to be more socioeconomically disadvantaged. So, lead exposure represents an insult at two levels: the biological and the social. As such, it helps highlight the interdependency between the environment and the brain. The notion that environmental factors influence biological states that in turn affect behavior should not be surprising, given that all stimulus-based learning and behavior is mediated by the brain, the development of which is driven in large part by the coding of our genes along with nutrition and sometimes injury and disease.[3] One litmus test of the biopsychosocial model is whether explanations of human behavior demonstrate greater predictive accuracy when the relevant biological, psychological, and social factors are all included in the model.

The hypothesis that predictions of antisocial behavior can be improved by including targeted brain metrics in risk models formed the basis of a series of peer-reviewed studies that together have lent support to this hypothesis. For example, in a sample of ninety-six adult offenders who engaged in an impulse control task while undergoing functional magnetic resonance imaging, brain activity within the ACC prospectively predicted being rearrested later (Aharoni, Vincent, et al., 2013). Offenders with relatively low ACC activity had roughly double the odds of getting rearrested for a violent or nonviolent crime as those with high activity, controlling for other known risk factors. Using advanced statistical techniques such as AUC, calibration, and discrimination, these neuropredictive models have demonstrated relatively high accuracy (e.g., the probability that a true positive was correctly classified exceeded 75 percent; Aharoni, Mallett, et al., 2014; for a review, see Poldrack et al., 2018). The findings comport with the ACC's known role in error processing and impulse control (e.g., Kerns et al., 2004).

In a test of the convergent validity of these methods, researchers have also examined the predictive utility of ACC-based models using machine learning techniques that attempt to classify predictive functional patterns from event-related potentials (ERPs) measured in members of the original sample of Aharoni and colleagues (Steele, Claus, et al., 2015). The best performing model included a brain component known as the P300, which was capable of predicting being rearrested with comparably high accuracy. Other studies have reported similar predictive effects using different models of brain function (Delfin et al., 2019; Kiehl et al., 2018; Pardini et al., 2014). Together, these studies provide preliminary evidence that brain factors such as ACC function may serve as candidate biomarker for antisocial behavior.

Finally, several studies of substance-dependent forensic populations have highlighted the potential value of neuroprediction for purposes of assessing the probability of drug relapse and treatment completion, which are themselves known predictors of the risk of reoffending (Camchong et al., 2012; Janes et al., 2010; Paulus et al., 2005; Sinha & Li, 2007; Steele, Fink, et al., 2014). For example, ERP components and functional connectivity associated with the ability to monitor response errors have separately predicted treatment completion with accuracy exceeding 80 percent (Steele, Fink, et al., 2014; Steele, Maurer, et al., 2017).

Neuroprediction is still very much a nascent research field. Much remains to be understood about how our biological makeup shapes our behavior. The observed predictive effects are not and will never be perfect. But as the science advances, models that ignore the influence of biology on behavior will inevitably be forced to compete with those that acknowledge such influence. So, it is critical to understand better the impact that biology has on antisocial behavior and to think more carefully about how such behavior should be managed by society.

6.4   Evaluating Common Objections to Actuarial Risk Assessment

Many of the concerns surrounding the usage of neuropredictive technologies apply more generally to all actuarial risk assessment methods. So, rather than focus more narrowly on neuroprediction, we will focus primarily on actuarial prediction (except when there are salient differences between the two). Given that actuarial risk assessment of all kinds has been the target of persistent objections, we think our approach is appropriate. In responding to critics of actuarial prediction, we identify four common types of concerns: statistical concerns and concerns about potential violations of norms of beneficence, justice, and respect for persons.

6.4.1 Concerns about Statistics

Critics have decried the low predictive accuracy among actuarial risk assessment tools (e.g., Starr, 2014). But this claim is misleading, as meta-analyses have demonstrated that the predictive value of many actuarial risk assessment instruments is superior to the traditional methods that would otherwise be used (e.g., Grove & Meehl, 1996). So, while we should continue to strive to increase the power of actuarial risk assessment, it already represents a significant step forward, even if it falls short of some critics' idealized standards. But from a legal perspective, its objective accuracy is beside the point. The Supreme Court has already made it clear that risk assessment is legally relevant. Thus, the only remaining legal questions are which risk assessment techniques we should incorporate and for which purposes.

A related and perhaps more compelling criticism is that the predictive accuracy of actuarial risk assessment holds only when making predictions about groups rather than individuals. This is referred to as the group-to-individual (G2i) inference problem. Indeed, scientists and practitioners must exercise great caution when considering whether and when to render diagnostic judgments based on group data (see Campbell & Eastman, 2014; Faigman, Monahan, & Slobogin, 2014; Heilbrun, Douglas, & Yasuhara, 2009). Predictable differences between groups do not imply that predictions about individuals within those groups will be similarly accurate (Buckholtz & Meyer-Lindenberg, 2012).

However, there are confusions about this objection that warrant clarification. A strong form of this objection states that, statistically, risk assessments may not be applied to individuals in principle because they are derived from group data. This stance cannot be true. By definition, the classification accuracy statistics employed by at least some of these instruments specifically describe an *individual's* degree of fit with a group, given certain shared attributes (Dawes et al., 1989). Thus, at least in principle, it is statistically justifiable to use these actuarial risk assessment techniques to make predictions about individuals.

Moreover, even if it weren't justifiable, this should be at least as problematic for clinical judgment. Like actuarial risk assessment, clinical judgment is an inductive process that relies on past observations. Whether explicitly or implicitly, clinicians reference attributes from past cases—either textbooks or personal experience—to inform their predictions. The assumption is that this offender will behave as similar offenders have behaved in the past. So, if the use of group-based inference renders actuarial risk assessment problematic on statistical grounds, then *all* risk assessment would thereby be problematized.

The weak form of the G2i objection states that statistical risk assessments are not yet accurate enough to be practicably applied to individual offenders, who might vary widely from the group average. However, this claim demands greater precision. In the medical field, the accuracy of some of the most common screening tools is often marginal. For example, the ten-year cumulative probability of a false-positive mammogram result (a diagnosis that cancer is present when, factually, it is not) from breast cancer screening may exceed 60 percent (Hubbard et al., 2011). Yet, it is part of the standard of care for individual patients. For a doctor screening for cancer, it makes sense to tolerate a lot of false-positives in exchange for a high true-positive rate because the purpose of the screening phase is usually not to deliver a major medical intervention but rather to gather more information. Here, the intended application matters. Likewise, in criminal law, it is important to scale the accuracy of the predictive tool, regardless of type, to the stakes of the decision. But given that risk assessment is already pervasive in the law, the question isn't whether risk assessment is accurate enough, but rather how to improve it and when to use it.

Given that risk assessment, despite its limitations, will continue to play a role in the law, it seems prudent to advocate for the most reliable and accurate methods available. Indeed, scholars have made a powerful case for why statistical tools that generate G2i inferences, despite their caveats, can be responsibly leveraged for risk assessment (Bedard, 2017; Faigman et al., 2014; Heilbrun et al., 2009). Therefore, when authors object to the use of actuarial risk assessment on grounds of problematic G2i inference, a charitable interpretation is that there is nothing inherently problematic about G2i inferences in principle, but its present-day accuracy levels fail to meet *normative* standards required by the particular applications of our justice system. In other words, the concern about G2i is not so much a scientific objection as it is a moral or legal one (Janus & Prentkey, 2003). These moral and legal objections are the subject of the next sections.

6.4.2 Concerns about Beneficence

One moral concern about prosecutorial uses of actuarial risk assessment is that classification errors could result in the gratuitous infliction of harm. For example, false-positive errors could support a longer sentence or revocation of privileges from a non-dangerous offender, and false-negative errors could result in an early release of a dangerous offender, placing other community members at risk of victimization. To avoid these errors, it may be tempting to invoke a simple rule to minimize unnecessary harm.

However, removing risk assessment does not reduce harm. Judges are obligated to make sanctioning decisions with consideration for public safety regardless of whether an actuarial risk assessment is employed. The only other alternative (clinical judgment) still yields classification errors that are likely to be substantially larger without actuarial risk assessment than with it (Monahan & Silver, 2003). So, revocation of actuarial risk assessment on grounds of its harmfulness perversely increases harm by increasing the number of low-risk offenders who will be needlessly sanctioned and also the number of community members who will become victims of a truly dangerous person who was misclassified.

Another concern is that actuarial risk assessment facilitates the erroneous view that offenders are incorrigible, depriving them of appropriate service opportunities and possibly triggering a self-fulfilling prophecy (Specker et al., 2018). Perceived and anticipated stigma has been shown to predict poorer social adjustment following offender release, and thus is an important factor when considering potential harms inherent to actuarial risk assessment (Moore, Stuewig, & Tangney, 2016). However, this is a problem for all forms of risk assessment, not just actuarial risk assessment. Even with traditional clinical assessment, offenders are labeled as high risk, low risk, and so on, which can be stigmatizing, depending on the context. In this respect, actuarial methods are once again superior to clinical methods, given that the former methods permit the definition of both dynamic (i.e., changeable) risk and protective factors, such as the presence of positive or negative peer influences, which help service providers to identify appropriate interventions based in part on established evidence of treatment responsivity. Because these factors can be formally built into the actuarial models, this is likely to result in less, not more, stigmatization compared to traditional tools.

6.4.3 Concerns about Justice

Perhaps the most common normative concern about offender risk assessment is that it violates principles of distributive justice or fairness. In this view, it is unfair to judge an individual based on group tendencies because classification errors are not randomly distributed across individuals. If classification errors are non-random, this could facilitate arbitrarily worse treatment for certain, often underprivileged, groups. Risk assessment using biomarkers may also medicalize criminality in a way that stigmatizes already marginalized groups (Berryessa, 2017; Coppola, 2018; Hall, Carter, & Yücel, 2014; Jurjako, Malatesti, & Brazil, 2018). In this view, we ought to judge people not as mere statistics but for who they are as individuals. But structured risk assessment tools, it is argued, erroneously target individuals based on group membership rather than their true dangerousness, thereby codifying and reinforcing our inherent social biases (e.g., Silver & Miller, 2002, p. 152; Starr, 2014, p. 807).

The argument that people have the right to be treated as individuals as opposed to members of a (usually underprivileged) group often invokes the Fourteenth Amendment's Equal Protection Clause, which mandates that the government cannot discriminate on the basis of differences that are not relevant to a legitimate governmental interest (Starr, 2014, p. 827). Here, it is suggested that by assuming that the individual is a member of a high-risk group, actuarial risk assessment deprives that individual of the presumption that their actions should be judged independently of the actions of other members of the groups to which they belong. They are said to be punished for what other members of their group do.

However, for this argument to be coherent, it cannot assume that individuals ought to be judged completely free of *any* group membership. That would imply complete subjectification of the law: an extreme and unfeasible requirement to apply different laws to each individual in accordance with some ostensible model of his or her "true" self. A more plausible interpretation of the fairness argument, then, is that actuarial risk assessment erroneously assumes that the individual is a member of the unfavorable higher-risk group instead of the more legally privileged lower-risk group.

Importantly, whether this group classification is actually erroneous depends on the accuracy of the prediction. If this prediction turns out to be correct, and the individual is actually of high risk, the individual forfeits the law's more favorable presumption (that they belong to the low-risk group).

However, even if the high-risk prediction turns out to be wrong, this does not mean the individual is entitled to be judged independently of all group membership. Rather, it means that he should be judged as a member of the more privileged lower-risk group. For this reason, we can't reject actuarial risk assessment on grounds that it treats individuals as members of a group because the only other coherent alternative assumes these individuals are just members of another group (see also Janus & Prentkey, 2003). The Supreme Court seemed to recognize this point when it argued that equal protection must "coexist" with the fact that most legislation engages in classification in a way that disadvantages some social groups (Hamilton, 2015).

The law has been fairly explicit about which group factors are appropriate targets of risk assessment and which are not. Under the Equal Protection Clause, the Supreme Court has delineated three levels of scrutiny of evidence: strict scrutiny, intermediate scrutiny, and the rational basis test—the lowest level of scrutiny. In risk assessment, race typically falls under strict scrutiny, gender under intermediate scrutiny, and age under the rational basis test (Monahan, 2014).[4] So, the use of race as a factor requires a much stronger justification than the use of age, for example, to be considered for admissibility. Why is race held to a higher standard? One reason might be that race has been the basis of ongoing discrimination in the U.S. In particular, African Americans are disproportionately represented in the criminal justice system due in part to arbitrary and unfair policing and adjudication practices, and so corrective measures are needed to offset this error. Another reason for holding racial status to a higher standard might be that race is not independently diagnostic of offending (e.g., although U.S. prison populations are disproportionately African American, most African Americans are not criminals). A third reason might be that unlike dynamic risk factors such as drug use, race is static: people cannot change it at will. However, all of these reasons also arguably apply to some lesser-scrutinized factors such as age. So, they do not easily explain the difference in standards.

Perhaps a better explanation for why race evidence is held to a higher standard than age is that whereas there are good theoretical reasons to expect that age (as a measure of maturation) might play a direct causal role in some criminal behavior, no respectable theory suggests that race itself causes crime. So, one key function of these legal standards could be to filter out evidence that is less causally relevant because the susceptibility of such factors to error and bias will tend to outweigh their probative value.

Conversely, the law's lower standard for age suggests at least a tacit recognition that while a given risk factor does not need to be independently diagnostic of recidivism to be relevant, it should at least bear a direct correspondence to a known causal contributor to recidivism (e.g., as age bears to maturation).

A related reason that age merits a lower evidentiary standard might be that it cannot easily be used as a marker for other persecuted groups. For example, knowing someone's age is not helpful in predicting their race. Conversely, risk factors such as race, zip code, and socioeconomic status could more easily be used to target already-persecuted groups disproportionately. The complication is that some of these factors—such as socioeconomic status—may also contain causally relevant, independently predictive information, while others—such as race—do not. In such cases, completely ignoring the role of such factors in actuarial models would only serve to obscure any biases in classification. A better solution would be to control for problematic factors (such as race) statistically so that any variance attributed to those factors can be subtracted from otherwise useful factors (such as socioeconomic status). To avoid misuse of surrogate factors, it is useful to code for those factors explicitly, not to pretend they don't exist (Aharoni, Anderson, et al., 2019).

Some scholars have suggested, on grounds of discrimination, that we shouldn't support any selection rule whose outcomes might disfavor certain social groups (e.g., Starr, 2014; Tonry, 2014). This argument seems to confuse at least two different definitions of discrimination: unfair *outcomes* and unfair *processes*. By analogy, consider the familiar social rule: "For sports team photos, taller people have to stand in the back row." For some sports, such a rule might incidentally disfavor a certain race (e.g., if African American team members happen to be taller than others on average), but whether this rule would be *discriminatory* in the legal sense is questionable because it is blind to race from a selection perspective (i.e., it's not optimized to remove opportunities for African Americans as a group). The racial disadvantage in this case is only a *corollary* of an otherwise useful selection factor, not as a direct *causally* relevant selection factor itself. This is important because causal selection factors will tend to have lower covariation with other extraneous factors than more distant corollaries do.

Arguably, discarding the height rule would introduce even greater discrimination against shorter people, including many with physical disabilities, whose faces could be even more obscured from the camera's view than the taller people were. In this mundane analogy, we do not mean to minimize the gravity of the issue of racial bias in the justice system. We use this analogy only to obviate the point that inequalities in *outcome* are often inescapable, but they do not necessarily indicate a need for a system overhaul. Discarding all rules that result in unequal outcomes among vulnerable groups would mean eliminating all risk assessment and either releasing all dangerous offenders into the community or locking up all non-dangerous ones. The fallout from either of these scenarios would be no consolation for those aiming to achieve equal outcomes. A humbler goal, and one that the Equal Protection Clause seems to support, is equality in legal process (Hamilton, 2015; Weiss, 2017).

What about risk factors that are more clearly biological, such as measures of brain activity? In his examination of the status of biological risk factors for violence risk assessment within the context of the Equal Protection Clause, Monahan (2014) concludes that as long as the risk factors in question are not being used as a surrogate for a more strictly regulated factor, biological risk factors would likely be admissible under a lower level of scrutiny. We agree with this point, provided that the biological evidence is demonstrated to be causally relevant. The broader lesson here is that while caution is warranted in deciding which risk factors merit inclusion in risk assessment instruments, these are tractable obstacles whose solutions can advance the utility of actuarial risk assessment beyond that of traditional clinical approaches.

The question at hand is not whether actuarial risk assessment is imperfect. It undoubtedly is. The question is whether the use of particular actuarial risk factors is any more problematic than the other alternatives (see also Dawes et al., 1989; Douglas et al., 2017; Nadelhoffer et al., 2012). Admittedly, the problem becomes more difficult when relatively accurate selection criteria are not available. So, under these conditions, it might indeed be justified to forgo the broader mission in order to protect minority groups. Our point is not to advocate all uses of actuarial risk assessment uncritically but instead to suggest that their justifiability must be judged relative to the ways in which they are to be used and what the other alternative options are.

On the broader point that actuarial risk assessment could codify society's social biases (Starr, 2014), we agree that these instruments might do so. It might ensure that some amount of bias is present in the predictions. Surely, for these tools to be effective and ethical, developers should always aim to minimize bias. But codification of bias is not the biggest problem faced by these tools. By design, actuarial risk assessment attempts to codify risk factors. This feature makes the process transparent and enables us to monitor and evaluate procedural mistakes and ultimately to minimize bias. For this reason, codification of bias can be understood more as a strength than as a flaw because it enables the assumptions of the predictive models to be explicit and subject to evaluation and revision.

That is not to say that all actuarial models will invariably codify bias. Whether and the degree to which a model is biased will depend on the predictor variables that it uses. For instance, while researchers could likely develop more accurate models if they included race as a predictor variable for violence—not because different races are naturally more or less violent but precisely because races are treated differently by the legal system—they virtually never do. The main reason is that if researchers used race as a predictor variable, then their actuarial models wouldn't be admissible for legal purposes. Do these models nevertheless use variables that are proxies for race, for example zip codes or socioeconomic status? Some do, but most of the main models do not (or at least not obviously so). Are some of the twelve items used by VRAG proxies for race? Here, again, it's possible, but it's not obvious either way. Yet, because VRAG uses specific variables transparently, we can look at these variables and determine whether some may be serving as surrogates for race. If so, we can remove these variables and see whether it affects the overall predictive validity of the model. If removing the item impacts accuracy, then we can and should have an open discussion concerning whether the trade-off is worth it. Depending on the legal context— for example, the death penalty versus bail—we might come down on one side of the issue or the other.

This level of transparency is not found in traditional clinical methods. As cognitive psychologists have long warned, when human beings make decisions in the absence of constraints, we overly rely on anecdotal information and discount true probability information (e.g., Kogut & Ritov, 2005). Importantly, we may still consult probability information, we just do it unconsciously, intuitively, and unchecked (Ryazanov et al., 2018), using our own flawed, selective memories. When it comes to unstructured decision making, we are *folk* statisticians (de Hevia & Spelke, 2009; Kahneman & Tversky, 1972; Kunda & Nisbett, 1986; Tajfel & Wilkes, 1963; Tversky & Kahneman, 1973). In light of this well-studied fact, actuarial risk assessment is no more problematic—indeed, it is less problematic—than traditional risk assessment. If we dispense with structured risk information in justice decisions, assessment bias does not go away, it just exerts its nasty effects behind a curtain of opacity.

6.4.4   Concerns about Respect for Persons

Regardless of whether risk assessment makes the offender the target of harm or discrimination, one might object to its use on grounds that it violates a basic obligation of respect for persons, including a requirement to protect the offender's autonomy—his natural freedom or right to make his own choices—and the privacy of his mental life. After being convicted, there are many ways in which the state already restricts the offender's autonomy and privacy, but at least prior to conviction and sentencing, decisions about how he lives his life should be up to him. Undergoing a neurological assessment of that person's dangerousness could violate his autonomy and privacy and therefore should be subject to his consent, just as it would be with any other citizen. Many offenders, of course, would not provide such consent, and even if they did, their consent may not be valid because the conditions under which it is solicited could be highly coercive. Bioethicist Paul Wolpe (2009) articulated this concern when concluding that "[t]he skull should be designated as a domain of absolute privacy," even in times of national security, on the grounds that bureaucratic intrusion into people's brains is to take away their "final freedom."

This concern about respect for persons raises a fair point. Indeed, this issue rings especially true when it comes to neuroprediction rather than actuarial prediction more generally. So, we will focus primarily on neuroprediction in this section. In addressing this issue, it is important to start by placing the importance of respecting offenders within the broader context of the other people whose interests are also at stake. After all, the offender's interests are not necessarily the only ones worth protecting; his rights to autonomy and privacy must be weighed alongside those of his potential victims. So, the relevant question is not whether the offender's rights have been violated, it is how many violations of the autonomy and privacy of the offender's potential victims equal the violation incurred by that offender. This is a difficult normative question, but it bears on legal judgments whether risk assessment is employed.

We have argued that the respect for persons is not unconditional, and the courts agree. In *Katz v. The United States* (389 U.S. 347 [1967]) the Supreme

Court ruled that in matters of public safety, some violations of privacy may be justified. For example, defendants may sometimes be required to undergo diagnostic tests and even medical surgeries in order to seize potentially probative evidence such as a bullet or ingested jewelry (Hanley, 2017). Courts have also compelled defendants to undergo psychiatric evaluations and other risk assessment measures. The courts recognize some limits to privacy, and they are not likely to hold potentially intrusive brain measures to a different standard than other types of intrusive measures (Nadelhoffer et al., 2012).

The respect for persons concern, however, presents another challenge, namely that neurotechnologies are different from other kinds of legal evidence because they reveal not just physical information but also information in the more privileged mental domain. To be sure, some neurotechnologies purport to read not just the brain but the mind. Examples of this so-called mind-reading technology include techniques for detecting lies (Schauer, 2010) or reconstructing a person's past visual experiences (Nishimoto et al., 2011). Using such techniques to draw inferences about a defendant's mental experiences has been the subject of strong scientific and ethical criticism (Ienca & Andorno, 2017), and, unsurprisingly, legal scholars have been cautious about them (Greely & Illes, 2007; cf. Schauer, 2010). Yet, even setting these criticisms aside, the question still remains of whether neurotechnologies can ever measure *mental* phenomena. If so, they might qualify as testimonial evidence (i.e., information that is evoked by questioning by the government), which might be more privileged than information that exists independently of prompting. This is important because the Fifth Amendment specifically protects defendants from being forced to testify against themselves (Ienca & Andorno, 2017; Nadelhoffer et al., 2012).

From a broad scientific perspective, neurotechnologies measure lots of different things, and many of these are better described as physical than mental properties, depending on what conclusions are to be drawn from them (Shen, 2013). The brain is responsible for myriad physical processes, such as metabolizing fat and regulating the heart, which aren't directly related to a person's subjective self. A scientist using neural markers to predict whether someone will develop Alzheimer's disease, for example, might be interested in future health outcomes without having any interest in that person's character traits or the content of her thoughts.

Appealing to a physical level of analysis, some scholars have suggested that brain information should not necessarily be privileged. People can be sources of information about themselves without necessarily implicating themselves. This distinction allows courts to use people's personal information to some degree without violating their Fifth Amendment rights (Farahany, 2012a). *Pennsylvania v. Muniz* (496 U.S. 582 [1990]) illustrates this point. The court decided that those suspected of driving under the influence of alcohol or drugs do not need to be read their Miranda rights before their sobriety is assessed because the relevant inference concerns "the physiological functioning of [their] brain," and the brain constitutes physical information (Shen, 2013). Similarly, in *Schmerber v. California* (384 U.S. 757 [1966]), the court used information from the defendant's blood test to determine his blood-alcohol level while he was hospitalized. He was too intoxicated to consent, but the court determined that the blood-alcohol evidence was admissible because that information existed independently of his testimonial capacities and did not need his participation or effort to be determined (Fox, 2009).

When it comes to the brain, however, others have argued that the information captured by neurobiological measures may in fact include mental information because the brain directly gives rise to the mind, including a person's sense of identity and self (Ienca & Andorno, 2017). If so, it may qualify as the more privileged testimonial evidence. To qualify as testimonial evidence under current precedent, the information must describe a conscious, intentional proposition, a characteristic that is privileged in part because it is relatively easy for a defendant to conceal (Fox, 2009). Defendants, for instance, may remain silent and refuse to cooperate with lawyers and even psychologists who attempt to extract guilty knowledge for court uses. So, if brain measures could potentially reveal information about this type of information (e.g., beliefs, feelings, memories), it is easy to see why neurotechnologies might be held to a higher standard than more clearly physical measures such as breathalyzer tests.

Though the law has distinguished physical evidence from testimonial evidence, advances in neurobiology have begun to blur the boundary. Using physical measures of brain function to draw inferences about mental states would seem to raise the testimonial flag. However, not all mental information is necessarily testimonial (or incriminating). Thus, some scholars have criticized the physical–testimonial dichotomy on the grounds that it doesn't sufficiently protect citizens from potential privacy intrusions (Farahany, 2012a; Pustilnik, 2013). Farahany offers an alternative model, distinguishing four types of brain–mind information, relayed here from least to most sacred: identifying information (using brain information to determine someone's identity, such as DNA profiling), automatic information (subconscious physiological, neurological, and reflexive patterns of behavior), memorialized information (memories determined by measures such as voice recognition tests, except those created under government coercion), and uttered information (conscious thoughts communicated through language).

Neuropredictive information, such as neural markers of behavioral traits such as impulsivity or emotional instability, is clearly not uttered. And unlike uttered information (or testimonial evidence more broadly), behavioral traits are not so easily veiled. On the other hand, such neuropredictive measures can be used to discern cognitive information beyond the subject's mere identity. This leaves automatic and memorialized information. In order to qualify as memorialized information, neuropredictive assessments would have to glean insight into the content or quality of a person's memories. But while significant strides have been made to capture neural signatures of episodic memories in the lab (e.g., Nishimoto et al., 2011; Rissman, Greely, & Wagner, 2010), in practice, neuropredictive assessments generally do not purport to probe subjective experiences. At best, they could provide indirect evidence of a behavioral trait. To qualify as memorialized information, the results of neuropredictive assessments would also need to be easily concealed. But again, while it might be possible for some individuals to refuse to cooperate or to learn to generate erroneous neural signals effortfully, the neural pattern of a typical cooperating subject would be a relatively reflexive, emergent, and cognitively impenetrable property of that person's neurocognitive functioning. Moreover, the behavioral information inferred by these tests, such as impulsivity, might already be somewhat discernible from public records, including academic, job, and criminal history, and from general interactions with others. Under this framework, neuropredictive measures that estimate risk of reoffending based on theoretical traits such as impulsivity would seem to qualify best as "automatic information." Thus, compared to some other kinds of evidence, they might be less eligible for protection. While people have a reasonable expectation of privacy of their brain function in their daily lives, if this information is classified as automatic, some of those expectations of privacy might plausibly be overridden in cases of public security.

Still, the use of neuropredictive measures to bypass people's conscious thoughts and preferences could raise new concerns, particularly if these measures are used to draw inferences about the content of the defendant's mental experiences or if they are physically intrusive (Farahany, 2012b). That said, it is not clear that functional imaging data could be collected from an unwilling offender in practice, since he could simply be noncompliant, refuse to be still in the scanner, and so on. But if such technologies could be used without the consent of the defendant (e.g., with some kind of sedatives or restraints) or without the aid of additional confirmatory/disconfirmatory evidence, great caution must be exercised to evaluate and regulate to what end the technology is to be used, since the particular conclusions that are drawn will likely vary widely from case to case as a function of the fact finder's specific question.

6.5   Conclusions and Future Directions for Risk Assessment

We began this chapter with a special interest in the ethical implications of neurobiological factors in offender risk assessments. However, we have learned that many of the concerns about such tools, such as concerns about justice and beneficence, apply to the use of offender risk assessment more generally. So, any attempt to evaluate the use of neurobiological risk assessment tools must first answer to those broader concerns. Much of the existing discourse on offender risk assessment has been focused on potential sentencing applications of actuarial risk assessment after a conviction. So, our present discussion has focused mainly on that case, where concerns about justice and beneficence seem justifiable in the abstract. But when legal actors carry an obligation to make a decision with real consequences for public safety—as is common for many risk assessment contexts—these ethical concerns apply at least as strongly to decision protocols that bypass the evidence about risk.

Despite the scholarly emphasis on sentencing applications of actuarial risk assessment, in our view, this emphasis is somewhat misdirected. The use of actuarial risk assessment for sentencing is an extreme case where a conflict of interest exists between the offender and society, delivery of a sentencing judgment is mandatory, and the stakes on both sides are high. This disproportionate emphasis in sentencing scholarship has overshadowed the variety of lower-conflict, lower-stakes legal contexts that could usefully implement actuarial risk assessment in more obvious ways (see Baum & Savulescu, 2013). These include decisions to reduce a charge; decisions to offer bail, probation, or parole in lieu of jail or prison time; decisions to place an inmate in a lower security level; decisions about early release from civil commitment; diversion from traditional court to drug court; or the provision of benign treatment services. In all of these cases, the default choice (e.g., neglecting to offer treatment) is generally less favorable to the offender, and the offender qualifies for the more favorable choice by demonstrating low risk. Certainly, concerns could still be raised in these cases, but these concerns would likely apply to an even greater extent when such choices are not offered. Scholars increasingly agree that more attention should be paid to the uses of risk assessment that serve rather than violate the offender's interests (Baum & Savulescu, 2013; Tonry, 2014).

Neurobiological tools may be especially valuable in such contexts. Although accurate prediction does not necessarily depend on knowing the causes of target behaviors, it can often help to illuminate such causes and thus could inform the development of tailored treatments that more effectively manage those behaviors (see also Barabas et al., 2018; Douglas et al., 2017; Gilligan & Lee, 2005; Latessa et al., 2010; Meynen, 2018). Two recent brain stimulation studies demonstrate a proof of this concept. In one study using transcranial direct-current stimulation, stimulation of the ACC resulted in improved performance on cognitive and affective attentional tasks (To et al., 2018). In another study using the same technology, a single stimulation session of the dorsolateral prefrontal cortex reduced aggressive criminal intentions twenty-four hours post stimulation (Choy, Raine, & Hamilton, 2018). As with any new treatment intervention, the value of such neurobiological approaches must meet high standards of validity and reliability (Large & Nielssen, 2017).

Whether brain measures are included in risk assessments or not, actuarial risk assessment cannot be judged using a monolithic ethical rule because its uses are heterogenous, reflecting plural often competing social values such as public safety versus civil liberties. Thus, its standards for evaluation must aspire to balance these competing values, and they must do so in a manner that is responsive to the particular context of their intended use.

When evaluating the justifiability of a given use of actuarial risk assessment, it would seem helpful to receive guidance from a formal ethical theory that offers ways to balance the many competing rights and values at play in risk assessment settings. Alas, no such universally accepted theory exists. Meanwhile, many shorter-term practical considerations remain for minimizing damage in the application of actuarial risk assessment.

First, to minimize classification errors, scientists and practitioners of actuarial risk assessment should demand large, well-matched validation samples (Douglas et al., 2017; for a review, see Poldrack et al., 2018). The larger and better matched the validation sample, the more predictive utility can be leveraged from smaller ontological categories. Relatedly, researchers should more heavily prioritize independent replication among prospective follow-up studies and randomized controlled trials whenever possible. In practice, such opportunities are often limited by funding constraints and policy regulations, but their value cannot be overstated.

Second, unlike traditional risk assessment methods, when an actuarial risk assessment classification error is detected, it is possible to track the likely source of the error so that the model can be strategically improved. This is a unique quality control advantage of actuarial risk assessment. Currently, many risk assessment tools and error rates are protected under rules of propriety. However, in criminal justice contexts where stakes are high, transparency of process is a necessary part of keeping these processes honest. Researchers and providers should be obligated to disclose all statistical and algorithmic information to practitioners and potentially the public (Hamilton, 2015; Wykstra, 2018).

Third, the use of dynamic factors in risk models should be upheld whenever possible (Wolpe, 2009) to increase self-efficacy in the offender's rehabilitative efforts while reducing stigmatization and perceptions that the offender is incorrigible.

Fourth, actuarial risk tools are only as good as their measures. Rather than ignore problematic factors such as race, developers of these tools should consider statistically controlling for such factors. Similarly, there are known problems with the use of official records as proxies for actual criminal histories. Developers should consider ways to supplement the official records of their validation samples with other measures such as anonymized self-reported criminal history.

Fifth, the criminal justice system needs clearly defined normative standards for classification accuracy. Appropriate standards will remain a moving target and will depend on a deliberate ongoing dialogue between policy makers, scientists, and practitioners. These standards must be sensitive to the different applications of actuarial risk assessment and should include contingencies for cases in which a risk assessment result is underpowered or inconclusive (see also Campbell & Eastman, 2014).

Finally, efforts to understand our collective and often conflicting normative attitudes toward different forms of risk assessment will benefit strongly from actuarial methods that help clarify the relative costs and benefits of each alternative. If we choose to exclude the evidence from our risk assessment practices, its harms do not go away—they merely operate under the cover of darkness.

Acknowledgments

We thank Amanda Haskell, Saskia Verkiel, Nicholas Alonso, Lyn Gaudet, Jason Kerkmans, Douglas Mossman, Felipe De Brigard, Walter Sinnott-Armstrong, Natalia Washington, and the participants of the Duke Summer Seminars on Neuroscience and Philosophy. Contributions to this chapter were supported in part by a grant from the National Science Foundation: #1829495.

Notes

1. Barefoot v. Estelle, *Id*. at § 901.

2. Several recent meta-analyses have further established the predictive validity of VRAG (see, e.g., Singh, Grann, & Fazel, 2011; Singh et al., 2014; Yang, Wong, & Coid, 2010). It is worth noting, however, that VRAG was designed using a male population. Some recent research suggests its use to assess the risk of female offenders may be problematic (Hastings et al., 2011).

3. For a recent review of the literature on the biopsychosocial model of violence, see Raine (2013).

4. It is worth pointing out that actuarial tools such as VRAG do not use predictor variables such as race precisely because the models would then be held to higher standards of legal scrutiny, thereby problematizing their use in legal contexts.

References

Aharoni, E., Anderson, N. E., Barnes, J. C., Allen, C. H., & Kiehl, K. A. (2019). Mind the gap: Toward an integrative science of the brain and crime. *BioSocieties, 14*(3), 463–468.

Aharoni, E., Mallett, J., Vincent, G. M., Harenski, C. L., Calhoun, V. D., Sinnott-Armstrong, W., . . . & Kiehl, K. A. (2014). Predictive accuracy in the neuroprediction of rearrest. *Social Neuroscience, 9*(4), 332–336.

Aharoni, E., Vincent, G. M., Harenski, C. L., Calhoun, V. D., Sinnott-Armstrong, W., Gazzaniga, M. S., & Kiehl, K. A. (2013). Neuroprediction of future rearrest. *Proceedings of the National Academy of Sciences of the United States of America, 110*(15), 6223–6228.

Andrews, D. A. (2006). Enhancing adherence to risk–need–responsivity: Making quality a matter of policy. *Criminology and Public Policy, 5*(3), 595–602.

Aos, S., Miller, M., & Drake, E. (2006). Evidence-based public policy options to reduce future prison construction, criminal justice costs, and crime rates. *Federal Sentencing Reporter, 19*(4), 275.

Banks, S., Robbins, P. C., Silver, E., Vesselinov, R., Steadman, H. J., Monahan, J., . . . & Roth, L. H. (2004). A multiple-models approach to violence risk assessment among people with mental disorder. *Criminal Justice and Behavior, 31*(3), 324–340.

Barabas, C., Virza, M., Dinakar, K., Ito, J., & Zittrain, J. (2018). Interventions over predictions: Reframing the ethical debate for actuarial risk assessment. *Proceedings of Machine Learning Research, 81*, 62–76.

Baum, M., & Savulescu, J. (2013). Behavioral biomarkers: What are they good for? In I. Singh, W. P. Sinnott-Armstrong, & J. Savulescu (Eds.), *Bioprediction, biomarkers, and bad behavior: Scientific, legal, and ethical challenges* (pp. 12–41). Oxford: Oxford University Press.

Bedard, H. L. (2017). The potential for bioprediction in criminal law. *The Columbia Science and Technology Law Review, XVIII*, 58.

Berns, G. S., & Moore, S. E. (2012). A neural predictor of cultural popularity. *Journal of Consumer Psychology, 22*(1), 154–160.

Berryessa, C. M. (2017). Jury-eligible public attitudes toward biological risk factors for the development of criminal behavior and implications for capital sentencing. *Criminal Justice and Behavior, 44*(8), 1073–1100.

Buckholtz, J., & Meyer-Lindenberg, A. (2012). Psychopathology and the human connectome: Toward a transdiagnostic model of risk for mental illness. *Neuron, 74*(6), 990–1004.

Camchong, J., Stenger, A., & Fein, G. (2012). Resting-state synchrony during early alcohol abstinence can predict subsequent relapse. *Cerebral Cortex, 23*(9), 2086–2099.

Campbell, C., & Eastman, N. (2014). The limits of legal use of neuroscience. In W. Sinnott-Armstrong, I. Singh, & J. Savulescu (Eds.), *Bioprediction, biomarkers, and bad behavior: Scientific, legal, and ethical challenges* (pp. 91–117). Oxford: Oxford University Press.

Cecil, K. M., Brubaker, C. J., Adler, C. M., Dietrich, K. N., Altaye, M., Egelhoff, J. C., . . . Lanphear, B. P. (2008). Decreased brain volume in adults with childhood lead exposure. *PLoS Medicine, 5*(5), e112.

Choy, O., Raine, A., & Hamilton, R. H. (2018). Stimulation of the prefrontal cortex reduces intentions to commit aggression: A randomized, double-blind, placebo-controlled, stratified, parallel-group trial. *Journal of Neuroscience, 38*(29), 6505–6512.

Coppola, F. (2018). Mapping the brain to predict antisocial behaviour: New frontiers in neurocriminology, "new" challenges for criminal justice. *UCL Journal of Law and Jurisprudence—Special Issue, 1*(1), 103–126.

Dawes, R. M., Faust, D., & Meehl, P. E. (1989). Clinical versus actuarial judgment. *Science, 243*(4899), 1668–1674.

de Hevia, M.-D., & Spelke, E. S. (2009). Spontaneous mapping of number and space in adults and young children. *Cognition, 110*(2), 198–207.

Delfin, C., Krona, H., Andiné, P., Ryding, E., Wallinius, M., & Hofvander, B. (2019). Prediction of recidivism in a long-term follow-up of forensic psychiatric patients: Incremental effects of neuroimaging data. *PLoS One, 14*(5), e0217127.

Douglas, T., Pugh, J., Singh, I., Savulescu, J., & Fazel, S. (2017). Risk assessment tools in criminal justice and forensic psychiatry: The need for better data. *European Psychiatry: The Journal of the Association of European Psychiatrists, 42*, 134–137.

Ellwood, C. A. (1912). Lombroso's theory of crime. *Journal of the American Institute of Criminal Law and Criminology, 2*(5), 716.

Ennis, G., & Litwack, R. (1974). Psychiatry and the presumption of expertise: Flipping coins in the courtroom. *California Law Review, 62*, 693–718.

Faigman, D. L., Monahan, J., & Slobogin, C. (2013). Group to individual (G2i) inference in scientific expert testimony. *The University of Chicago Law Review, 81*(2), 417–480.

Farahany, N. A. (2012a). Incriminating thoughts. *Stanford Law Review, 64*, 351–408.

Farahany, N. A. (2012b). Searching secrets. *University of Pennsylvania Law Review, 160*, 1239.

Fox, D. (2009). The right to silence as protecting mental control. *Akron Law Review, 42*(3), 763.

Gilligan, J., & Lee, B. (2005). The Resolve to Stop the Violence Project: Reducing violence in the community through a jail-based initiative. *Journal of Public Health, 27*(2), 143–148.

Greely, H. T., & Illes, J. (2007). Neuroscience-based lie detection: The urgent need for regulation. *American Journal of Law and Medicine, 33*(2–3), 377–431.

Grove, W. M., & Meehl, P. E. (1996). Comparative efficiency of informal (subjective, impressionistic) and formal (mechanical, algorithmic) prediction procedures: The clinical-statistical controversy. *Psychology, Public Policy, and Law, 2*(2), 293–323.

Hall, W. D., Carter, A., & Yücel, M. (2014). Ethical issues in the neuroprediction of addiction risk and treatment response. In J. Clausen & N. Levy (Eds.), *Handbook of neuroethics* (pp. 1025–1044). Dordrecht, Netherlands: Springer.

Hamilton, M. (2015). Risk-needs assessment: Constitutional and ethical challenges. *American Criminal Law Review, 52*, 61.

Hanley, B. J. (2017). In search of evidence against a criminal defendant: The constitutionality of judicially ordered surgery. *The Catholic Lawyer, 22*(4), 6.

Hastings, M. E., Krishnan, S., Tangney, J. P., & Stuewig, J. (2011). Predictive and incremental validity of the Violence Risk Assessment Guide scores with male and female inmates. *Psychological Assessment, 23*(1), 174–183.

Heider, F. (1958). *The psychology of interpersonal relations*. New York: John Wiley.

Heilbrun, K., Douglas, K. S., & Yasuhara, K. (2009). Violence risk assessment: Core controversies. In J. L. Skeem, K. S. Douglas, & S. O. Lilienfeld (Eds.), *Psychological science in the courtroom: Consensus and controversy* (pp. 333–357). New York: Guilford Press.

Hubbard, R. A., Kerlikowske, K., Flowers, C. I., Yankaskas, B. C., Zhu, W., & Miglioretti, D. L. (2011). Cumulative probability of false-positive recall or biopsy recommendation after 10 years of screening mammography. *Annals of Internal Medicine, 155*(8), 481.

Ienca, M., & Andorno, R. (2017). Towards new human rights in the age of neuroscience and neurotechnology. *Life Sciences, Society and Policy, 13*(1), 5.

Janes, A. C., Pizzagalli, D. A., Richardt, S., Frederick, B. de B., Chuzi, S., Pachas, G., . . . Kaufman, M. J. (2010). Brain reactivity to smoking cues prior to smoking cessation predicts ability to maintain tobacco abstinence. *Biological Psychiatry, 67*(8), 722–729.

Janus, E. S., & Prentky, R. A. (2003). Forensic use of actuarial risk assessment with sex offenders: Accuracy, admissibility and accountability. *American Criminal Law Review, 40*, 1443.

Jurjako, M., Malatesti, L., & Brazil, I. A. (2018). Some ethical considerations about the use of biomarkers for the classification of adult antisocial individuals. *International Journal of Forensic Mental Health, 18*(3), 228–242.

Just, M. A., Pan, L., Cherkassky, V. L., McMakin, D. L., Cha, C., Nock, M. K., & Brent, D. (2017). Machine learning of neural representations of suicide and emotion concepts identifies suicidal youth. *Nature Human Behaviour, 1*(12), 911.

Kahneman, D., & Tversky, A. (1972). Subjective probability: A judgment of representativeness. *Cognitive Psychology, 3*(3), 430–454.

Kerns, J. G., Cohen, J. D., MacDonald, A. W., Cho, R. Y., Stenger, V. A., & Carter, C. S. (2004). Anterior cingulate conflict monitoring and adjustments in control. *Science, 303*(5660), 1023–1026.

Kiehl, K. A., Anderson, N. E., Aharoni, E., Maurer, J. M., Harenski, K. A., Rao, V., . . . & Kosson, D. (2018). Age of gray matters: Neuroprediction of recidivism. *NeuroImage: Clinical, 19,* 813–823.

Kogut, T., & Ritov, I. (2005). The "identified victim" effect: An identified group, or just a single individual? *Journal of Behavioral Decision Making, 18*(3), 157–167.

Krauss, D. A., & Sales, B. D. (2001). The effects of clinical and scientific expert testimony on juror decision making in capital sentencing. *Psychology, Public Policy, and Law, 7*(2), 267–310.

Kunda, Z., & Nisbett, R. E. (1986). The psychometrics of everyday life. *Cognitive Psychology, 18*(2), 195–224.

Large, M., & Nielssen, O. (2017). The limitations and future of violence risk assessment. *World Psychiatry, 16*(1), 25–26.

Latessa, E. J., Lovins, L. B., & Smith, P. (2010). *Follow-up evaluation of Ohio's community based correctional facility and halfway house programs: Program characteristics supplemental report.* Cincinnati, OH: University of Cincinnati.

Lidz, C. W., Mulvey, E. P., & Gardner, W. (1993). The accuracy of predictions of violence to others. *JAMA, 269*(8), 1007.

Liu, J., Xu, X., Wu, K., Piao, Z., Huang, J., Guo, Y., . . . & Huo, X. (2011). Association between lead exposure from electronic waste recycling and child temperament alterations. *Neurotoxicology, 32*(4), 458–464.

Lombroso, C. (2006). Criminal literature. In M. Gibson & N. H. Rafter (Trans.), *Criminal man* (pp. 79–80). Durham, NC: Duke University Press.

MacKenzie, D. L. (2006). *What works in corrections: Reducing the criminal activities of offenders and delinquents.* Cambridge: Cambridge University Press.

Meynen, G. (2018). Forensic psychiatry and neurolaw: Description, developments, and debates. *International Journal of Law and Psychiatry, 65,* 101345.

Monahan, J. (1995). *The clinical prediction of violent behavior.* Lanham, MD: Jason Aronson, Inc. (Originally published in 1981 as *Predicting violent behavior: An assessment of clinical techniques* by Sage).

Monahan, J. (2014). The inclusion of biological risk factors in violence risk assessments. In I. Singh, W. P. Sinnott-Armstrong, & J. Savulescu (Eds.), *Bioprediction, biomarkers, and bad behavior: Scientific, legal, and ethical challenges* (pp. 57–76). Oxford: Oxford University Press.

Monahan, J., Brodsky, S. L., & Shan, S. A. (1981). *Predicting violent behavior: An assessment of clinical techniques.* Beverly Hills, CA: Sage.

Monahan, J., Steadman, H., Silver, E., Applebaum, P. S., Clark-Robbins, A., Mulvey, E. P., . . . Banks, S. (2001). *Rethinking risk assessment: The MacArthur Study of mental disorder and violence.* Oxford: Oxford University Press.

Monahan, J., & Silver, E. (2003). Judicial decision thresholds for violence risk management. *International Journal of Forensic Mental Health, 2*(1), 1–6.

Moore, K. E., Stuewig, J. B., & Tangney, J. P. (2016). The effect of stigma on criminal offenders' functioning: A longitudinal mediational model. *Deviant Behavior, 37*(2), 196–218.

Nadelhoffer, T., Bibas, S., Grafton, S., Kiehl, K. A., Mansfield, A., Sinnott-Armstrong, W., & Gazzaniga, M. (2012). Neuroprediction, violence, and the law: Setting the stage. *Neuroethics, 5*(1), 67–99.

Nishimoto, S., Vu, A. T., Naselaris, T., Benjamini, Y., Yu, B., & Gallant, J. L. (2011). Reconstructing visual experiences from brain activity evoked by natural movies. *Current Biology, 21*(19), 1641–1646.

Pardini, D. A., Raine, A., Erickson, K., & Loeber, R. (2014). Lower amygdala volume in men is associated with childhood aggression, early psychopathic traits, and future violence. *Biological Psychiatry, 75*(1), 73–80.

Paulus, M. P., Tapert, S. F., & Schuckit, M. A. (2005). Neural activation patterns of methamphetamine-dependent subjects during decision making predict relapse. *Archives of General Psychiatry, 62*(7), 761.

Pinker, S. (2016). *The blank slate: The modern denial of human nature.* New York: Viking.

Poldrack, R. A., Monahan, J., Imrey, P. B., Reyna, V., Raichle, M. E., Faigman, D., & Buckholtz, J. W. (2018). Predicting violent behavior: What can neuroscience add? *Trends in Cognitive Sciences, 22*(2), 111–123.

Pustilnik, A. C. (2013). Neurotechnologies at the intersection of criminal procedure and constitutional law. In J. T. Parry & L. S. Richardson (Eds.), *The constitution and the future of criminal justice in America* (pp. 109–134). Cambridge: Cambridge University Press.

Quinsey, V. L., Harris, G. E., Rice, M. E., & Cormier, C. (1998). *Violent offenders: Appraising and managing risk.* Washington, DC: American Psychological Association.

Raine, A. (2013). *The anatomy of violence: The biological roots of crime.* New York: Pantheon Books.

Rissman, J., Greely, H. T., & Wagner, A. D. (2010). Detecting individual memories through the neural decoding of memory states and past experience. *Proceedings of the National Academy of Sciences of the United States of America, 107,* 9849–9854.

Ryazanov, A. A., Knutzen, J., Rickless, S. C., Christenfeld, N. J. S., & Nelkin, D. K. (2018). Intuitive probabilities and the limitation of moral imagination. *Cognitive Science, 42*(Suppl. 1), 38–68.

Sanders, T., Liu, Y., Buchner, V., & Tchounwou, P. B. (2009). Neurotoxic effects and biomarkers of lead exposure: A review. *Reviews on Environmental Health, 24*(1), 15–45.

Schauer, F. (2010). Neuroscience, lie-detection, and the law: Contrary to the prevailing view, the suitability of brain-based lie-detection for courtroom or forensic use should be determined according to legal and not scientific standards. *Trends in Cognitive Sciences, 14*(3), 101–103.

Shen, F. X. (2013). Neuroscience, mental privacy, and the law. *Harvard Journal of Law and Public Policy, 36*(2), 653–713.

Silver, E., & Miller, L. L. (2002). A cautionary note on the use of actuarial risk assessment tools for social control. *Crime and Delinquency, 48*(1), 138–161.

Singh, J. P., Desmarais, S. L., Hurducas, C., Arbach-Lucioni, K., Condemarin, C., Dean, K., . . . & Otto, R. K. (2014). International perspective on the practical application of violence risk assessment: A global survey of 44 countries. *International Journal of Forensic Mental Health, 13*, 193–206.

Singh, J. P., Grann, M., & Fazel, S. (2011). A comparative study of violence risk assessment tools: A systematic review and metaregression analysis of 68 studies involving 25,980 participants. *Clinical Psychology Review, 31*, 499–513.

Sinha, R., & Li, C. S. R. (2007). Imaging stress-and cue-induced drug and alcohol craving: Association with relapse and clinical implications. *Drug and Alcohol Review, 26*(1), 25–31.

Specker, J., Focquaert, F., Sterckx, S., & Schermer, M. H. N. (2018). Forensic practitioners' expectations and moral views regarding neurobiological interventions in offenders with mental disorders. *BioSocieties, 13*, 304–321.

Starr, S. (2014). Evidence-based sentencing and the scientific rationalization of discrimination. *Stanford Law Review, 66*(4), 803–872.

Steadman, H. J., Silver, E., Monahan, J., Appelbaum, P. S., Clark Robbins, P., Mulvey, E. P., . . . & Banks, S. (2000). A classification tree approach to the development of actuarial violence risk assessment tools. *Law and Human Behavior, 24*(1), 83–100.

Steele, V. R., Claus, E. D., Aharoni, E., Vincent, G. M., Calhoun, V. D., & Kiehl, K. A. (2015). Multimodal imaging measures predict rearrest. *Frontiers in Human Neuroscience, 9*, 425.

Steele, V. R., Fink, B. C., Maurer, J. M., Arbabshirani, M. R., Wilber, C. H., Jaffe, A. J., . . . & Kiehl, K. A. (2014). Brain potentials measured during a go/nogo task predict completion of substance abuse treatment. *Biological Psychiatry, 76*(1), 75–83.

Steele, V. R., Maurer, J. M., Arbabshirani, M. R., Claus, E. D., Fink, B. C., Rao, V., . . . & Kiehl, K. A. (2017). Machine learning of functional magnetic resonance imaging network connectivity predicts substance abuse treatment completion. *Biological Psychiatry: Cognitive Neuroscience and Neuroimaging, 3*(2), 141–149.

Tajfel, H., & Wilkes, A. L. (1963). Classification and quantitative judgement. *British Journal of Psychology, 54*(2), 101–114.

Taxman, F. S. (2002). Supervision—Exploring the dimensions of effectiveness. *Federal Probation Journal, 66*, 14.

To, W. T., Eroh, J., Hart, J., & Vanneste, S. (2018). Exploring the effects of anodal and cathodal high definition transcranial direct current stimulation targeting the dorsal anterior cingulate cortex. *Scientific Reports, 8*(1), 4454.

Tonry, M. (2014). Remodeling American sentencing: A ten-step blueprint for moving past mass incarceration. *Criminology and Public Policy, 13*(4), 503–533.

Tversky, A., & Kahneman, D. (1973). Availability: A heuristic for judging frequency and probability. *Cognitive Psychology, 5*(2), 207–232.

Verplaetse, J. (2009). *Localizing the moral sense: Neuroscience and the search for the cerebral seat of morality, 1800–1930*. Dordrecht, Netherlands: Springer Science & Business Media.

Weiss, B. (2017, April 1). Jonathan Haidt on the cultural roots of campus rage. *Wall Street Journal*. Retrieved from https://www.wsj.com/articles/jonathan-haidt-on-the -cultural-roots-of-campus-rage-1491000676

Wolpe, P. (2009). Is my mind mine? *Forbes*. Retrieved from https://www.forbes.com /2009/10/09/neuroimaging-neuroscience-mind-reading-opinions-contributors-paul -root-wolpe.html?sh=12d8f7936147

Wykstra, S. (2018, July). Just how transparent can a criminal justice algorithm be? *Slate Magazine*. Retrieved from https://slate.com/technology/2018/07/pennsylvania -commission-on-sentencing-is-trying-to-make-its-algorithm-transparent.html

Yang, M., Wong, S. C. P., & Coid, J. (2010). The efficacy of violence prediction: A meta-analytic comparison of nine risk assessment tools. *Psychological Bulletin, 136*, 740–767.

7 Ethical Issues Raised by Recent Developments in Neuroscience: The Case of Optogenetics

Gidon Felsen and Jennifer Blumenthal-Barby

7.1 Introduction

New tools for recording and manipulating neuronal activity in humans and animals are developing at a rapid pace. Driven in part by public and private funding for ambitious neuroscience research (Insel, Landis, & Collins, 2013; Bargmann & Newsome, 2014; Brose, 2016; Yuste & Bargmann, 2017; Musk & Neuralink, 2019), these tools offer the promise of transforming our understanding of the brain in health and disease. Discussion of the ethical implications of these technological advances, however, has often lagged, despite increasing recognition of these ethical implications by the emerging field of neuroethics (Farah, 2010; Illes & Hossain, 2017). In order to maximize the benefits and minimize the risks of applying these novel methods, it is critical to identify and address the ethical issues early (Goering & Yuste, 2016; Greely, Ramos, & Grady, 2016; Ramos et al., 2019). We focus here on the ethical implications associated with the clinical application of optogenetics—an approach that allows neural activity to be controlled with light (Bi et al., 2006; Boyden et al., 2005; Deisseroth, 2015). While optogenetics is only one of several recent transformative neurotechnological developments, it is perhaps among the closest to clinical application. Indeed, clinical trials using optogenetics to treat retinitis pigmentosa, in which opsins (light-sensitive proteins) are expressed in the retina to restore vision, are underway (Simunovic et al., 2019), and there is great interest in applying optogenetics to a range of psychiatric and neurological disorders (Deisseroth, 2012; Delbeke et al., 2017). In this chapter, after briefly describing optogenetics and its potential clinical applicability, we focus on the ethical issues associated with its clinical use.

To appreciate the significance of optogenetics, it may be helpful for some readers to review very briefly here some fundamentals of neuroscience

(neuroscience experts may wish to skip ahead). Neurons communicate information via action potentials, which occur when positively charged ions enter the neuron. Ions cross the cell membrane passively through open channels, which can be opened or closed by factors such as chemical compounds, temperature, and light, and can be actively pumped across the membrane as well. Light-gated channels and pumps are found in a variety of microbial species, in which they promote survival (e.g., by allowing the organism to navigate toward a food source). Advances in molecular biology and genetics have allowed researchers to clone the genes encoding these light-sensitive channels and pumps (among many other proteins) and to express them in cells in which they are not normally found. Expressing light-sensitive ion channels in the cell membrane of a neuron, for example, allows the flow of ions across the membrane to be controlled by light, which can be delivered to the brain via an optical fiber. In this way, optogenetics allows for the manipulation of activity of genetically specified populations of neurons with light, providing unprecedented control of specific neural circuits (Deisseroth, 2015). Depending on which light-sensitive protein is expressed, and in which types of neurons, specific groups of neurons can be activated or inactivated with high temporal resolution. For example, neuronal expression of channelrhodopsin-2, a blue-light sensitive cation channel protein found in some algae (Boyden et al., 2005), allows these neurons to be activated by blue light. Light causes the channel to open, allowing cations to enter the neuron, resulting in an increased rate of action potentials. When archaerhodopsin-3, a proton pump found in the single-celled organism archaea (Chow et al., 2010), is expressed in neurons and exposed to yellow light, protons are pumped out of the neuron, ultimately decreasing the rate of action potentials. Channelrhodopsin-2 and archaerhodopsin-3 therefore allow the activity (the rate of action potentials) of neurons to be increased or decreased, respectively, and are only two of a rapidly expanding set of optogenetic proteins for controlling cellular processes.

Additional molecular tools allow for the expression of optogenetic proteins restricted to specific types of neurons based on, for example, what neurotransmitter they release or what downstream brain regions they project to (i.e., send axons; Luo, Callaway, & Svoboda, 2018). For example, suppose one wanted to activate only those neurons in one brain area (say, area A) that project to another brain area (say, area B). There are several approaches by which channelrhodopsin-2 can be expressed in only those area A neurons that project to area B, and not in other area A neurons. Delivering blue light to area A (e.g., via an optical fiber) then activates only that subset of area B–projecting neurons (Gradinaru, Zhang, et al., 2010). This level of specificity is critical because neurons in any given brain area frequently project to many different brain areas, and each projection may have a specific function. Similarly, neurons in any given area may release different neurotransmitters, which may also mediate specific functions. Finally, because light delivery can be controlled with high temporal and spatial precision, neural activity in specific brain regions can be modulated at a relevant timescale for the processing that occurs in neural circuits (i.e., milliseconds).

By providing such cell-type specificity and spatiotemporal specificity, optogenetics has enabled researchers to examine whether activity in specific cell types is necessary and/or sufficient for a range of functions, from stimulus representation to neural circuit dynamics to complex social behavior, transforming how basic neuroscience is performed as well as our understanding of neural function (Yizhar, 2012; Yoshihara & Yoshihara, 2018). Given the circuit-based nature of many psychiatric and neurological disorders, optogenetics offers the potential to transform clinical care as well in ways that are not possible with current standard treatment options. Pharmaceuticals, for example, have themselves transformed psychiatric treatment over the last several decades via the development of drugs that up- or downregulate neurotransmission (Ban, 2006). However, the relatively slow time course of drug effects—on the order of hours—as well as the frequency of off-target effects, limits the efficacy of this approach. Direct current delivery, on the other hand, can provide millisecond-level modulation of neural activity, but affects all nearby neurons rather than only the desired type (Henderson, Federici, & Boulis, 2009; Nowak et al., 2010; Delbeke et al., 2017). Thus, these and other current approaches lack the necessary combination of temporal resolution and cell-type specificity to treat circuit-level disorders optimally.

Deep brain stimulation (DBS), a treatment for Parkinson's disease (PD) and other movement disorders approved by the Food and Drug Administration (FDA), provides an illustrative example of the shortcomings of traditional approaches and the potential clinical application of optogenetics (Lin, Deisseroth, & Henderson, 2011; Delbeke et al., 2017). In DBS for PD, current is delivered to a particular brain region (typically the subthalamic nucleus [STN], a component of the basal ganglia) in order to normalize its pathological activity and mitigate parkinsonian motor symptoms

(Kringelbach et al., 2007). We will subsequently refer to this therapy as "electrical DBS" (eDBS) to differentiate it from optogenetic DBS discussed below. Although eDBS typically relieves some parkinsonian symptoms, its mechanism of action is surprisingly poorly understood (Herrington, Cheng, & Eskandar, 2015), hindering improvements to the therapy. Standard models of basal ganglia function predict that the parkinsonian STN is hyperactive, suggesting that eDBS works by reducing STN activity. Typically, however, delivering current is thought to increase, not decrease, activity. One possible explanation is that eDBS somehow modulates the activity of axons projecting to the STN rather than STN neurons themselves, even though the delivered current presumably affects all nearby tissue similarly. By allowing specific neurons (e.g., STN neurons) to be manipulated in a known direction (e.g., excited), optogenetics avoids these problems, offering the potential to understand the mechanism of action of eDBS and identifying beneficial stimulation parameters, both of which would improve eDBS therapy (Gradinaru, Mogri, et al., 2009; Kravitz et al., 2010).

There is thus intense interest in developing and applying DBS protocols that use optogenetics (which we will call "oDBS"). To be sure, there are many practical barriers to overcome before oDBS can be a viable treatment option, including the delivery of the genes encoding the opsin, achieving stable long-term opsin expression, and safe light delivery. While these methods are standard in animal models, establishing them in humans presents particular challenges (some of which are discussed below). Conceptual barriers also exist. While we understand a great deal about PD and its treatment at the circuit level (McGregor & Nelson, 2019), our current understanding is insufficient to determine exactly how activity should be altered to relieve parkinsonian symptoms (i.e., what pattern of activity in which cell types would be required, and how frequently the stimulation protocol might need to be revisited in response to plasticity in the affected circuits). However, our technological capabilities and conceptual understanding are advancing rapidly. For example, it was recently demonstrated that "upconversion" nanoparticles, which absorb light of one wavelength and emit light of another, can be used to activate opsins (Chen et al., 2018). This is an important advance with translational relevance, as it would allow opsins normally sensitive to short-wavelength (e.g., blue) light to be activated by long-wavelength (e.g., red) light, which can penetrate through the skull and relatively deep into the brain and therefore be delivered noninvasively

(Feliu, Neher, & Parak, 2018). Thus, while there remain obstacles to over-come, it is critical to examine the ethical issues associated with oDBS (Gil-bert et al., 2014). We next describe the normative ethical issues associated with clinical trials for oDBS and with the potential for widespread adoption of optogenetics in clinical care, and briefly discuss how these issues can and should be informed by empirical data.

7.2   Normative Ethical Issues Associated with Clinical Optogenetics

Clinical optogenetics requires genetic engineering in order to express opsins in neurons, as well as the direct modulation of neural activity. As such, several of the associated ethical issues are similar to those of gene therapy and eDBS. We review these briefly here, highlighting differences specific to optogenetics, before focusing on novel ethical issues. While we concentrate here on clinical applications of optogenetics, it is possible that optogenet-ics could at some point be leveraged to enhance cognitive function, fol-lowing the familiar shift from restoration to enhancement seen with other therapeutic approaches. Such use would involve many of the same ethical issues as would clinical applications, while also introducing additional issues discussed in the cognitive enhancement literature (e.g., access and equity; Farah et al., 2004).

7.2.1   Ethics of Opsin Expression in Patients

Expressing opsins in neurons requires introducing DNA encoding the opsin protein to the neurons of interest, which can be accomplished in several ways. Critically, these methods require only somatic editing and not the more ethically fraught germline editing (in which the exogenous DNA is passed to future generations). Genetically modifying somatic cells to treat diseases is generally considered potentially ethical as long as standard issues are addressed (i.e., safety, risk–benefit calculation, protection of vulner-able subjects, informed consent, patient monitoring, and equity of access; National Academies of Sciences, 2017; Cavaliere, 2019; Coller, 2019), includ-ing by the public (Condit, 2010; Robillard et al., 2014). To the extent that introducing opsin DNA is seen as therapeutic, we would expect the same normative ethical issues and empirical attitudes.

However, in contrast to cases in which gene therapy is used to repair faulty genetic information (i.e., resulting from genetic disorders), in the case of optogenetics, the opsin is not itself therapeutic but instead provides a *tool* for therapy by providing the neurons with an enhanced functionality to be controlled by light. There does not appear to be a deep normative distinction between these cases, given that the ultimate purpose of the opsin expression is therapeutic. However, one might be concerned about some of the morally relevant consequences differentiating a therapy itself and a tool that could be used for therapy in that the tool (in this case, gene editing for opsin expression) could be used for other purposes (see below). Moreover, attitudes toward gene therapy differ for restoring, as opposed to enhancing, function (Robillard et al., 2014). So, it is possible that the gene therapy required for optogenetics would be met with some resistance that is worth understanding and addressing. Empirical data addressing this question would therefore be useful.

7.2.2   Ethics of Directly Manipulating Brain Activity in Patients

oDBS involves the direct manipulation of brain activity. As such, many of the ethical issues raised by eDBS apply to oDBS as well (and some are not applicable; e.g., oDBS may not require an invasive implant for light delivery). These issues include balancing the benefits of therapy with the risks of unintended effects of stimulation, informed consent in a patient population in which medical decision making may be impaired, the potential for enhancement rather than restoration of function, and potential changes to personality and agency (Schermer, 2011; Skorburg & Sinnott-Armstrong, 2020). In the case of closed-loop eDBS, in which stimulation parameters are adapted based on recorded brain activity, privacy is an additional concern (Zuk et al., 2018). We do not identify any ethical issues related to the direct manipulation of brain activity that are novel to oDBS. On the contrary, given that oDBS (including closed-loop oDBS) has the potential to be more effective than eDBS due to its cell-type specificity, some of the above concerns may be mitigated. For example, unintended effects, including personality change, may be less likely to occur.

7.3   Novel Ethical Issues with Optogenetics: First-in-Human Clinical Trials

Perhaps the nearest clinical application of optogenetics to a neurological disorder is oDBS (e.g., for movement disorders such as PD). In light of this, a novel ethical issue raised by optogenetics involves considering the ethical requirements for clinical trials of oDBS (Gilbert et al., 2014). Following the framework of Emanuel, Wendler, and Grady (2000), we focus on aspects unique to oDBS of several criteria that such trials would need to meet.

7.3.1 Fair Subject Selection

Which patients would be most appropriate for first-in-human oDBS trials? While for the reasons described above oDBS holds high potential for improvement over eDBS, the latter is already sufficiently effective and safe, has been approved by the FDA for certain conditions, and is in widespread clinical practice. Thus, the rationale for subjecting a particular patient to oDBS requires a potential relative benefit, or at least no detriment, of oDBS over eDBS for that patient. One promising population would be those for whom eDBS would otherwise be indicated but is not possible. For example, patients with structural basal ganglia abnormalities present a problem for targeting specific brain regions with the stimulating electrode (Kocabicak et al., 2015). This particular problem would be mitigated by oDBS, in which, as long as opsin expression can be sufficiently restricted to a particular type of neuron (an active area of research showing steady advancement), neither the viral injection nor the light delivery needs to be as spatially restricted to achieve therapeutic success.

Although trials testing the therapeutic efficacy of oDBS would need to be performed in patients, there may be advantages to performing parallel trials, in separate groups of subjects, to test the safety of opsin expression and the safety of light delivery. In principle, opsins can be activated by light of sufficiently long wavelength to be deliverable noninvasively through the skull (e.g., via LEDs on the scalp). Assessing the effects of increasing light power on brain tissue (absent opsin expression) can therefore be performed relatively noninvasively in healthy human subjects and will be able to build on work in preclinical studies in animal models. To assess the safety of opsin expression alone, although trials would require an invasive intracranial injection, they would not need to be paired with light delivery. Alternatively, if safety-related trials are performed in patients only, it may be advantageous to also test for efficacy in the same patients (Gilbert et al., 2014).

An important aspect of fair subject selection involves ensuring that vulnerable individuals are not targeted for risky research, and also that the rich and socially advantaged are not favored for potentially beneficial research (Emanuel et al., 2000). Subjects should be selected based on the scientific goals of the study. In the context of oDBS, it is unlikely that vulnerable or stigmatized individuals would be specifically targeted (e.g., because of convenience). However, it may be true that vulnerabilities exist that should be managed during trial design and execution, and that the rich or socially advantaged may be favored as subjects, given that they may be presumed to be more open to advanced neurotechnologies such as oDBS. Thus, care should be taken for researchers and Institutional Review Boards to use an analytic framework for identifying and addressing vulnerabilities, such as the integrative-functional account (Racine & Bracken-Roche, 2019). This framework provides a toolkit that can be applied to any study protocol. Care should also be taken to recruit subjects from all socioeconomic groups so that the privileged are not prioritized and benefits and risks are fairly distributed.

7.3.2 Favorable Risk–Benefit Ratio

In addition to the risks and benefits associated with eDBS and gene therapy, a unique risk of oDBS is the potential for covert manipulation. Once the neurons of a patient are expressing the opsin, their activity is susceptible, by design, to being manipulated by light. As noted above, in the preferred clinical scenario, the light would be deliverable noninvasively, likely via long-wavelength light capable of penetrating deeper into the brain (Feliu et al., 2018). However, this means that other light sources of sufficient power could, in principle, activate these neurons as well. This leaves open the possibility that patients' brain activity could be manipulated covertly with imperceptible light that has no effect on individuals who are not expressing the opsin. Such manipulation could induce patterns of activity in opsin-expressing neurons that cause harmful motor or cognitive effects. The importance of this concern would grow as more patients express a particular opsin, which would be expected to occur if oDBS were to attain clinical viability.

This risk echoes more traditional concerns about the widespread ability to manipulate brain activity of groups of individuals (e.g., via chemical agents delivered to the water supply). In this case, however, the manipulating agent (light) can be delivered indiscriminately while only having an effect on the known targeted population, making this population particularly vulnerable (Racine & Bracken-Roche, 2019). Thus, ethical clinical trials for oDBS should include plans to minimize these risks while still maximizing the potential benefits at the subject and societal level. For example, if covert manipulation with external light sources is deemed too great a risk, opsin expression levels can be titrated such that only sufficiently high-intensity light (i.e., that emitted by the implanted light source) is capable of manipulating neural activity.

7.3.3   Scientific Validity

In general, much more preclinical work must be performed in animal models before scientifically valid clinical trials can be performed. For example, despite a recent National Institutes of Health requirement to examine sex differences in preclinical studies (Clayton & Collins, 2014), to our knowledge there have been no studies on whether any oDBS-related variables (e.g., strength of opsin expression) depend on sex. One issue unique to oDBS, which may not be addressable in preclinical studies, is its long-term viability. oDBS requires that opsins remain stably expressed in the desired cell types, and that light be continually delivered to the brain with a constant (known) effect on neural activity. Typically, this approach would be expected to last for several years, similar to the time course of eDBS therapy. Little is known, however, about the stability of opsin expression over these timescales. The duration of the vast majority of studies in animal models is limited to months at most, and even at this timescale, age-related morphological changes in opsin-expressing neurons have been reported (Miyashita et al., 2013). While expression stability has begun to be examined in the retina (Ameline et al., 2017; Simunovic et al., 2019), to our knowledge, no studies have addressed this question in the brain structures required for oDBS (e.g., the STN). Beyond the stability of opsin expression alone, it is unknown whether there are long-term effects of chronically activating opsins, both on the health of the neurons themselves and on the circuits being manipulated, which may undergo activity-dependent plasticity (Zucker, 1999).

It is worth noting that in addition to the three ethical requirements discussed above (i.e., fair subject selection, favorable risk–benefit ratio, and scientific validity), Emanuel and colleagues' (2000) framework for ethical research includes four other requirements: social or scientific value (that the research improve health and well-being or knowledge), independent review (that the research be evaluated by individuals unaffiliated with the research), informed consent (that subjects are provided with information about the purpose of the research, the procedures, its perceived risks and benefits and alternatives, and that the subject understands this and makes a voluntary decision), and respect for enrolled subjects (including permitting withdrawal, protecting privacy, maintaining welfare, keeping them apprised of changes in risk–benefit ratio, and informing them of research results). Although any research study using optogenetics should ensure that all seven ethical requirements are met, we do not discuss these four in depth because optogenetics research does not raise any unique challenges or questions in these domains.

However, it is perhaps worth making a few remarks about informed consent because one might wonder whether the conditions being treated in oDBS prevent patients from understanding the information sufficiently to give informed consent. As we indicated, the most likely first use (and trials) would be for oDBS for movement disorders such as PD (perhaps with patients with structural basal ganglia abnormalities that present problems for the targeting of specific brain regions required by eDBS). PD is known to produce minor cognitive impairment, such as difficulty in focusing and memory impairment. Informed consent processes should be developed that account for these sequelae, but there is no reason to think that patients will be unable to provide informed consent—and oDBS does not present unique issues here. Even if oDBS is used for psychiatric conditions such as depression or OCD, the normative stance is the same: while capacity for consent should be assessed, it should not be assumed that patients with these conditions are incapable of providing informed consent to participate in oDBS research, nor is it true that oDBS presents unique issues compared to other types of research with these populations. Empirical work can be helpful in understanding where potential gaps or weaknesses in understanding about oDBS research exist so that consent processes can be designed to address them (see below).

Some of the ethical concerns discussed above are limited not to clinical trials themselves but to the continued use of oDBS, should it ultimately succeed in clinical trials and receive regulatory approval. For example, the long-term effectiveness of oDBS must continue to be monitored after approval (i.e., in Phase IV trials). In addition, as noted above, the concern about covert manipulation would only grow with increased acceptance and adoption of optogenetics in clinical settings, and perhaps even in non-clinical settings in which manipulating brain activity is used for enhancement of normal function.

7.4   Empirical Data

Empirical data have the potential to inform some of the normative ethical questions discussed above (Sugarman & Sulmasy, 2010). In particular, how do relevant stakeholders—patients, medical caregivers, and family members—perceive the potential risks and benefits of clinical optogenetics, particularly compared to the risks and benefits of eDBS? For example, eDBS requires an invasive permanent implant. oDBS would require a one-time invasive surgery (to inject the viral vector), but light could, in principle, then be delivered noninvasively. Is the invasiveness of eDBS and oDBS therefore perceived differently? Understanding these attitudes would help with the assessment of favorable risk–benefit ratio and design of informed consent processes in a way that includes the patients' and subjects' perspectives and not only the researchers' and ethicists' perspectives. While we may gain some insight into these questions by considering studies that have examined attitudes about eDBS and about gene therapy, to our knowledge, no empirical studies have focused on attitudes and understanding about clinical optogenetics. While it would be necessary for surveys and qualitative studies first to provide sufficient objective information about the "basics" of optogenetics to stakeholders before assessing their attitudes, concerns, and understanding (as is the case with any novel therapy), such studies could help to evaluate the ethics of clinical optogenetics.

7.5   Conclusions

Optogenetics offers the potential to manipulate neural circuit activity precisely, and there is increasing interest in harnessing this potential to treat a broad range of neurological and psychiatric disorders. Identifying and addressing the ethical issues associated with the first clinical trials and with the potential for widespread adoption of the technology, as we have attempted to do in this chapter, is a necessary step for realizing this potential.

References

Ameline, B., Tshilenge, K.-T., Weber, M., Biget, M., Libeau, L., Caplette, R., . . . Isiegas, C. (2017). Long-term expression of melanopsin and channelrhodopsin causes no gross alterations in the dystrophic dog retina. *Gene Therapy, 24*, 735–741.

Ban, T. A. (2006). Academic psychiatry and the pharmaceutical industry. *Progress in Neuro-Psychopharmacology and Biological Psychiatry, 30*, 429–441.

Bargmann, C. I., & Newsome, W. T. (2014). The Brain Research Through Advancing Innovative Neurotechnologies (BRAIN) Initiative and neurology. *JAMA Neurology, 71*(6), 675–676.

Bi, A., Cui, J., Ma, Y.-P., Olshevskaya, E., Pu, M., Dizhoor, A. M., & Pan, Z.-H. (2006). Ectopic expression of a microbial-type rhodopsin restores visual responses in mice with photoreceptor degeneration. *Neuron, 50,* 23–33.

Boyden, E. S., Zhang, F., Bamberg, E., Nagel, G., & Deisseroth, K. (2005). Millisecond-timescale, genetically targeted optical control of neural activity. *Nature Neuroscience, 8,* 1263–1268.

Brose, K. (2016). Global neuroscience. *Neuron, 92,* 557–558.

Cavaliere, G. (2019). *The ethics of human genome editing.* Geneva: World Health Organization.

Chen, S., Weitemier, A. Z., Zeng, X., He, L., Wang, X., Tao, Y., . . . McHugh, T. J. (2018). Near-infrared deep brain stimulation via upconversion nanoparticle-mediated optogenetics. *Science, 359,* 679–684.

Chow, B. Y., Han, X., Dobry, A. S., Qian, X., Chuong, A. S., Li, M., . . . Boyden, E. S. (2010). High-performance genetically targetable optical neural silencing by light-driven proton pumps. *Nature, 463,* 98–102.

Clayton, J. A., & Collins, F. S. (2014). Policy: NIH to balance sex in cell and animal studies. *Nature, 509,* 282–283.

Coller, B. S. (2019). Ethics of human genome editing. *Annual Review of Medicine, 70,* 289–305.

Condit, C. M. (2010). Public attitudes and beliefs about genetics. *Annual Review of Genomics and Human Genetics, 11,* 339–359.

Deisseroth, K. (2012). Optogenetics and psychiatry: Applications, challenges, and opportunities. *Biological Psychiatry, 71,* 1030–1032.

Deisseroth, K. (2015). Optogenetics: 10 years of microbial opsins in neuroscience. *Nature Neuroscience, 18,* 1213–1225.

Delbeke, J., Hoffman, L., Mols, K., Braeken, D., & Prodanov, D. (2017). And then there was light: Perspectives of optogenetics for deep brain stimulation and neuro-modulation. *Frontiers in Neuroscience, 11,* 663.

Emanuel, E. J., Wendler, D., & Grady, C. (2000). What makes clinical research ethical? *JAMA, 283,* 2701–2711.

Farah, M. J. (2010). *Neuroethics: An introduction with readings.* Cambridge, MA: MIT Press.

Farah, M. J., Illes, J., Cook-Deegan, R., Gardner, H., Kandel, E., King, P., . . . Wolpe, P. R. (2004). Neurocognitive enhancement: What can we do and what should we do? *Nature Reviews Neuroscience, 5,* 421–425.

Feliu, N., Neher, E., & Parak, W. J. (2018). Toward an optically controlled brain. *Science, 359,* 633–634.

Gilbert, F., Harris, A. R., & Kapsa, R. M. I. (2014). Controlling brain cells with light: Ethical considerations for optogenetic clinical trials. *AJOB Neuroscience, 5*, 3–11.

Goering, S., & Yuste, R. (2016). On the necessity of ethical guidelines for novel neurotechnologies. *Cell, 167*, 882–885.

Gradinaru, V., Mogri, M., Thompson, K. R., Henderson, J. M., & Deisseroth, K. (2009). Optical deconstruction of parkinsonian neural circuitry. *Science, 324*, 354–359.

Gradinaru, V., Zhang, F., Ramakrishnan, C., Mattis, J., Prakash, R., Diester, I., . . . Deisseroth, K. (2010). Molecular and cellular approaches for diversifying and extending optogenetics. *Cell, 141*, 154–165.

Greely, H. T., Ramos, K. M., & Grady, C. (2016). Neuroethics in the age of brain projects. *Neuron, 92*, 637–641.

Henderson, J. M., Federici, T., & Boulis, N. (2009). Optogenetic neuromodulation. *Neurosurgery, 64*, 796–804.

Herrington, T. M., Cheng, J. J., & Eskandar, E. N. (2015). Mechanisms of deep brain stimulation. *Journal of Neurophysiology, 115*, 19–38.

Illes, J., & Hossain, S. (Eds.). (2017). *Neuroethics: Anticipating the future.* Oxford: Oxford University Press.

Insel, T. R., Landis, S. C., & Collins, F. S. (2013). The NIH BRAIN Initiative. *Science, 340*, 687–688.

Kocabicak, E., Temel, Y., Höllig, A., Falkenburger, B., & Tan, S. K. (2015). Current perspectives on deep brain stimulation for severe neurological and psychiatric disorders. *Neuropsychiatric Disease and Treatment, 11*, 1051–1066.

Kravitz, A. V., Freeze, B. S., Parker, P. R. L., Kay, K., Thwin, M. T., Deisseroth, K., & Kreitzer, A. C. (2010). Regulation of parkinsonian motor behaviours by optogenetic control of basal ganglia circuitry. *Nature, 466*, 622–626.

Kringelbach, M. L., Jenkinson, N., Owen, S. L. F., & Aziz, T. Z. (2007). Translational principles of deep brain stimulation. *Nature Reviews Neuroscience, 8*, 623–635.

Lin, S.-C., Deisseroth, K., & Henderson, J. M. (2011). Optogenetics: Background and concepts for neurosurgery. *Neurosurgery, 69*, 1–3.

Luo, L., Callaway, E. M., & Svoboda, K. (2018). Genetic dissection of neural circuits: A decade of progress. *Neuron, 98*, 256–281.

McGregor, M. M., & Nelson, A. B. (2019). Circuit mechanisms of Parkinson's disease. *Neuron, 101*, 1042–1056.

Miyashita, T., Shao, Y. R., Chung, J., Pourzia, O., & Feldman, D. (2013). Long-term channelrhodopsin-2 (ChR2) expression can induce abnormal axonal morphology and targeting in cerebral cortex. *Frontiers in Neural Circuits, 7*, 8.

Musk, E., & Neuralink. (2019). An integrated brain–machine interface platform with thousands of channels. *Journal of Medical Internet Research, 21,* e16194.

National Academies of Sciences, Committee on Human Gene Editing. (2017). *Somatic genome editing.* Washington, DC: National Academies Press.

Nowak, V. A., Pereira, E. A. C., Green, A. L., & Aziz, T. Z. (2010). Optogenetics—Shining light on neurosurgical conditions. *British Journal of Neurosurgery, 24,* 618–624.

Racine, E., & Bracken-Roche, D. (2019). Enriching the concept of vulnerability in research ethics: An integrative and functional account. *Bioethics, 33,* 19–34.

Ramos, K. M., Grady, C., Greely, H. T., Chiong, W., Eberwine, J., Farahany, N. A., . . . Koroshetz, W. J. (2019). The NIH BRAIN Initiative: Integrating neuroethics and neuroscience. *Neuron, 101,* 394–398.

Robillard, J. M., Roskams-Edris, D., Kuzeljevic, B., & Illes, J. (2014). Prevailing public perceptions of the ethics of gene therapy. *Human Gene Therapy, 25,* 740–746.

Schermer, M. (2011). Ethical issues in deep brain stimulation. *Frontiers in Integrative Neuroscience, 5,* 17.

Simunovic, M. P., Shen, W., Lin, J. Y., Protti, D. A., Lisowski, L., & Gillies, M. C. (2019). Optogenetic approaches to vision restoration. *Experimental Eye Research, 178,* 15–26.

Skorburg, J. A., & Sinnott-Armstrong, W. (2020). Some ethics of deep brain stimulation. In D. J. Stein & I. Singh (Eds.), *Global mental health and neuroethics* (pp. 117–132). Cambridge, MA: Academic Press.

Sugarman, J., & Sulmasy, D. P. (2010). *Methods in medical ethics.* Washington, DC: Georgetown University Press.

Yizhar, O. (2012). Optogenetic insights into social behavior function. *Biological Psychiatry, 71,* 1075–1080.

Yoshihara, M., & Yoshihara, M. (2018). "Necessary and sufficient" in biology is not necessarily necessary—Confusions and erroneous conclusions resulting from misapplied logic in the field of biology, especially neuroscience. *Journal of Neurogenetics, 32,* 53–64.

Yuste, R., & Bargmann, C. (2017). Toward a global BRAIN Initiative. *Cell, 168,* 956–959.

Zucker, R. S. (1999). Calcium- and activity-dependent synaptic plasticity. *Current Opinion in Neurobiology, 9,* 305–313.

Zuk, P., Torgerson, L., Sierra-Mercado, D., & Lázaro-Muñoz, G. (2018). Neuroethics of neuromodulation: An update. *Current Opinion in Biomedical Engineering, 8,* 45–50.

II Cognitive Neuroscience and Philosophy

8   Touch and Other Somatosensory Senses

Tony Cheng and Antonio Cataldo

8.1   Introduction: The Somatosensory World

In 1925, David Katz published an influential monograph on touch, *Der Aufbau der Tastwelt*, which was translated into English in 1989. Although it is called "the world of touch," it also discusses the thermal and the nociceptive senses, albeit briefly. In this chapter, we will follow this approach, but we will speak about "somatosensory senses" in general in order to remind ourselves that perception of temperature and pain should also be considered together in this context.

Many human interactions with the environment rely on the somatosensory or *bodily* senses. Imagine this simple scenario: at night, you stand outside the door of your house, and—without looking—you easily find the keys in your backpack. You open the door, feeling the cold, smooth surface of the handle, and, still in the dark, you grope around with your hand and reach for the light switch. While you put the kettle on, your shoulder starts itching, and your hand unerringly reaches the exact point where the sensation originated. You pour your favorite infusion into a mug, and a warm sensation pleasantly spreads from the mug to your fingers. Abruptly, your mobile phone vibrates in your pocket. You spill a drop of hot water on your hand, and automatically start rubbing it to alleviate the sting. In each of these—and many other— examples, the intricate stream of signals arising from the skin, muscles, tendons, and internal organs constantly inform us about the events occurring on the surface and inside of our body, enabling us to select the actions that are more likely to increase the chance and quality of our survival.

Somatosensation, the umbrella term describing such a stream of bodily information, entails several sensory functions that are conventionally defined by contrast with the other sensory systems. For example, in a classical definition by Vernon Mountcastle (1918–2015), often referred as the father of neuroscience (Sejnowski, 2015; Andersen, 2015), somatosensation is any sensory activity originating elsewhere than in the specialized sense organs (eyes and ears), and conveying information about the state of the body and its immediate environment (Mountcastle, 1982). It would therefore be an overly ambitious goal for this chapter to provide a comprehensive account of the somatosensory system; for this, we refer to more extensive works (e.g., Roberts, 2002; Gallace & Spence, 2014). Rather, following the example of previous influential works (e.g., Katz, 1925/1989), we will focus mainly on tactile perception, discussing the thermal and nociceptive senses only in passing.

This chapter has two parts. In part 1, we briefly review some key issues in the psychology and neuroscience of touch. In part 2, we discuss some key issues in the philosophy of perception as they apply to touch. At the end, we conclude by briefly considering the interface problem (Bermúdez, 2005), that is, how different levels of explanation can possibly interface with one another, as it offers the opportunity to discuss how neuroscience, psychology, and philosophy can interact with one another.

8.2   The Psychology and Neuroscience of Touch

It is a commonplace to regard physics, chemistry, biology, psychology, and so on as dealing with different levels of nature, with each discipline involving distinct explanatory resources and goals. The question as to how these different disciplines connect with each other is still a matter of controversy (e.g., Cartwright, 1999; Heil, 2003). Since early on, a lot of work in behavioral neuroscience described in Lashley (1932), Tolman (1949), Hebb (1949), and Penfield (1975), for example, clearly demonstrates the influence of neuroscience on psychology, even if psychology was understood in behaviorists' terms. In 1976, Michael Gazzaniga and George Armitage Miller usefully coined the term "cognitive neuroscience" to refer to studies in which cognitive psychology and neuroscience coupled with each other to produce meaningful research, thus formalizing a cross-disciplinary field of research (Gazzaniga, 1984). Given the historical intertwinement of these fields, understanding the neuroscience of touch requires some background in its psychology as well.

8.2.1   A Brief History of Empirical Research on Touch

Cognitive psychology took off in the 1960s, after varieties of behaviorism finally declined (e.g., Chomsky, 1959). Ulric Neisser (1967) made popular the label "cognitive psychology" to characterize the discipline that studies mental processes such as attention, memory, perception, metacognition, language, and so on. Perception, too, was central to cognitive psychology, as it interfaces with those cognitive processes. However, most studies on perception are confined to vision, which has dominated research on perception in both psychology and neuroscience since their very beginning. This kind of "visuocentrism" exists not only in the sciences but also in philosophy (O'Callaghan, 2008), and might reflect various contingent reasons, such as the fact that vision tends to be the last sense one wishes to lose (Robles-De-La-Torre, 2006; Goldstein & Brockmole, 2016). The complex organization of the somatosensory system, relative to the other sensory modalities, has undoubtedly played an important role in slowing down empirical research on this topic while simultaneously nourishing philosophical debates about the ontological properties of this sense. For example, while other senses—vision, audition, olfaction, and gustation—have clearly localized sensory organs (eyes, ears, nose, mouth/tongue), the somatic senses gather information from the entire body, including skin, muscles, tendons, lips, and internal organs. Thus, while vision or audition receive inputs from only two kinds of receptors each (rods and cones for vision, inner and outer hairy cells for audition), the somatosensory system processes information coming from up to thirteen different afferent receptors.

That being said, studies of somatosensation have been steadily increasing since the 1960s. Around the 1990s, about 2,000 studies were published per year. In contrast, more than 10,000 studies on somatosensation were published in 2019 only (cf. Gallace & Spence, 2014; see figure 8.1 for the trends).

Research on touch owes this current momentum, in part, to efforts starting from more than half a century earlier. In 1960, for example, an important conference "Symposium on Cutaneous Sensitivity" was held at Fort Knox in the U.S. (Hawkes, 1960). Soon after that, tactile visual substitution systems appeared and stimulated several studies in the 1970s (e.g., Bach-y-Rita, 1972). In the 1980s, research on touch kept growing steadily, though it still largely focused on unisensory studies (e.g., Barlow & Mollon, 1982). In the 1990s, more applied studies appeared, such as human factor studies on tactile feedbacks in soldiers and fighter pilots (Sklar & Sarter, 1999). Since the 2000s,

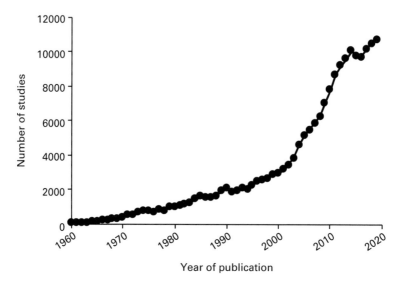

Figure 8.1
Number of studies published on the somatic senses from 1960 to 2019. Search performed on Scopus on July 2020, using "touch," "tactile," "haptic," and "somatosens*" as search terms in the title, abstract, or keywords.

there has been an explosion of research on touch, with new and different emphases including multisensory processing, integration, and related issues (e.g., Stein, 2012; Bremner, Lewkowicz, & Spence, 2012; Deroy, 2017).

8.2.2 The Physiology of Touch

Human skin has complicated physiological structures with a large variety of receptors; it is the surface of the interface between oneself and the external environment. Before entering into much detail, one only needs to consider how many different elements can be involved when it comes to touch: temperature, pressure, pain, muscle and joint sensations, and movements (Iggo, 1977; McGlone & Reilly, 2010; see table 8.1). It is also constructive to consider how many properties touch can take in: shapes, sizes, structures, and locations (macro-geometric; Spence & Driver, 2004), on the one hand, and texture, roughness, stickiness, and spatial density (Bensmaia & Hollins, 2005), on the other. Different kinds of receptors work together to produce complex somatosensory sensations, or "sensory blends" (Gallace & Spence, 2014; Saal & Bensmaia, 2014), such as wetness (Filingeri et al., 2014), tickle (Selden, 2004), and tingling (Cataldo, Hagura, et al., 2021).

The skin is the largest and heaviest sensory organ of the human body, covering an area of about 1.8 square meters and weighing around four kilograms. Divided in two primary layers, epidermis and dermis, it contains the large majority of somatosensory receptors (~2,500 per square centimeter on the fingertips). The skin can also be distinguished between glabrous and hairy parts. The former is thicker and less permeable, and can be found on the palms, the soles, and parts of the genitals. The latter covers the rest of the body, with different hair densities. The lips are special transitional regions between the two (Gallace & Spence, 2014).

Most tactile information is registered by four different mechanoreceptors, each sensitive to a different type skin deformation: Meissner corpuscles are preferentially activated by light touches and changes in textures, Merkel disks are for sustained touches and pressures, Pacinian corpuscles are sensitive to high frequency mechanical and vibratory pressures, and Ruffini's end organs respond to tensions and stretches of skins and sustained pressures (see table 8.1). In addition to these receptors, C-tactile

Table 8.1
Organization of the somatosensory system

Somatosensory Sub-Modality	Corresponding Sensations	Afferent Receptors/ Fibers	Somatosensory Pathway
Pain	Sharp cutting dull burning Deep aching	Aδ-fibers C-fibers	Neospinothalamic Paleospinothalamic Archispinothalamic
Temperature	Warm/hot Cool/cold	C-warm C-cold	Paleospinothalamic Neospinothalamic
Touch	Itch/tickle Touch Pressure Flutter Vibration	C-tactile Merkel corpuscles Hair follicle Meissner corpuscles Pacini corpuscles	Medial lemniscal
Proprioception and kinesthesia	Muscle length Muscle tension Joint angle Joint pressure	Muscle spindle Golgi tendon Golgi joint Ruffini joint Pacini joint	Medial lemniscal

The somatosensory system is conventionally subdivided into four sub-modalities (leftmost column), each responsible for specific bodily sensations, such as pain, tickle, touch, warmth, etc. Signals within each somatosensory sub-modality are transduced by dedicated afferent receptors (third column). Finally, signals ascending from different afferent fibers are organized into distinct somatosensory pathways.

fibers are responsible for itch/tickle sensations. Each mechanoreceptor transduces its adequate tactile stimulus (i.e., a stimulus impinging the skin with the right type and amount of mechanical energy necessary to activate that receptor) into neural signals that ascend the somatosensory pathways for further processing. While the exact transduction mechanisms by which mechanoreceptors convey electric signals to the brain are not yet fully understood (Lumpkin & Caterina, 2007; Kung, 2005), most tactile sensations originate when ion channels in the mechanoreceptor's membrane are directly activated by the stretching or deformation of the skin (Jiang et al., 2002; Lumpkin, Marshall, & Nelson, 2010). The opening of the gates leads to an increased concentration of cations (Na^+ and Ca^{2+}) in the receptor neuron, which in turn induces the depolarization of the cell and the generation of an action potential. Free nerve endings are unspecialized afferent nerve fibers that are responsive to contacts, pressures, and stretches. Hair follicle receptors respond to stroking, sending and receiving nerve impulses to and from the brain when the hairs move. Thermoreceptors are responsible for the non-painful range of temperatures (Schepers & Ringkamp, 2010), while nociceptors respond to potentially damaging stimuli, including thermal, mechanical, and chemical ones (Dubin & Patapoutian, 2010).

The action potentials generated by the cutaneous receptors are transmitted to the central nervous system through different types of neural fibers. Importantly, the conduction velocity of each type of fiber (i.e., the maximal traveling speed of a neural signal within that fiber) depends on the diameter of the fiber itself, with larger fibers conducting faster signals than smaller fibers. Thus, for example, mechanical and proprioceptive stimuli transmitted by large (Aα: ~16 μm diameter) and medium (Aβ: ~9 μm diameter) fibers show a maximal conduction velocity of 120 and 72 meters per second, respectively. Conversely, noxious and thermal signals are transmitted by small Aδ (~3.5 μm diameter) and C fibers (~0.8 μm diameter), and their maximal conduction velocity only reaches around 36 and 2 meters per second, respectively. Finally, afferent somatosensory signals reach the brain through three main pathways: the dorsal column–medial lemniscus (for discriminative tactile, vibratory, and proprioceptive stimuli); the anterolateral system (for thermal, nociceptive and crude touch); and the somatosensory pathways to the cerebellum (for proprioceptive signals that do not contribute to consciousness; Gallace & Spence, 2014; see table 8.1).

Primary afferent fibers of dorsal root ganglia and the trigeminal sensory neurons then conduct signals to the ventral posterior lateral and medial nuclei of the thalamus. The majority of these neurons project to the contralateral primary somatosensory cortex, or S1, which is located in the postcentral gyrus (Blatow et al., 2007). As shown by the pioneering work of Penfield and Boldrey (1937), electrical stimulation applied to different parts of S1 elicits tactile sensations in different parts of the body. The somatotopic representation of the body in S1 is classically illustrated with the famous somatosensory homunculus (Penfield & Boldrey, 1937; Penfield & Rasmussen, 1950; see figure 8.2). Notably, brain regions corresponding to the hands and the lips are much bigger than other parts, underlying the greater tactile sensitivity of those areas (Nakamura et al., 1998). Adjacent to S1 is the secondary somatosensory cortex, or S2, which is reciprocally connected to ipsilateral S1 through cortico-cortical connections (Burton, 1986). It has been shown through macaque studies that there are sixty-two pathways linking thirteen cortical areas in the somatosensory and motor systems (Friedman et al., 1986). Tactile signals are also processed in higher areas such as the supplementary motor area, the primary motor area, and the granular and agranular insula (Felleman & van Essen, 1991). The richness and complexity of such intracortical connections may underlie the interesting clinical observation that patients with S1 damage might still be able to process tactile information from the dorsal column–lemniscal system via the ventral posterolateral nucleus of the thalamus (Brochier, Habib, & Brouchon, 1994).

8.2.3   The Psychology of Touch

The physiological studies presented in the previous section have contributed to build a satisfactory—though far from complete—understanding of how we detect, localize, and identify single somatosensory events. However, physiology alone is unable to explain the complexity and richness of our "somatosensory scene." While the tactile stimuli commonly used in psychophysical studies are isolated and well-controlled, our daily experience is crowded with a constant stream of multiple somatosensory patterns spread across the body. Yet, rather than being baffled by this "great blooming, buzzing, confusion," our brain seems able to effortlessly generate uniform and coherent percepts. For example, while holding a mug of coffee in our hand, we do not perceive five independent tactile sensations arising from our fingertips, nor we can clearly segregate the feeling of warmth from

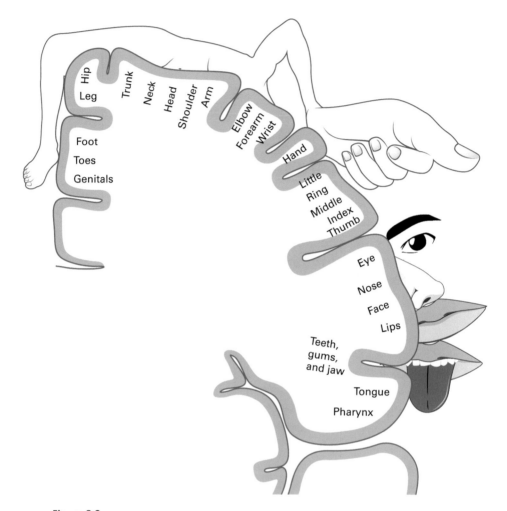

Figure 8.2
Somatosensory and motor homunculus. The figure shows a distorted representation of the human body based on the cortical representation of each body part in the primary somatosensory (S1; left) and motor (M1; right) cortices. (OpenStax College, CC BY 3.0, via Wikimedia Commons.)

the mechanical contact. Instead, we have an immediate and "holistic" perception of a single, warm mug.

The apparent paradox between the complexity of somatosensory inputs and our simple and coherent phenomenological experience has puzzled philosophers, psychologists, and neuroscientists since the dawn of their disciplines. In what follows, we survey some of the most relevant psychological phenomena and theories that may explain how the somatosensory system provides us with a rich and coherent perception of the world.[1]

Gestalt grouping and the "tactile field" "Gestalt" is a German word for "form" or "shape." In psychology, it means an organized holistic shape that is perceived as more than the sum of its parts. In vision science, there has been a debate concerning the stage at which gestalt grouping happens during visual information processing (e.g., Palmer, Brooks, & Nelson, 2003). Despite "gestaltian grouping principles" having been studied almost exclusively in vision, recent studies have investigated whether similar principles also apply to the tactile system (for a review, see Gallace and Spence, 2011; 2014). For example, Cataldo, Ferrè, di Pellegrino et al.(2016) have recently shown that the perceived intensity of complex tactile or thermal (Walsh et al., 2016) patterns does not correspond to the simple sum of the individual events, and that integration of multiple simultaneous somatosensory stimuli relies on nonlinear summation mechanisms driven by the salience of the stimuli. Notably, gestaltian groupings can be relative to different spatial frames of reference, including retinotopic, spatiotopic (external space), and object based (Spence & Driver, 2004).

In the case of touch, it has been proposed that the skin can be considered analogous to the retina in this regard as the source of a "skin space" (Cheng & Haggard, 2018; Cheng, 2019) with elementary grouping functions. Furthermore, it has been argued that "tactile fields" as psychological constructs are sustained by the skin space (Haggard & Giovagnoli, 2011; Fardo, Beck, et al., 2018; see also Cataldo, Cheng, Schwenkler, et al. "Sensory Field in Touch: Insights from Tactile Change Blindness and Gestalt Structures, unpublished). The "tactile field" hypothesis is also supported by recent studies on self-touch showing that the spatial properties of the somatosensory system are—to a certain extent—independent from motor signals (Cataldo, Dupin, Gomi, et al., 2020; Cataldo, Dupin, Dempsey-Jones, et al., unpublished; Moscatelli et al., 2019). While many philosophical accounts of space perception give motor signals an underpinning role (Lotze, 1852; von Helmholtz, 2013; Hatfield, 1990; for a review, see Rose, 1999), these studies show that tactile input influences extent perception of movement (Cataldo, Dupin, Gomi, et al., 2020) and motor control (Moscatelli et al., 2019), thus ruling out strong motor-based theories of space perception. By contrast, in the case of thermal perception

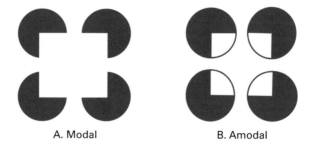

A. Modal   B. Amodal

Figure 8.3
A comparison between amodal and modal completion. From Ringach and Shapley (1996).

and nociception, is not clear if such field-like organization exists (Marotta, Ferrè, & Haggard, 2015; Mancini et al., 2015; Cataldo, Ferrè, di Pellegrino, et al., 2016; Cataldo, Ferrè, & Haggard, 2019).[2]

Perceptual completion   A related phenomenon is that people have a tendency to perceive complete figures, even when some of the information is missing. Researchers have distinguished between two kinds of completion: *amodal* completion and *modal* completion. Amodal completion occurs when parts of one object are occluded by other objects, but the target in question is still perceived as a single entity (Michotte, Thines, & Crabbe, 1964/1991). This is one of the examples of "perceptual presence," as Alva Noë (2005, 2015) calls it. For example, we have a visual sense in which things being occluded or the backs of objects somehow show up in our perceptual phenomenology. This phenomenon has also been discovered in audition (e.g., O'Callaghan, 2008). Modal completion refers to cases in which people can apparently perceive complete shapes out of separate inducing elements (see figure 8.3 for a comparison between amodal and modal completion). The famous Kanizsa triangle is one such example (Kanizsa, 1979; Kanizsa & Gerbino, 1982). Although there have been fewer studies on tactile completion, there are still some examples. For instance, Kennedy and Juricevic (2006) conducted interesting studies on tactile completion in blind populations and found that "the directions of objects from a vantage point in touch converge much as they do in vision" (p. 506). Here is an easy way to appreciate tactile completion: imagine holding a bottle with your fingers spreading apart. And now ask yourself: "Why do I not feel holes in the bottle where the spaces are between my fingers" (MacKay, 1967)? Tactile completion appears to be the answer.

Change/inattentional blindness   These are two phenomena that are different from each other but similar in interesting ways. There are many kinds of visual change blindness, but typically it occurs when two scenes are presented consecutively and are exactly the same except for a few elements. Very often, the elements are quite salient with regard to locations, colors, shapes, or sizes. However, people tend to find it difficult to identify the difference. Changes can occur during saccades (Grimes, 1996), eye blinks (O'Regan et al., 2000), or a number of other temporal scales, and despite some changes being massive, participants show no awareness (or at least no report) that anything has changed. For tactile change blindness, it has been shown that people often cannot detect identity and positional changes between simple vibrotactile patterns consecutively presented on their skin (Gallace, Tan, & Spence, 2006). Now compare this with inattentional blindness, where people's attention is directed away from the target and therefore fail to detect it (Mack & Rock, 1998). Even fewer studies have been done for touch in this regard, but for a recent effort, see Murphy and Dalton (2016) where they invoke visual attentional load to induce "inattentional numbness"—"awareness of a tactile event is reduced when attention is focused elsewhere" (p. 763).[3]

Attention and memory   The discussion concerning inattentional blindness and numbness naturally brings us to attention and, relatedly, memory. These two topics have occupied central places in cognitive psychology since the very early days. Very generally, attention can be understood as selection, access, and modulation of information (Treisman, 1964; Huang & Pashler, 2007; Carrasco et al., 2013). There are many theories of attention, including the load theory that we briefly touched on above (e.g., Lavie and Tsal, 1994), attention as cognitive unison (Mole, 2010), attention as selection for action (Wu, 2011), and attention as structuring of the stream of consciousness (Watzl, 2017), to name just a few. When it comes to touch, it has been shown that attention can be directed specifically to this modality: people respond more quickly to tactile targets when attention is directed toward them (Spence, Nicholls, and Driver, 2001). It has also been shown that tactile attention occurs at a very early stage of perceptual processing (Spence, Shore, & Klein, 2001). Relatedly, memory has been central in cognitive psychology because when targets are attended and processed, such information needs to be stored for further usage. According to a standard taxonomy, there are (at least) four kinds of memories: iconic or sensory memory, which occurs right after people briefly perceive the targets (Neisser, 1967); fragile short-term memory, which has a larger capacity and can last for several seconds (Vandenbroucke, Sligte, & Lamme, 2011; Lamme, 2003); working memory, which tides closely to reportability and action (Block, 2007); and long-term memory, which can last for days or even years. Like in other domains, analogous studies in touch are not abundant, but researchers have studied tactile short-term memory (Gallace et al., 2008) and its neural correlates (Gallace & Spence, 2009). The gist of these findings is that "short-lasting tactile representations can be accessed using partial report procedures similar to those used previously in visual studies" (Gallace et al., 2008, p. 132). It has also been shown that, unsurprisingly, visually impaired people are better at recognizing tactile objects than sighted people are (Saito et al., 2007).

Blind-touch, neglect, and extinction   Blindsight patients are individuals who have suffered occipital lesions and as a result developed cortical blindness. Although they are blind on a portion of their visual field, they tend to perform much better than chance when forced to guess about what visual information is presented in their blind hemifield (Kentridge, Heywood, & Weiskrantz, 1999; for skepticism concerning this "blindness" claim, see Phillips, 2020). For touch, it was discovered that some patients are able to point to the locations of tactile stimulations, though without being able to report them verbally (Paillard, Michel, & Stelmach, 1983). There are some crucial differences between sight and touch here, to be sure. For example, in the tactile case, some patients are able to point to the locations of tactile stimulations spontaneously (i.e., without the forced-choice paradigm), while patients with blindsight can perform the task successfully only with forced choice. That is to say, "super-blindsight" (i.e., those who can spontaneously form beliefs on the basis of unconscious information gained from the blind field) is only a conceptual possibility at best (Block, 1995). Relatedly, patients who suffer unilateral spatial neglect often fail to detect and act on targets presented in regions contralateral to the relevant brain lesions (Vallar, 1997, 1998). Spatial extinction patients cannot simultaneously perceive multiple stimuli of the same type (Bisiach, 1991). It has been found that unilateral spatial neglect and spatial extinction affect tactile stimuli as well (Vallar et al., 1994).

Rubber hand and other multisensory illusions   The rubber hand illusion (RHI) can be induced in neurotypical subjects. To do so, a person positions one of their hands out of sight, while seeing a lifelike rubber hand in roughly the same location as where the hidden hand would normally be. When the hidden hand and the rubber hand are stroked synchronously in homologous points, most participants report that the rubber hand feels like their own hand (Botvinick & Cohen, 1998)—an instance of illusory body ownership (for a review, see Blanke, 2012). Novel somatosensory (Ehrsson, Holmes, & Passingham, 2005), motoric (Hara et al., 2015), and even "haptic" (Cataldo, di Luca, & Hayward, 2019) and oculo-tactile (Cataldo, di Luca, Deroy, et al., unpublished) versions of the RHI have been recently described, expanding our ability to test how different aspects of somatosensation affect our bodily self-consciousness. Bodily illusions involving touch have been extensively studied recently, for example out-of-body illusion (Ehrsson, 2007, p. 1048), "the experience in which a person who is awake sees his or her body from a location outside the physical body," and body-swap illusion (Petkova & Ehrsson, 2008; Cataldo, Garofalo, & di Pellegrino, unpublished). Before these illusions were discovered, the most famous tactile illusion was probably the cutaneous rabbit illusion, in which the perceived location of a target on the skin is shifted toward the location where rapid, subsequent stimulation is delivered (Geldard & Sherrick, 1972; Flach & Haggard, 2006). There are too many somatosensory illusions or hallucinations to be listed exhaustively here. Some examples include thermal illusions such as the thermal grill illusion, in which an interlaced warm–cold–warm setup can induce burning sensation in the middle (Thunberg, 1896; Fardo, Finnerup, & Haggard, 2018), and similarly thermal referral illusion, where the middle stimulus is with a neutral temperature (Green, 1977; Cataldo, Ferrè, di Pellegrino, et al., 2016).

8.3   The Philosophy of Touch

The philosophy of perception is one of the oldest branches of philosophy, although the label is relatively new. From the very beginning of documented philosophical thinking, at least in the Western tradition, reflections on perception have been central (Kalderon, 2015). Intuitively, perception is a primitive relation (i.e., more basic than thought) between the subjective and the objective. In what follows, we discuss three topics in this general area, focusing in particular on how studies in psychology and neuroscience can potentially shed light on traditional issues in philosophy of touch or tactile perception.

8.3.1 Direct versus Indirect Perception

Under a common construal, perception is an epistemological notion: it refers to how subjects gain information, and thus get to know about the external world. This includes the so-called non-epistemic seeing, where we can visually differentiate objects from their immediate surroundings (Dretske, 1969, 1981). Traditionally, this notion of perception is coupled with a metaphysical realism about the world (i.e., the world exists independent of us), which conforms to our commonsensical understanding. Anti-realism denies the mind-independence of the "external world," and together with other assumptions, it can be developed into idealism (Berkeley, 1710/1999; roughly, objects are collections of ideas) or phenomenalism (Mill, 1865/2015; roughly, objects are collections of perceptual phenomena). For our purposes, let us focus on realism for the moment. Assuming the world exists independently of us, there seems to be a further epistemological question: Do we perceive the world directly or indirectly? That is, is there a "veil of perception" between the perceiver and the world? This question arose from various arguments, such as the argument from illusion and the argument from hallucination, which are often attributed to British Empiricism (Robinson, 1994, p. 33), and can be summarized as follows. Very roughly, when one undergoes a visual illusion, one seems to see a certain *property* that is not instantiated by the visual target in one's visual field. When one undergoes a visual hallucination, one seems to see a certain *object* that is not in one's visual field. Some have argued that in illusory and hallucinatory cases, the subject visually senses something—call it "sense-datum"—that instantiates the illusory property in question. This view seems to imply that at least in illusory cases, one *directly* sees sense-data and *indirectly* sees the external visual targets. Similar inferences can be conducted with hallucinations and some similar cases.[4]

Just like in the sciences, visuocentrism has also been dominant in philosophy. It is often assumed that at least for the above issues concerning perceptual objects and the external world, what can be said about vision can be easily generalized to other senses. For example, if one sees sense-data directly, one also touches sense-data directly (for exceptions, see e.g., O'Shaughnessy, 1989; more on this below). Reflecting on the science of somatosensory sensations, however, will show that this generalization is questionable at best.

One natural entry point for approaching the above generalization idea is the various illusions we touched on above: Can they offer insights into the argument from illusion? The tentative answer is that it is not obvious.

Consider what the argument needs. It needs a property that is sensed by the subject, but also that that very property does not instantiate in anything in front of or nearby the subject. Consider the RHI. Here, subjects would misattribute the sensations to the wrong object (i.e., the rubber hand). But it is not as if there is a sensed property that is not instantiated by any nearby object. Rather, it is that the property is instantiated nearby but is misattributed to the wrong object. Similar things can be said about many illusions found in psychology. This is why when it comes to the argument from illusion, we often hear that the illusions in question are philosophers' illusions. For the argument to work, it was even claimed that the illusions do not need to be actual; possible illusions will do (A. D. Smith, 2002). Therefore, insights from psychology and neuroscience of somatosensory sensations need to be found elsewhere when it comes to their connections to traditional issues in philosophy.

Another attempt of approaching the above generalization idea comes from philosophy itself and extends to psychology and neuroscience. Above, we mentioned that visuocentrism also operates in philosophy. Examples for the argument from illusion are often visual, but its conclusion is often taken to be generalizable to the other senses. Philosophers such as Brian O'Shaughnessy (1989), M. G. F. Martin (1992), Louise Richardson (2010), and Matthew Soteriou (2013) have been arguing that sight and touch have some essentially different structural features. For example, Soteriou (2013) writes:

> [T]he structural feature of normal visual experience that accounts for the existence of its spatial sensory field is *lacking* in the form of bodily awareness involved when one feels a located bodily sensation. (p. 120; emphasis added)

O'Shaughnessy (1989) puts a similar point in a different way:

> There is in touch *no* analogue of the visual field of visual sensations—in touch a body investigates bodies as one body amongst others. (p. 38)[5]

Now, recall the first example we went through: gestalt grouping. We mentioned that some studies suggest that there might be a tactile field that is sustained by the skin space. Here is a succinct summary of the view:

> Taken as a whole, these studies provide good evidence for the existence of a tactile field supporting perception of tactile spatial patterns. That is, tactile processing includes the capacity to represent spatial relations between locations of individual stimuli over relatively large distances. We showed that tactile pattern judgements depend on secondary factors over and above local tactile perceptual ability at the stimulated locations. (Haggard & Giovagnoli, 2011, p. 73)

They also state that "[o]verall, the results suggest that the human sense of touch involves a tactile field, *analogous* to the visual field. The tactile field supports computation of spatial relations between individual stimulus locations, and thus underlies tactile pattern perception" (Haggard & Giovagnoli, 2011, p. 65).[6] Now, the debate between the Haggard group and the philosophers above is ongoing, and there are still many open questions. For example, do these studies really support the Tactile Field Hypothesis (Cheng, 2019)? If so, can we make sure that the two sides of the debate are not talking past each other? Whatever it turns out to be, the comparison between sight, touch, and other sense modalities can shed light, albeit indirectly, on whether the argument from illusion, even if sound, can generalize its conclusion across the board for various sense modalities.

8.3.2 The Body and the Self

The mind–body problem is often said to be the most canonical topic in philosophy of mind, and various major textbooks can testify to this (e.g., Heil, 2019; Kim, 2010). In psychology and neuroscience, versions of the identity theory or material monism are often assumed. Even if we do not question such assumptions, there is a different though related question concerning the body and the self. One way of seeing what this is about is to compare "bodily awareness" and "bodily *self*-awareness." Bodily awareness is a common topic between philosophy and cognitive neuroscience, as proprioception, kinesthesia, and interoception are phenomena under this umbrella (Bermúdez, Marcel, & Eilan, 1995; De Vignemont & Alsmith, 2018; Tsakiris & De Preester, 2018). However, controversies arise when we consider whether we have bodily *self*-awareness for free when we have bodily awareness. Psychologists and neuroscientists tend to use these two terms interchangeably (e.g., Tsakiris et al., 2007; Blanke, 2012; Aspell, Palluel, & Blanke, 2012; Gallace & Spence, 2014, p. 54). This is understandable, since if I am bodily aware of some limbs, of course I am aware of *my own* limbs. This natural thought can be challenged, however. It is possible, for example, that through certain configurations I am actually aware of *someone else's* limbs (cf. Dennett, 1978). An interesting actual case where self-awareness and bodily awareness do not go hand in hand is depersonalization (DP)—an intriguing form of altered subjective experience in which people report feelings of unreality and detachment from their sense of self. In a recent study (Farmer et al., 2020), we have investigated self-bias in visual tactile integration in non-clinical participants reporting high and low levels of DP. Interestingly, the high DP group showed no self-face facilitation, instead showing a greater visuotactile integration when observing the face of another person. In addition, across all participants, self-bias was negatively predicted by the occurrence of anomalous body experiences. These results indicate disrupted integration of tactile and visual representations of the bodily self in those experiencing high levels of DP.

Yet, even if we consider only normal cases, that is, what I am aware of now are my own limbs, it seems to be a further claim that what I am aware of has a self-component. This passage nicely summarizes this bodily-self view in question:

> [I]n bodily awareness, one is not simply aware of one's body as one's body, but one is aware of one's body *as oneself*. That is, when I attend to the object of bodily awareness I am presented not just with my body, but with my "bodily self." (J. Smith, 2006, p. 49)

J. Smith (2006) then argues that "imagining being Napoleon having a pain in the left foot will not contain *me* as an object. The only person in the content of this imagining is Napoleon . . . Thus, when I simply imagine a pain, but without specifying *whose* pain, the imagined experience is not first personal" (p. 57, emphasis added). This is a complicated issue that we cannot detail here (Cheng, 2018). Our goal is just to indicate that both the metaphysical question concerning the relation between the body and the self and the epistemological question concerning bodily self-awareness are yet another set of questions where psychology, neuroscience, and philosophy should work together and make progress. To give one example, consider this passage again from Haggard and Giovagnoli (2011):

> [A] substantive representation of one's own body as a volumetric object mediates spatial judgements on the body surface. Tactile pattern perception involves representing oneself both as a source of sensory experiences, but also as a physical object with a characteristic body structure, and therefore having spatial attributes analogous to other objects. In touch, then, the linkage between primary experiences and self-consciousness seems stronger than in vision. The linkage shows that the body is a physical as well as a psychological object. In this sense, tactile pattern perception presupposes a self that is an object embedded in the world, rather than simply a viewpoint on the world (Merleau-Ponty, 1962). (p. 76)

There is much to say concerning this passage. For example, it moves from representation to consciousness, which is a substantial and controversial move. Also, it does not say enough about how the *self* comes into the picture. More positively, it draws our attention to the relation between tactile pattern perception and body structure, and gestures toward a certain way to understand embodiment and embeddedness. Moreover, it is no doubt an interesting attempt to develop the "duality of touch" idea (Katz, 1925/1989). It is also interesting to observe that the phenomenological tradition is made relevant here in an intelligible way.

8.3.3 Individuating the Senses

How many senses do we have, and by which criteria do we individuate them? Aristotle (circa 350 BC) famously distinguished the five senses with different proper objects: colors for vision, sounds for audition, and so on. But he also pointed out a crucial difficulty with this approach: for touch, there is no obvious unique proper object.[7] On the one hand, touch can register not only hardness, but also humidity, texture, vibration, weight, and many other properties. On the other hand, touch can take in shapes and sizes, but these properties are also visible. Since then, philosophers have been trying to solve this puzzle. Major proposals include the representational criterion, the phenomenal character criterion, the proximal stimulus criterion, and the sense-organ criterion. Each criterion seems to have putative counterexamples, and there is no consensus to date (Macpherson, 2011). The existence of multisensory integration, which generates percepts jointly through supposedly different senses, and crossmodal illusions, which are possible only due to interactions between at least two senses (Stein, Stanford, & Rowland, 2014), further complicate the issues. Famous cases include the McGurk effect, the ventriloquist illusion, and synesthesia, to name just a few (Deroy, 2017).

There are two levels of difficulties here: first, there is a general question about how to individuate the senses; second, there is a more specific question concerning the somatosensory system. As indicated above, touch can register not only hardness but also humidity, texture, and many other properties, and touch can take in shapes and sizes, but these properties are also visible. In these considerations, Aristotle did not even include pains and thermal sensations explicitly. How should we proceed here? Do the psychology and neuroscience of touch shed any light on this question?

We believe the answer is yes. Above, we saw that a tactile field should be postulated in the case of neutral touch. We also mentioned, when reviewing the relevant psychology, that in the case of thermal perception and nociception, there is no such field-like organization (Marotta et al., 2015; Mancini et al., 2015). If the nociceptive, the thermal, and the tactile were to be grouped into one and the same sense, it is hard if not impossible to make sense of the differences just stated. Between the nociceptive and the thermal, it is striking to see that the existence of the thermal grill illusion described above is responsive to external spatiotopic space (Marotta et al., 2015). Between the thermal and the tactile, it is likewise striking to see that the thermal grill illusion can still be robust without contact (Ferrè et al., 2018). Between the tactile and the other two, it is further striking to see that only the tactile has a field organization that facilitates object (thing) perception. Notice that this consideration is different from a less interesting view that simply refers to different physiologies: physiological makeups are indeed important in many regards, but various considerations in Macpherson (2011) have shown that simply relying on physiological considerations cannot be satisfactory. Just to give one example, it is unclear which sense organ it is for touch. In the case of human, we have tactile experiences within our bodies, but in those cases there is no skin involved. To be sure, the functional differences identified above are not themselves decisive reasons for distinguishing the nociceptive, the thermal, and the tactile into three senses. Still, they provide prima facie reasons for doing so. For those who disagree with this view, they must explain how their alternatives can square with the significant functional differences between the nociceptive, the thermal, and the tactile. For example, some have argued that given the diversity of touch, there is no satisfying answer yet (Ratcliffe, 2012). Others have argued that the unity of touch can be achieved through a proper understanding of explorative actions (Fulkerson, 2014). These theorists will need to come up with a response to the current perspective.[8] Again, whatever this turns out to be, it is instructive to look into the psychology and neuroscience of somatosensory sensations to renew traditional issues in philosophy.

8.4   Conclusion: The Interface Problem

In this chapter, we have gone through some elements and examples in the psychology, neuroscience, and philosophy of somatosensation, and how they can work together to yield a comprehensive picture of touch and somatosensation. Psychology needs to be incorporated because without it, the gap between philosophy and neuroscience seems to be too huge: it would be like doing biology and trying to bridge it to physics without the help with chemistry. Now, even with all the levels needed, there still lies the interface problem, as characterized by José Luis Bermúdez (2005):

> How does commonsense psychological explanation interface with the explanations of cognition and mental operations given by scientific psychology, cognitive science, cognitive neuroscience and the other levels in the explanatory hierarchy? (p. 35; also see Marr, 1982)

There are four broad responses to this problem: the autonomous mind (Davidson, 1974; Hornsby, 1980–1981; McDowell, 1985), the functional mind (Lewis, 1972; Cummings, 2000), the representational mind (Fodor, 1975), and the neurocomputational mind (Churchland, 1981, 1986). The above discussions have not responded to this problem directly; we have not taken a *theoretical* stand with regard to this problem. We have, however, offered examples to show how *practically* neuroscience, psychology, and philosophy can possibly mesh with one another. Ned Block (2007) has provided yet another good example how this can be done. In the paper title, he mentions only psychology and neuroscience, but as a philosopher, he has demonstrated how philosophy can take active roles in ongoing empirical debates—in this case, the complicated relation between attention and consciousness. We hope this brief review can serve as a guide for those who are interested in such projects, in particular with regard to the somatosensory senses.

Notes

1. This review is not intended to be comprehensive. Many topics we have not covered here are worthy of discussions, e.g., tactile figure–ground segregation (Tsakiris, Constantini, & Haggard, 2008; MacCumhaill, 2018).

2. A possible precursor of tactile field can be found in Husserl (1939/1997): fields of sensations.

3. For load theory of attention, see various works led by Nilli Lavie (e.g., Lavie, 1995).

4. The argument from illusion and the argument from hallucination have generated an enormous literature in the past several hundred years. A. D. Smith (2002) offers a very good review on these matters. For skepticism concerning philosophical discussions of direct perception, see Snowdon (1992). For an empirical defense of indirect perception, see Rock (1997).

5. These points have roots in Berkeley (1709/2017) and Strawson (1959). This quote, among other things, also indicates that O'Shaughnessy, as a sense-datum theorist, might have different views for different sense modalities. Cheng (2020) argues that

sight and touch actually share crucial structural features, and this points to the positive answer for Molyneux's question.

6. For a fuller review of these matters and updated studies, see Haggard and colleagues (2018).

7. De Vignemont and Massin (2015) propose that pressure and tension are proper and primary objects of touch. This line seems to be promising, but the proposal needs to answer potential difficult cases, such as the ears being able to sense *pressure waves* and generate audition but not touch.

8. For more on these issues, see Cheng (2015, forthcoming).

References

Andersen, R. A. (2015). Vernon B. Mountcastle (1918–2015). *Current Biology, 25*(8), R310–313.

Aspell, J. E., Palluel, E., & Blanke, O. (2012). Early and late activity in somatosensory cortex reflects changes in bodily self-consciousness: An evoked potential study. *Neuroscience, 216*, 110–122.

Bach-y-Rita, P. (1972). *Brain mechanisms in sensory substitution.* Cambridge, MA: Academic Press.

Barlow, H., & Mollon, J. (1982). *The senses.* Cambridge: Cambridge University Press.

Bensmaia, S. J., & Hollins, M. (2005). Pacinian representations of fine surface texture. *Perception and Psychophysics, 67*, 842–854.

Berkeley, G. (1709/2017). *An essay towards a new theory of vision.* Scotts Valley, CA: CreateSpace.

Berkeley, G. (1710/1999). *Principles of human knowledge and three dialogues.* Oxford: Oxford University Press.

Bermúdez, J. L. (2005). *Philosophy of psychology: A contemporary introduction.* London: Routledge.

Bermúdez, J. L., Marcel, A., & Eilan, N. (1995). *The body and the self.* Cambridge, MA: MIT Press.

Bisiach, E. (1991). Extinction and neglect: Same or different? In J. Paillard (Ed.), *Brain and space* (pp. 251–257). Oxford: Oxford University Press.

Blanke, O. (2012). Multisensory brain mechanisms of bodily self-consciousness. *Nature Reviews Neurosciences, 13*, 556–571.

Blatow, M., Nennig, E., Durst, A., Sartor, K., & Stippich, C. (2007). fMRI reflects functional connectivity of human somatosensory cortex. *NeuroImage, 37*, 927–936.

Block, N. (1995). On a confusion about a function of consciousness. *Behavioral and Brain Sciences, 18*(2), 227–247.

Block, N. (2007). Consciousness, accessibility, and the mesh between psychology and neuroscience. *Behavioral and Brain Sciences, 30*(5), 481–548.

Botvinick, M., & Cohen, J. (1998). Rubber hands "feel" touch that eyes see. *Nature, 391*(6669), 756–756.

Bremner, A., Lewkowicz, D., & Spence, C. (Eds.). (2012). *Multisensory development.* Oxford: Oxford University Press.

Brochier, T., Habib, M., & Brouchon, M. (1994). Covert processing of information in hemianesthesia: A case report. *Cortex, 30*, 135–144.

Burton, H. (1986). Second somatosensory cortex and related areas. In E. G. Jones & A. Peters (Eds.), *Cerebral cortex, sensory-motor areas and aspects of cortical connectivity* (Vol. 5, pp. 31–98). New York: Plenum.

Carrasco, M., Eckstein, M., Krauzlis, R., & Verghese, P. (2013). Attention modulation: Target selection, active search and cognitive processing. *Vision Research, 85*, 1–4.

Cartwright, N. (1999). *The dappled world: A study of the boundaries of science.* Cambridge: Cambridge University Press.

Cataldo, A., Cheng, T., Schwenkler, J., & Haggard, P. (in preparation). Sensory field in touch: Insights from tactile change blindness and gestalt structures.

Cataldo, A., di Luca, M., & Hayward, V. (2019). The "haptic" rubber hand illusion: Limb ownership through active touch of a fake hand. Demo at IEEE World-Haptics Conference—Tokyo 2019. Retrieved from https://osf.io/z65tu/?view_only =9685bfb5398f43fc

Cataldo, A., Dupin, L., Gomi, H., & Haggard, P. (2020). Sensorimotor signals underlying space perception: an investigation based on self-touch. *Neuropsychologia,* 107729.

Cataldo, A., Ferrè, E. R., di Pellegrino, G., & Haggard, P. (2016). Thermal referral: Evidence for a thermoceptive uniformity illusion without touch. *Scientific Reports, 6*, 35286.

Cataldo, A., Ferrè, E. R., & Haggard, P. (2019). Thermonociceptive interaction: Interchannel pain modulation occurs before intrachannel convergence of warmth. *Journal of Neurophysiology, 121*(5), 1798–1808.

Cataldo, A., Hagura, N., Hyder, Y., & Haggard, P. (2021). Touch inhibits touch: sanshool-induced paradoxical tingling reveals perceptual interaction between somatosensory submodalities. *Proceedings of the Royal Society B: Biological Sciences, 288*, 20202914.

Cheng, T. (2015). Book review: The first sense. *Frontiers in Psychology, 6*, 1196.

Cheng, T. (2018). *Sense, space, and self.* [Unpublished doctoral dissertation]. University College London.

Cheng, T. (2019). On the very idea of a tactile field. In T. Cheng, O. Deroy, & C. Spence (Eds.), *Spatial senses: Philosophy of perception in an age of science* (pp. 226–247). London: Routledge.

Cheng, T. (2020). Molyneux's question and somatosensory spaces. In G. Ferretti and B. Glenney (Eds.), *Molyneux's question and the history of philosophy*. London: Routledge.

Cheng, T. (forthcoming). Perception. In B. D. Young, & C. D. Jennings (Eds.), *Mind, cognition, and neuroscience: A philosophical introduction*. London: Routledge.

Cheng, T., & Haggard, P. (2018). The recurrent model of bodily spatial phenomenology. *Journal of Consciousness Studies, 25*(3–4), 55–70.

Chomsky, N. (1959). A review of B. F. Skinner's verbal behavior. *Language, 35*(1), 26–58.

Churchland, P. M. (1981). Eliminative materialism and propositional attitudes. *The Journal of Philosophy, 78*(2), 67–90.

Churchland, P. S. (1986). *Neurophilosophy: Toward a unified science of the mind-brain*. Cambridge, MA: MIT Press.

Cummings, R. (2000). How does it work? vs. What are the laws? Two conceptions of psychological explanation. In F. Keil & R. Wilson (Eds.), *Explanation and cognition* (pp. 117–145). Cambridge, MA: MIT Press.

Davidson, D. (1974). Psychology as philosophy. In S. Brown (Ed.), *Philosophy of psychology* (pp. 41–52). New York: Macmillan.

Dennett, D. C. (1978). Where am I? In *Brainstorms: Philosophical essays on mind and psychology* (pp. 310–323). Cambridge, MA: MIT Press.

Deroy, O. (2017). *Sensory blending: On synaesthesia and related phenomena*. Oxford: Oxford University Press.

De Vignemont, F., & Alsmith, A. (2018). *The subject's matter: Self-consciousness and the body*. Cambridge, MA: MIT Press.

De Vignemont, F., & Massin, O. (2015). Touch. In M. Matthen (Ed.), *The Oxford handbook of the philosophy of perception* (pp. 294–313). Oxford: Oxford University Press.

Dretske, F. (1969). *Seeing and knowing*. Chicago: University of Chicago Press.

Dretske, F. (1981). *Knowledge and the flow of information*. Cambridge, MA: MIT Press.

Dubin, A. E., & Patapoutian, A. (2010). Nociceptors: The sensors of the pain pathway. *Journal of Clinical Investigation, 120*, 3760–3772.

Ehrsson, H. H. (2007). The experimental induction of out-of-body experiences. *Science, 317*, 1048.

Ehrsson, H. H., Holmes, N. P., & Passingham, R. E. (2005). Touching a rubber hand: Feeling of body ownership is associated with activity in multisensory brain areas. *Journal of Neuroscience, 25*, 10564–10573.

Fardo, F., Beck, B., Cheng, T., & Haggard, P. (2018). A mechanism for spatial perception on human skin. *Cognition, 178,* 236–243.

Fardo, F., Finnerup, N. B., & Haggard, P. (2018). Organization of the thermal grill illusion by spinal segments. *Annals of Neurology, 84*(3), 463–472.

Farmer, H., Cataldo, A., Adel, N., Wignall, E., Gallese, V., Deroy, O., . . . & Ciaunica, A. (2020). The detached self: Investigating the effect of depersonalisation on self-bias in the visual remapping of touch. *Multisensory Research, 1*(aop), 1–22.

Felleman, D. J., & van Essen, D. C. (1991). Distributed hierarchical processing in primate cerebral cortex. *Cerebral Cortex, 1,* 1–47.

Ferrè, E. R., Iannetti, G. D., van Dijk, J. A., and Haggard, P. (2018). Ineffectiveness of tactile gating shows cortical basis of nociceptive signaling in the thermal grill illusion. *Scientific Reports, 8*(1), 6584.

Filingeri, D., Fournet, D., Hodder, S., & Havenith, G. (2014). Why wet feels wet? A neurophysiological model of human cutaneous wetness sensitivity. *Journal of Neurophysiology, 112*(6), 1457–1469.

Flach, R., & Haggard, P. (2006). The cutaneous rabbit revisited. *Journal of Experimental Psychology: Human Perception and Performance, 32,* 717–732.

Fodor, J. A. (1975). *The language of thought.* New York: Crowell.

Friedman, D. P., Murrar, E. A., O'Neill, J. B., & Mishkin, M. (1986). Cortical connections of the somatosensory fields of the lateral sulcus of macaques: Evidence for a corticolimbic pathway for touch. *Journal of Comparative Neurology, 252,* 323–347.

Fulkerson, M. (2014). *The first sense: A philosophical study of human touch.* Cambridge, MA: MIT Press.

Gallace, A., & Spence, C. (2009). The cognitive and neural correlates of tactile memory. *Psychological Bulletin, 135,* 380–406.

Gallace, A., & Spence, C. (2011). To what extent do Gestalt grouping principles influence tactile perception?. *Psychological Bulletin, 137*(4), 538.

Gallace, A., & Spence, C. (2014). *In touch with the future: The sense of touch from cognitive neuroscience to virtual reality.* Oxford: Oxford University Press.

Gallace, A., Tan, H. Z., Haggard, P., & Spence, C. (2008). Short term memory for tactile stimuli. *Brain Research, 1190,* 132–142.

Gallace, A., Tan, H. Z., & Spence, C. (2006). The failure to detect tactile change: A tactile analog of visual change blindness. *Psychonomic Bulletin and Review, 13,* 300–303.

Gazzaniga, M. (Ed.). (1984). *Handbook of cognitive neuroscience.* New York: Springer.

Geldard, F. A., & Sherrick, C. E. (1972). The cutaneous "rabbit": A perceptual illusion. *Science, 178,* 178–179.

Goldstein, E. B., & Brockmole, J. R. (2016). *Sensation and perception* (10th ed.). Boston, MA: Cengage Learning.

Green, B. G. (1977). Localization of thermal sensation: An illusion and synthetic heat. *Perception and Psychophysics, 22*(4), 331–337.

Grimes, J. (1996). On the failure to detect changes in scenes across saccades. In K. Akins (Ed.), *Perception* (Vancouver Studies in Cognitive Science, Vol. 2, pp. 89–110). New York: Oxford University Press.

Haggard, P., Beck, B., Cheng, T., & Fardo, F. (2018). Spatial perception and the sense of touch. In F. D. Vignemont & A. Alsmith (Eds.), *The subject's matter: Self-consciousness and the body* (pp. 97–114). Cambridge, MA: MIT Press.

Haggard, P., & Giovagnoli, G. (2011). Spatial patterns in tactile perception: Is there a tactile field? *Acta Psychologica, 137*, 65–75.

Hara, M., Pozeg, P., Rognini, G., Higuchi, T., Fukuhara, K., Yamamoto, A., . . . & Salomon, R. (2015). Voluntary self-touch increases body ownership. *Frontiers in Psychology, 6*, 1509.

Hatfield, G. C. (1990). *The natural and the normative: Theories of spatial perception from Kant to Helmholtz*. Cambridge, MA: MIT Press.

Hawkes, G. R. (1960). *Symposium on cutaneous sensitivity* (Medical Research Laboratories Report No. 424). Fort Knox, KY: U.S. Army Medical Research Laboratory.

Hebb, D. O. (1949). *The organization of behavior: A neuropsychological theory*. New York: John Wiley.

Heil, J. (2003). *From an ontological point of view*. Oxford: Clarendon Press.

Heil, J. (2019). *Philosophy of mind* (4th ed.). London: Routledge.

Hornsby, J. (1980–1981). Which mental events are physical events? *Proceedings of the Aristotelian Society, 81*, 73–92.

Huang, L., & Pashler, H. (2007). A Boolean map theory of visual attention. *Psychological Review, 114*(3), 599–631.

Husserl, E. (1939/1997). *Experience and judgment* (J. S. Churchill & K. Ameriks, Trans.). Evanston, IL: Northwestern University Press.

Iggo, A. (1977). Cutaneous and subcutaneous sense organs. *British Medical Bulletin, 33*, 97–102.

Jiang, Y., Lee, A., Chen, J., Cadene, M., Chait, B. T., & MacKinnon, R. (2002). Crystal structure and mechanism of a calcium-gated potassium channel. *Nature, 417*(6888), 515–522.

Kalderon, M. E. (2015). *Form without matter: Empedocles and Aristotle on color perception*. Oxford: Oxford University Press.

Kanizsa, G. (1979). *Organization in vision*. New York: Praeger.

Kanizsa, G., & Gerbino, W. (1982). Amodal completion: Seeing or thinking? In J. Beck (Ed.), *Organization and representation in perception* (pp. 167–190). Mahwah, NJ: Lawrence Erlbaum Associates.

Katz, D. (1925/1989). *The world of touch* (L. E. Krueger, Trans.). Mahwah, NJ: Lawrence Erlbaum Associates.

Kennedy, J. M., & Juricevic, I. (2006). Form, projection, and pictures for the blind. In M. A. Heller & S. Ballesteros (Eds.), *Touch and blindness: Psychology and neuroscience* (pp. 73–93). Mahwah, NJ: Lawrence Erlbaum Associates.

Kentridge, R., Heywood, C. A., & Weiskrantz, L. (1999). Attention without awareness in blindsight. *Proceedings of the Royal Society B: Biological Sciences, 266*(1430), 1805–1811.

Kim, J. (2010). *Philosophy of mind* (3rd ed.). New York: Westview Press.

Kung, C. (2005). A possible unifying principle for mechanosensation. *Nature, 436*(7051), 647–654.

Lamme, V. (2003). Why visual attention and awareness are different. *Trends in Cognitive Sciences, 7*(1), 12–18.

Lashley, K. (1932). *Studies in the dynamics of behavior*. Chicago: Chicago University Press.

Lavie, N. (1995). Perceptual load as a necessary condition for selective attention. *Journal of Experimental Psychology: Human Perception and Performance, 21*(3), 451–468.

Lavie, N., & Tsal, Y. (1994). Perceptual load as a major determinant of the locus of selection in visual attention. *Perception and Psychophysics, 56*(2), 183–197.

Lewis, D. (1972). Psychophysical and theoretical identifications. *Australasian Journal of Philosophy, 50*, 249–258.

Lotze, H. (1852). *Medicinische Psychologie oder Psychologie der Seele*. Leipzig: Weidmann.

Lumpkin, E. A., & Caterina, M. J. (2007). Mechanisms of sensory transduction in the skin. *Nature, 445*(7130), 858–865.

Lumpkin, E. A., Marshall, K. L., & Nelson, A. M. (2010). The cell biology of touch. *Journal of Cell Biology, 191*, 237–248.

MacCumhaill, C. (2018). Absential locations and the figureless ground. *Sartre Studies International, 24*(1), 34–47.

Mack, A., & Rock, I. (1998). *Inattentional blindness*. Cambridge, MA: MIT Press.

MacKay, D. M. (1967). Ways of looking at perception. In W. Wathen-Dunn (Ed.), *Models for the perception of speech and visual form* (pp. 25–43). Cambridge, MA: MIT Press.

Macpherson, F. (Ed.). (2011). *The senses: Classic and contemporary philosophical perspective*. New York: Oxford University Press.

Mancini, F., Stainitz, H., Steckelmacher, J., Iannetti, G. D., & Haggard, P. (2015). Poor judgment of distance between nociceptive stimuli. *Cognition, 143*, 41–47.

Marotta, A., Ferrè, E. R., & Haggard, P. (2015). Transforming the thermal grill effect by crossing the fingers. *Current Biology, 25*(8), 1069–1073.

Marr, D. (1982). *Vision: A computational investigation into the human representation and processing of visual information*. San Francisco, CA: W. H. Freeman.

Martin, M. G. F. (1992). Sight and touch. In T. Crane (Ed.), *The contents of experience: Essays on perception* (pp. 196–215). Cambridge: Cambridge University Press.

McDowell, J. (1985). Functionalism and anomalous monism. In E. Lepore & B. P. McLaughlin (Eds.), *Action and events: Perspectives on the philosophy of Donald Davidson* (pp. 387–398). Oxford: Blackwell.

McGlone, F., & Reilly, D. (2010). The cutaneous sensory system. *Neuroscience and Behavioral Review, 34*, 148–159.

Merleau-Ponty, M. (1945/1962/2013). *Phenomenology of perception* (D. A. Landes, Trans.). London: Routledge.

Michotte, A., Thines, G., & Crabbe, G. (1964/1991). Amodal completion of perceptual structures. In G. Thines, A. Costall, & G. Butterworth (Eds.), *Michotte's experimental phenomenology of perception* (pp. 140–167). Mahwah, NJ: Lawrence Erlbaum Associates.

Mill, J. S. (1865/2015). *An examination of Sir William Hamilton's philosophy and of the principal philosophical questions discussed in his writing* (Vol. 2). Toronto: University of Toronto Press.

Mole, C. (2010). *Attention is cognitive unison*. New York: Oxford University Press.

Moscatelli, A., Bianchi, M., Ciotti, S., Bettelani, G. C., Parise, C. V., Lacquaniti, F., & Bicchi, A. (2019). Touch as an auxiliary proprioceptive cue for movement control. *Science Advances, 5*(6), eaaw3121.

Mountcastle, V. B. (1982). Somatoestesia. In *Enciclopedia del Novecento* (Vol. 6, pp. 997–1038). Milan, Italy: Istituto dell'Enciclopedia Italiana.

Murphy, S., & Dalton, P. (2016). Out of touch? Visual load induces inattentional numbness. *Journal of Experimental Psychology: Human Perception and Performance, 42*(6), 761–765.

Nakamura, A., Yamada, T., Goto, A., Kato, T., Ito, K., Abe, Y., . . . Kakigi, R. (1998). Somatosensory homunculus as drawn by MEG. *NeuroImage, 74*, 377–386.

Neisser, U. (1967). *Cognitive psychology*. New York: Appleton-Century-Crofts.

Noë, A. (2005). *Action in perception*. Cambridge, MA: MIT Press.

Noë, A. (2015). *Varieties of presence*. Cambridge, MA: Harvard University Press.

O'Callaghan, C. (2008). *Sounds: A philosophical theory*. Oxford: Oxford University Press.

O'Regan, J. K., Deubel, H., Clark, J. J., & Rensink, R. A. (2000). Picture changes during blinks: Looking without seeing and seeing without looking. *Visual Cognition, 7*, 191–212.

O'Shaughnessy, B. (1989). The sense of touch. *Australasian Journal of Philosophy, 67*(1), 37–58.

Paillard, J., Michel, F., & Stelmach, G. (1983). Localisation without content: A tactile analogue of "blind-sight." *Archives of Neurology, 40*, 548–551.

Palmer, S. E., Brooks, J. L., & Nelson, R. (2003). When does grouping happen? *Acta Psychologica, 114*, 311–330.

Penfield, W. (1975). *Mystery of the mind: A critical study of consciousness and the human brain*. Princeton, NJ: Princeton University Press.

Penfield, W., & Boldrey, E. (1937). Somatic motor and sensory representation in the cerebral cortex of man as studied by electrical stimulation. *Brain, 60*, 389–443.

Penfield, W., & Rasmussen, T. L. (1950). *The cerebral cortex of man: A clinical study of localization of function*. New York: Macmillan.

Petkova, V. I., & Ehrsson, H. H. (2008). If I were you: Perceptual illusion of body swapping. *PLoS One, 3*, e3832.

Phillips, I. (2020). Blindsight is qualitatively degraded conscious vision. *Psychology Review*. Advance online publication. doi:10.1037/rev0000254.

Ratcliffe, M. (2012). What is touch? *Australasian Journal of Philosophy, 90*(3), 413–432.

Richardson, L. (2010). Seeing empty space. *European Journal of Philosophy, 18*(2), 227–243.

Ringach, D. L., & Shapley, R. (1996). Spatial and temporal properties of illusory contours and amodal boundary completion. *Vision Research, 36*(19), 3037–3050.

Roberts, D. (Ed.). (2002). *Signals and perception: The fundamentals of human sensation*. Basingstoke, UK: Palgrave.

Robinson, H. (1994). *Perception*. London: Routledge.

Robles-De-La-Torre, G. (2006). The importance of the sense of touch in virtual and real environments. *IEEE Multimedia, 13*(3), 24–30.

Rock, I. (1997). *Indirect perception*. Cambridge, MA: Harvard University Press.

Rose, D. (1999). The historical roots of the theories of local signs and labelled lines. *Perception, 28*, 675–685.

Saal, H. P., & Bensmaia, S. J. (2014). Touch is a team effort: Interplay of submodalities in cutaneous sensibility. *Trends in Neurosciences, 37*(12), 689–697.

Saito, D. N., Okada, T., Honda, M., Yonekura, Y., & Sadato, N. (2007). Practice makes perfect: The neural substrates of tactile discrimination by Mah-Jong experts include the primary visual cortex. *BMC Neuroscience, 7*(1), 79.

Schepers, R. J., & Ringkamp, M. (2010). Thermoreceptors and thermosensitive afferents. *Neuroscience and Biobehavioral Reviews, 34*, 177–184.

Sejnowski, T. J. (2015). Vernon Mountcastle: Father of neuroscience. *Proceedings of the National Academy of Sciences of the United States of America, 112*(21), 6523–6524.

Selden, S. T. (2004). Tickle. *Journal of the American Academy of Dermatology, 50*, 93–97.

Serino, A., Giovagnoli, G., De Vignemont, F., & Haggard, P. (2008). Spatial organisation in passive tactile perception: Is there a tactile field? *Acta Psychologica, 128*(2), 355–360. Sklar, A. E., & Sarter, N. B. (1999). Good vibrations: Tactile feedback in support of attention allocation and human-automation condition in event-driven domains. *Human Factors, 41*, 543–552.

Smith, A. D. (2002). *The problem of perception.* Cambridge, MA: Harvard University Press.

Smith, J. (2006). Bodily awareness, imagination, and the self. *European Journal of Philosophy, 14*(1), 49–68.

Snowdon, P. F. (1992). How to interpret direct perception. In T. Crane (Ed.), *The contents of experience: Essays on perception* (pp. 48–78). Cambridge: Cambridge University Press.

Soteriou, M. (2013). *The mind's construction: The ontology of mind and mental action.* Oxford: Oxford University Press.

Spence, C., & Driver, J. (Eds.). (2004). *Crossmodal space and crossmodal attention.* Oxford: Oxford University Press.

Spence, C., Nicholls, M. E. R., & Driver, J. (2001). The cost of expecting events in the wrong sensory modality. *Perception and Psychophysics, 63*, 330–336.

Spence, C., Shore, D. L., & Klein, R. M. (2001). Multimodal prior entry. *Journal of Experimental Psychology: General, 130*, 799–832.

Stein, B. E. (Ed.). (2012). *The new handbook of multisensory processing.* Cambridge, MA: MIT Press.

Stein, B. E., Stanford, T. R., & Rowland, B. A. (2014). Development of multisensory integration from the perspective of the individual neuron. *Nature Reviews Neuroscience, 15*(8), 520–535.

Strawson, P. F. (1959). *Individuals: An essay in descriptive metaphysics.* London: Routledge.

Thunberg, T. (1896). Förnimmelserne vid till samma ställe lokaliserad, samtidigt pågående köld-och värmeretning. *Uppsala Läkareförenings Förh, 1,* 489–495.

Tolman, E. C. (1949). *Purposive behavior in animals and men.* Berkeley: University of California Press.

Treisman, A. (1964). Selective attention in man. *British Medical Bulletin, 20*(1), 12–16.

Tsakiris, M., Costantini, M., & Haggard, P. (2008). The role of the right temporo-parietal junction in maintaining a coherent sense of one's body. *Neuropsychologia, 46*(12), 3014–3018.

Tsakiris, M., & De Preester, H. (2018). *The interoceptive mind: From homeostasis to awareness.* Oxford: Oxford University Press.

Tsakiris, M., Hesse, M. D., Boy, C., Haggard, P., & Fink, G. R. (2007). Neural signatures of body ownership: A sensory network for bodily self-consciousness. *Cerebral Cortex, 17,* 2235–2244.

Vallar, G. (1997). Spatial frames of reference and somatosensory processing: A neuropsychological perspective. *Philosophical Transactions of the Royal Society B: Biological Sciences, 352,* 1401–1409.

Vallar, G. (1998). Spatial hemineglect in humans. *Trends in Cognitive Sciences, 2,* 87–97.

Vallar, G., Rusconi, M. L., Bignamini, L., Geminiani, G., & Perani, D. (1994). Anatomical correlates of visual and tactile extinction in humans: A clinical CT scan study. *Journal of Neurology, Neurosurgery, and Psychiatry, 57,* 464–470.

Vandenbroucke, A. R., Sligte, I. G., & Lamme, V. (2011). Manipulating of attention dissociate fragile visual short-term memory from visual working memory. *Neuropsychologia, 49*(6), 1559–1568.

von Helmholtz, H. (2013). *Treatise on physiological optics.* North Chelmsford, MA: Courier Corporation.

Walsh, L., Critchlow, J., Beck, B., Cataldo, A., de Boer, L., & Haggard, P. (2016). Salience-driven overestimation of total somatosensory stimulation. *Cognition, 154,* 118–129.

Watzl, S. (2017). *Structuring mind: The nature of attention and how it shapes consciousness.* Oxford: Oxford University Press.

Wu, W. (2011). Attention as selection for action. In C. Mole, D. Smithies, & W. Wu (Eds.), *Attention: Philosophical and psychological essays* (pp. 97–116). Oxford: Oxford University Press.

9 What Do Models of Visual Perception Tell Us about Visual Phenomenology?

Rachel N. Denison, Ned Block, and Jason Samaha

9.1 Introduction

The science of visual perception has a long history of developing quantitative, computational models to explain and predict visual performance on a variety of tasks. These models have typically been developed to account for objective visual performance, such as observer's accuracy, reaction time, or perceptual sensitivity in discriminating different visual stimuli. Much less examined is how these models relate to the subjective appearance of a visual stimulus—that is, the observer's phenomenal character of seeing the stimulus. The goal of this chapter is to examine that second link—between models and phenomenal experience.

9.1.1 What Is Phenomenology?

By "phenomenology" or "phenomenal experience," we mean the first-person, subjective, conscious experience an observer has of a visual stimulus. We are interested here in the experience of the properties of that stimulus, for example its contrast, color, or shape. That particular visual phenomenal content is to be distinguished from the mere fact of the observer's being conscious (i.e., awake vs. in a dreamless sleep), from the totality of one's conscious experience (which includes non-visual content such as thoughts, feelings, and non-visual perceptual experiences), and from the experience of visual content unrelated to the stimulus of interest (e.g., content in the periphery of one's visual field). In short, an observer's phenomenal experience of the properties of a stimulus is a matter of what a particular stimulus looks like to them. Here, we ask whether current vision models can make predictions about this kind of subjective visual experience.

9.1.2   Key Concepts in Visual Neuroscience

First, we introduce a few key concepts in visual neuroscience that will be relevant for understanding the models. Much of what we know about the response properties of visual neurons comes from studying early sensory brain areas such as the primary visual cortex (V1) while presenting relatively simple stimuli such as oriented luminance gratings (striped patches, as in figure 9.4). The properties of V1 neurons described below have informed the models discussed in this chapter, as well as attempts to link model components to neural responses. However, the models we consider are meant to be general, and could be used to describe the encoding and decision-making mechanisms for diverse stimulus properties and sensory and decision-related brain regions.

Contrast sensitivity   The contrast of a stimulus refers to its variation in luminance across space—the differences between the lightest and darkest regions of the image. Visual neurons tend to respond to luminance changes rather than absolute luminance. So, the response of a neuron typically increases with higher contrast.

Orientation selectivity   Many neurons in V1, and at later stages of visual processing, are selective for the orientation of edges in an image. For example, a neuron may respond strongly to a region of an image with a vertical edge but respond very little to a horizontal edge. The response profile of a neuron to different orientations (0–180°) is called an orientation tuning curve. Many vision science experiments use grating stimuli with different orientations and contrasts because extensive physiological experiments have related the responses of visual neurons to such stimuli, providing a theoretical foundation for linking neural activity and behavior.

Response variability   Neurons have ongoing spontaneous activity, and they don't respond exactly the same way to repeated presentations of a stimulus. On one trial, a neuron may respond a bit more, and on another trial a bit less to the same stimulus. This trial-to-trial response variability is sometimes called internal noise. It creates uncertainty in the mapping between a stimulus and a neural response. Such response variability of sensory neurons is considered one of the sources of variability in human behavior.

9.1.3   What Is a Model?

Here, we define a model as a compact mathematical explanation of some phenomenon. Visual perception is a complicated phenomenon. So, we make models to capture and test our understanding of how it comes about—that is, we use models to explain and to predict. First, we might like to *explain* observers' behavior in a specific visual task. Why did they report seeing the things they reported seeing? That is, what are the underlying computations or processes that produced their behavior? Second, we might like to *predict* observers' behavior on another session of the same task or on a somewhat different task. How well can we predict what they will report seeing? Predictions allow us to test a model quantitatively, compare the performance of alternative models, and identify mismatches between predictions and data, which point to how the model might be improved.

Models can be specified at different levels of abstraction (Marr, 1982), allowing different kinds of explanations and predictions, depending on one's scientific question. For example, a more abstract model—at what has been called the computational level—can be specified with variables and equations that correspond to cognitive processes but bear little connection to neurobiology. A less abstract model—at the implementational level— could involve simulating large numbers of spiking neurons. The kinds of components that make up a model are quite flexible.

Here, we are concerned with observer models (Geisler, 2011; Rahnev & Denison, 2018; Kupers, Carrasco, & Winawer, 2019), which take as input stimuli and task goals and produce as output behavior and/or neural responses (figure 9.1). The output of an observer model can therefore be compared to human behavioral and neural data for the same stimulus and task. What goes inside the model can vary, but all the models we will discuss can be thought of as having two stages: a sensory stage and a decision stage. The sensory stage encodes stimulus features. The decision stage reads out, or decodes, the responses at the sensory stage to make a perceptual decision. At the sensory stage, a fine-grained representation of the stimulus feature may exist, but at the decision stage, this representation is lost. For example, in an orientation discrimination task, the sensory stage might encode the stimulus orientation as a continuous variable, whereas the decision stage converts that graded sensory response to a binary choice about whether the orientation is clockwise or counterclockwise of a decision boundary. The decision stage is more integral to models designed to

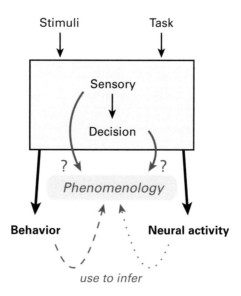

Figure 9.1
Observer models and their relation to phenomenal experience. An observer model
(gray box) takes as input stimuli and task goals. The models we consider here can
include sensory processing and decision stages. Traditionally, observer models are
designed to output measurable quantities (shown in bold), including behavioral and/
or neural responses. Here, we are interested in understanding how observer models
could predict the contents of phenomenal visual experience (*phenomenology*) whose
properties must be inferred from behavior and neural activity. (We use a sparser
arrow to indicate inferences from neural activity because such inferences require
established relations between neural activity and phenomenology in the absence of
a behavioral report—an active research topic in itself.) Specifically, we ask how sensory and decision components of standard vision models could map to phenomenal
experience.

produce categorical choice behavior than to models focused on describing
sensory representations.

9.1.4   Models of Visual Perception and Decision Making
Here, we consider four widely used vision models: (1) signal detection theory
(SDT), (2) drift diffusion models (DDM), (3) probabilistic population codes
(PPC), and (4) sampling models. These models can be organized along two
dimensions: (1) whether they are static or dynamic, and (2) whether they
use point representations or probabilistic representations of stimulus features (table 9.1). Static models (SDT and PPC) describe sensory and decision

Table 9.1
Classification of models

	Point Representation	Probabilistic Representation
Static	SDT	PPC
Dynamic	Drift diffusion	Sampling

SDT: signal detection theory; PPC: probabilistic population codes.

variables at a single time point, or some steady state, whereas dynamic models (DDM and sampling) unfold over time (usually, for perceptual decisions, timescales of hundreds of milliseconds to seconds).[1] SDT and DDM models represent some stimulus feature as a single continuous-valued number or *point estimate*. (This term is used to mean a point representation.) This number could represent, for example, the strength of evidence for one choice versus another in a two-choice perceptual decision. In contrast, models with probabilistic representations (PPC and sampling) also contain distributional information about the uncertainty of the stimulus feature.[2] In coding a sensory feature, probabilistic models represent not only a point estimate (e.g., the distance is 5.3 meters), but also the certainty of the estimate or even the probabilities associated with all possible feature values (e.g., the distance is 5.3 meters with probability x, 5.4 meters with probability y, and so forth).

9.1.5 Modeling Phenomenal Experience

How does phenomenal experience relate to the components of these models? Here, we find a gap. These models have been developed largely to predict and explain objective behavioral performance and neural activity, not phenomenal experience. At first glance, this gap is surprising—a major aspect of visual perception is our conscious experience of the way things look. So, why hasn't visual phenomenology been a modeling focus so far? Here are three possible reasons:

First, there is good reason for a scientific strategy that does not focus on phenomenal experience. As experimenters, we can devise models that predict and explain neural and behavioral variables without probing observers' phenomenal experience per se. And whereas behavior and neural activity can be measured directly, phenomenal experience must be inferred from behavioral reports and/or neural responses (figure 9.1). As a result, ascertaining observers' phenomenal experience—for example, whether they consciously perceived a stimulus—is fraught with conundrums. Should we use subjective measures of conscious awareness, such as asking observers if they saw something, or objective ones, such as above-chance performance? What is the relation between metacognition (e.g., observers' appraisal of their perceptual decisions) and phenomenal experience itself? Is consciousness all or nothing—or does it come in degrees? Confronting these issues is challenging and can reduce the attractiveness of phenomenal experience as a research target. While this argument reflects a real challenge in the empirical study of phenomenal experience, we believe there is still much room for progress on this front. We will return to this topic in section 9.3.2.

Second, objective and subjective measures of visual perception are often tightly coupled, which might suggest that models that explain objective reports will also straightforwardly explain subjective reports and phenomenology. For example, if an observer is asked to report whether a stimulus was one of two possibilities (an *objective report* because it has accuracy conditions with respect to the physical stimulus) and then report their confidence that their choice was correct (a *subjective report* because it only refers to the observer's internal state), correct trials will typically receive higher confidence ratings than incorrect trials. However, although objective and subjective measures are often correlated, they can dissociate—sometimes dramatically, as in blindsight (Weiskrantz, 1996; see chapter 10 of this volume, which describes such dissociations in detail). Such dissociations indicate that objective reports are not a perfect proxy for phenomenology.

Third, many perceptual phenomena—for example, rivalry and perceptual constancies—depend on unconscious processing. So, it is not surprising that models of visual perception often focus on aspects of visual processing that are not specific to visual phenomenology.

Therefore, although it's understandable why computational models of visual perception have mainly focused on capturing objective performance, there is also a need for computational models that predict the properties of phenomenal experience specifically. We think it is surprising that current standard approaches to modeling visual perception have not addressed this challenge more directly.

The goal of this chapter is therefore not to describe the precise links between current vision models and phenomenal experience, as such links have not been fully developed. Rather, to work toward bridging this gap, in section 9.2, we introduce each model according to the following structure. First, we introduce the model and describe its sensory and decision stages. Second, we discuss empirical studies, philosophical work, and theoretical considerations that bear on how the model components could be mapped to aspects of phenomenology. Under the umbrella of phenomenology, we consider both awareness (whether some stimulus property is consciously experienced) and appearance (how exactly a stimulus looks, given that it is experienced). Finally, we summarize various options for possible mappings between model components and phenomenal experience. We view these mappings as alternative hypotheses about how model components could relate to phenomenal experience. In some cases, existing evidence can help adjudicate among these alternatives, but the matter is not yet settled. Although at this point, we cannot decide which option is best, each model provides a formal structure through which to articulate different options, which itself is a theoretical advance. In section 9.3, we discuss scientific and philosophical research directions that follow from this approach. Although the models we discuss have primarily been used to explain and predict objective performance and neural activity, they could also be applied and extended to explain and predict a variety of subjective reports.

9.1.6   Clarifications

Before we launch into the models, here are a few clarifications:

1. Are we realist or as-if (Chakravartty, 2017) about these models? We think that the computations represented by these models could be instantiated as neural processes, though the mechanics of how these computations are implemented by the brain could look very different from the model's mechanics. Note that some models (such as PPC and sampling) are explicitly neural, so the realist mapping is more direct, whereas others (SDT and DDM) are more abstract and allow different neural implementations.

2. What do we mean by a *mapping* between model components and phenomenal experience? We mean a correspondence relation with a particular phenomenal experience. One formulation of this idea is that the information represented by the model component matches the information represented in phenomenal experience. Note that when we say that a model component maps to phenomenal experience, that does not mean that the component is sufficient for consciousness. We assume these components are embedded in a functioning human brain that already meets the conditions for consciousness. Rather, we are interested in how a model component could relate to a particular phenomenal

content. Ultimately, we are interested in the neural basis of a conscious content—what philosophers call a core realizer,[3] which would require understanding the neural implementation of model components.

3. Does specifying a mapping in this way mean that conscious states are epiphenomenal—that is, they have no causal/functional properties? No. The brain activity that realizes a conscious state has causal/functional interactions within the physical system.

4. The sensory stage representations of all the models discussed here assume that the visual system has feature detectors (e.g., for orientation, motion, faces, etc.; see section 9.1.2) that respond automatically to the feature of interest. More complex or novel feature combinations (e.g., a pink elephant) may not have dedicated detectors. Such detectors may have to be learned by the brain for these models to apply to complex perceptual decisions.

9.2   Specific Models of Visual Perception and Decision Making and their Possible Mappings to Conscious Perception

9.2.1   Signal Detection Theory

SDT is perhaps the most widely applied model of perceptual decision making in psychology and neuroscience. Its popularity stems from the fact that it can elegantly explain a wide range of perceptual behaviors, can tease apart the perceptual sensitivity of an observer from response biases they may have, and provides a straightforward framework in which to model the variability of perception and behavior.

Model characterization   Although the theory has been extended and built upon in many different ways, the core idea of SDT is as follows (figure 9.2). Neural detectors representing stimulus features (e.g., a horizontal line at a particular location in the visual field) respond when their preferred stimulus is present. The response magnitude indicates the amount of evidence that the particular feature is present in the environment. A detector tuned to horizontal lines at the center of gaze would respond maximally when there is a clear horizontal line present at the center of gaze, and would, on average, respond less when the image is less clear (e.g., if it were lower contrast) or when the image deviates from the detector's preference (e.g., if the image were a diagonal line or in the periphery). In this way, a single detector presented with an image outputs a single number that reflects the amount of evidence that a particular feature is present on that trial. This quantity is sometimes called the internal response. The observer must determine: How much evidence do I require to decide that the stimulus feature was present? This amount (or threshold) is known as the criterion.

Going forward, we use the word "detector" to mean the part of the sensory stage that carries information about the stimulus feature. This could correspond to an elementary or a highly processed neural signal. For example, the neural signal corresponding to the detector's activity could be purely feedforward or could incorporate feedback. In short, the activity of the detector corresponds to the sensory representation at the computational level.

Sensory stage   A core idea of SDT is that the internal response of a detector on any given trial is not determined by the stimulus properties alone but is also subject to stochastic variation (noise). This means that the same physical stimulus can produce different responses in the detector due to myriad other factors (e.g., the number of photons absorbed by the retina, the amount of neurotransmitter uptake at a synapse, the attentional state of the observer). The noise is often modeled as Gaussian with an independent mean and standard deviation. Note that talk of Gaussian distributions, in the context of SDT, refers to distributions built up over many repeated trials of an experiment. All that is represented at the sensory stage on a given trial is a single value (conceptualized as being drawn from a distribution) reflecting the amount of evidence that a particular feature is present.

Importantly, the same detector, again because of noise, will not be completely silent when its preferred stimulus feature is absent—that is, there will be another distribution from which the detector's response is drawn when the stimulus is absent. For example, many of us have confidently experienced our phone vibrating in our pocket despite no actual phone vibration. This experience can be interpreted as arising from noise in a hypothetical "vibration detector" that, due to random variability, is strong enough to be noticed. The stimulus-absent distribution is often modeled as having the same standard deviation (but lower mean) as the stimulus-present distribution, though this is a parameter that can be adjusted. Crucially, if the noise of one or both distributions is very large, or if the means of the two distributions are close, it will not be obvious which distribution the internal response on a given trial came from—that is, whether the signal was truly present or absent on that trial. In other words, the amount of overlap between distributions determines the difficulty of the decision. This overlap can be estimated by analyzing behavioral responses to repeated presentations of stimulus-absent and stimulus-present trials and is called d'. As the overlap between internal responses on stimulus-present and stimulus-absent trials increases, the detection task becomes harder and d' decreases. Thus, d' reflects the sensitivity of an observer's perceptual system, which is corrupted by noise.

Decision stage Given a single sample from one of these distributions, which response should be chosen? This is the question addressed in the decision stage. SDT posits that the observer's choice is determined by whether the internal response of the detector falls above or below some criterion level. For instance, internal responses above the criterion would result in a stimulus-present response, and responses below would result in a stimulus-absent response.

A key feature of SDT is that the response criterion is independent of the sensory distributions. Thus, the amount of evidence needed to commit to a decision is theoretically separate from the sensitivity of the perceptual system. The value of d' could be low or high, but the observer may still require a lot of evidence to decide stimulus presence. The independence of criterion and sensitivity allows an SDT-based analysis to separate out an observer's overall propensity to give a certain response from the actual discrimination sensitivity of their perceptual system.

A point worth bearing in mind when evaluating empirical evidence is that the SDT criterion, as measured from behavior, is defined relative to the distributions themselves. So, a change to the SDT criterion can be produced by either a change in the amount of internal response needed to commit to a decision or by shifting the means of both sensory distributions while keeping the same absolute criterion (Ko & Lau, 2012; Iemi et al., 2017; Denison et al., 2018; figure 9.2B).

When modeling perceptual decisions with SDT, the criterion is understood to be a flexible decision boundary that the observer can adjust according to a number of non-perceptual factors, including the probability the stimulus is present and the rewards or costs associated with detecting or missing stimuli. We refer to this standard SDT criterion as a *decisional criterion*, and we take it as obvious that deliberate or strategic shifts in decisional criteria do not necessarily affect conscious awareness. Here, however, our aim is to map SDT to phenomenal experience. Therefore, in the

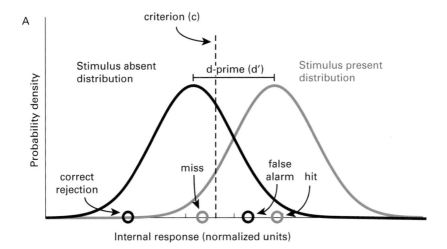

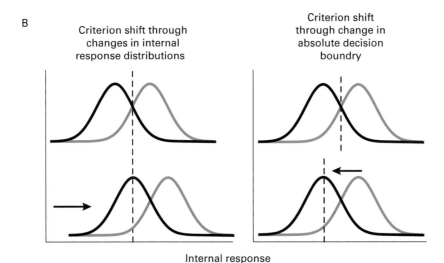

Figure 9.2

Signal detection theory (SDT) illustrated for a detection task. (A) The sensory stage of SDT is a single internal response, generated on every trial (open circles on the x-axis represent four example trials). When no stimulus is presented, the internal response is drawn from a Gaussian probability distribution (stimulus-absent distribution). When the stimulus is presented, the internal response is drawn from another Gaussian distribution with a higher mean (stimulus-present distribution). When the internal responses are normalized by the standard deviation of the Gaussian, the discriminability of the two distributions (d') is the difference in distribution means. The criterion represents the decision stage and is independent of d'. It defines the next section, we pose the question of whether there is a *phenomenal criterion*, such that only when the internal response crosses this criterion is a conscious percept produced (see Wixted, 2019, for a related discussion of threshold theories). That is, if an observer could directly report their conscious experience, does conscious perception follow the sensory stage representation, or does it depend on a criterion crossing?

Where is conscious perception in SDT? Given SDT models' single-trial point representations of sensory evidence along with the two possible interpretations of the criterion (phenomenal and decisional), where in these models could conscious perception fit? Now, we detail possible mappings from SDT components to phenomenology, followed by empirical evidence that bears on various mappings.

Options for mapping model components to phenomenology

Option 1: Conscious perception follows the representation at the sensory stage. That is, phenomenal experience is reflected in the activity of the detector that responds, with variability, to the stimulus. Crossing the criterion (or not) only determines the post-perceptual decision. This option would imply that activity in the detector is always associated with a conscious experience that has a graded strength.

Option 2: Conscious perception follows the representation at the sensory stage but only occurs once the internal response surpasses the phenomenal

Figure 9.2 (continued)

magnitude of internal response needed to report that the stimulus was present. When the stimulus is absent, an internal response below the criterion (report absent) gives a correct rejection, and an internal response above the criterion (report present) gives a false alarm (open circles). When the stimulus is present, an internal response below the criterion gives a miss, and an internal response above gives a hit (open circles). (B) The SDT criterion computed from behavior is a relative measure: it indicates how much evidence is needed to make a decision relative to the two distributions. However, when relating behavior to neural activity, we may be interested in the absolute decision boundary—for example, the neural response magnitude required to report that the stimulus was present. This example illustrates a potential difficulty in inferring an absolute decision boundary from the SDT criterion in a detection task. The criterion estimated from behavior could change due to either a shift in the internal response distributions with no change in absolute decision boundary (left) or a shift in the absolute decision boundary with no change in internal response distributions (right). See color plate 1.

criterion. This option would imply that only responses that surpass the criterion produce graded, conscious strengths, whereas all responses below the criterion are equally unconscious. This situation would allow for unconscious perception when the internal response is above zero but below the criterion.

Option 3: Conscious perception depends on a phenomenal criterion and follows the (binary) representation at the decision stage. This option would imply that phenomenal experience has an all-or-none categorical character, such that an observer either does or does not experience a certain stimulus property. Information about the strength of evidence for the stimulus property (i.e., internal response magnitude) would be lacking in phenomenal experience.

Options 2 and 3 would also involve the existence of a separate decisional criterion (in addition to the phenomenal criterion) that could be flexibly adjusted to adapt behaviors to task contingencies (e.g., reward structure). Note that options 1 and 2 predict something about the quality of one's subjective percept—namely, the strength or intensity of that percept—whereas option 3 predicts only whether the stimulus was seen.

Evidence for mappings   Recent experiments examining neural responses in signal detection tasks have started to shed light on what role a threshold-crossing process could play in whether a stimulus is consciously experienced. According to options 2 and 3, any signal that does not surpass the phenomenal criterion is unconscious. If we assume that confidence reports reflect conscious percepts, these options make predictions about metacognitive sensitivity, or how well confidence reports track performance accuracy. (An accurate response is one that matches the stimulus.) Specifically, options 2 and 3 predict that variability in an observer's confidence ratings for stimulus-absent responses should be unrelated to the accuracy of those responses[4] (Wixted, 2019). If a stimulus does not exceed the phenomenal criterion, then there should be no conscious access to how far away from the criterion the stimulus was. So, confidence will be unrelated to the internal response magnitude on these trials. Accuracy, on the other hand, should still relate statistically to the internal response magnitude.

This prediction has been tested by examining whether confidence or visibility judgments are higher on correct versus incorrect trials, even when the stimulus was reportedly not seen. Whereas abundant evidence makes clear that, for seen trials, the more visible a stimulus is rated, the higher confidence observers report (and the higher their metacognitive sensitivity;

Galvin et al., 2003), it is less obvious that the same is true of misses and false alarms. Koeing and Hofer (2011) used a near-threshold detection task with a rating scale that afforded a low and high confidence stimulus-absent response. Even though both responses correspond to a no-stimulus report (thus, below the criterion), high-confidence "no" responses were significantly more accurate than low-confidence "no" responses. However, this result is not always obtained. Using several manipulations of stimulus visibility, Kanai and colleagues (2010) asked whether confidence differed for misses and correct rejections. Confidence did not differ with perceptual manipulations of stimulus visibility such as masking. Notably, task accuracy could reach 70 percent before confidence ratings began to distinguish between misses and correct rejections. Interestingly, when perceptual reports were manipulated via attentional load (e.g., cueing or rapid visual presentation), there was a clear relationship between confidence and accuracy for unseen trials. In general, metacognitive sensitivity is worse for stimulus-absent than for stimulus-present responses, even when metacognitive sensitivity is above chance for both (Meuwese et al., 2014; Mazor, Friston, & Fleming, 2020). Lower variance of the stimulus-absent compared to the stimulus-present SDT distribution could also contribute to poorer metacognitive sensitivity for stimulus-absent trials, but it would not predict zero sensitivity (Lau, 2019).

Taking these findings together, it seems that observers at least sometimes have cognitive access to signals from stimuli they report not seeing. These findings can be interpreted in three alternative ways. (1) Conscious perception exists below the criterion for report, violating the predictions of options 2 and 3. (2) Metacognitive processes are sensitive to unconscious information, such that sensory responses below the criterion are unconscious but still capable of influencing confidence ratings. (3) Reports of stimulus presence or absence do not perfectly reflect phenomenology, and observers sometimes report a stimulus as absent when in fact a weak conscious sensation occurred. On this account, metacognitive ability is a product of a weak but nevertheless still conscious experience. The first two alternatives assume that the measured SDT criterion is the phenomenal criterion, whereas the third does not.

To the extent that the sensory stage representation in SDT can be mapped onto activity in early sensory brain areas, a number of neuroimaging and electrophysiology experiments also shed light on the model-phenomenology mapping. Using functional magnetic resonance imaging (fMRI) in a contrast detection experiment, Ress and Heeger (2003) showed that BOLD responses in the visual cortical area V1 tracked observers' reported perceptions rather than physical stimulus properties. That is, V1 responses were not detectable when the target stimulus was present yet undetected by the observer. In addition, when the observer reported seeing a target that was not present (a false alarm), V1 responded above baseline. Thus, on the basis of these results (and many similar findings from other paradigms; Polonsky et al., 2000; Lee, Blake, & Heeger, 2005; Wunderlich, Schneider, & Kastner, 2005), the sensory response seems to track conscious perception rather well,[5] in favor of option 1.

However, recent work has shown that stimuli that are subjectively invisible and do not afford above-chance discrimination accuracy can nevertheless be decoded from fMRI patterns in V1 (though not in V2 or V3; Haynes & Rees, 2005), suggesting that a mere representation of the stimulus in V1 is not sufficient for reporting awareness. Likewise, recent results from multi-unit recordings in the visual cortex demonstrate that high-contrast targets that are undetected by the animal still elicit large V1 and V4 responses, which are even larger than responses to low-contrast targets that are detected (van Vugt et al., 2018). One interpretation is that there is no threshold level of activity that V1 or V4 could produce that would result in a conscious sensation, suggesting that a mechanism beyond evaluating the response magnitude in visual areas is needed to trigger a report that the stimulus was present. However, the same behavior could be explained by assuming that the animal used a different decision criterion for high- versus low-contrast stimuli, or that the animal's present/absent reports do not perfectly track their phenomenal experience.

SDT has also been used to explain the phenomenon of blindsight, whereby an individual with damage to the primary visual cortex reports blindness in part of the visual field but can still discriminate certain visual stimuli in the "blind" field at above-chance levels (Cowey & Stoerig, 1991). Ko and Lau (2012) argue that in blindsight, the sensory distribution on stimulus-present trials is reduced to just barely above the noise distribution, but the absolute phenomenal criterion remains at higher pre-lesion levels. Thus, blindsight patients almost never report awareness (no threshold crossing) but have some preserved sensitivity (because the post-lesion sensory distribution is slightly above the noise distribution). Under this account, then, the setting of the criterion, and not the sensory response itself, is what is critical for conscious perception (as in options 2 and 3). This interpretation is concordant with *higher-order* theories of perception in philosophy, which argue that mere sensory representation is not sufficient for consciousness (Brown, Lau, & LeDoux, 2019). Alternatively, it has been argued that blindsight patients are not blind in the sense of lacking phenomenal visual experience, but rather just have severely degraded, albeit still conscious, visual perception (Overgaard et al., 2008), which is below the decisional criterion for report (though see Azzopardi & Cowey, 1997, for an attempt to address the issue of response bias in blindsight). This interpretation is more in line with *first-order* theories in philosophy, which maintain that sensory representation of a particular kind *is* sufficient for conscious perception (Block, 2011), more akin to option 1.

Thus, evidence to date does not definitively rule out any of the possible mappings between SDT components and phenomenology. Later, in section 9.3, we outline future directions that could aid in adjudicating between models and options.

9.2.2 The Drift Diffusion Model

In its simplicity, SDT fails to capture a salient aspect of perceptual decision making, namely that decisions unfold over time. Whereas SDT posits a single internal response on each trial, the DDM assumes that internal responses occur across time and are accumulated until a threshold is reached (Ratcliff et al., 2016). At that point, a decision is made and, importantly, a decision time is specified. That is, DDM has the advantage of providing quantitative predictions of not only the sensitivity and bias of an observer, but also the time it takes to commit to a particular decision. Additionally, recent developments to the DDM allow for estimations of confidence (Kiani & Shadlen, 2009).

Model characterization   In the DDM (figure 9.3), an internal response is generated at each moment during stimulus processing (this is sometimes called the momentary evidence). For instance, one can think of the firing rate of a visual neuron in some short time window as the momentary evidence for that neuron's preferred stimulus. A secondary neuron then accumulates that evidence by integrating it across time (perhaps with some leak) until the amount of accumulated evidence surpasses a threshold, whereby a decision is made. The DDM has gained prominence, in part, due to the discovery of neurons in frontal and parietal cortices that ramp up like evidence accumulators when their activity is averaged across trials (Gold &

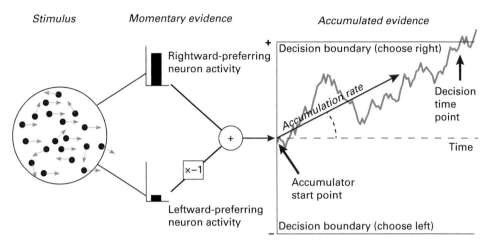

Figure 9.3
Drift diffusion model (DDM) illustrated for a motion-direction discrimination task. Assume a choice between net leftward or rightward motion of some moving dots, where the majority of dots are moving rightward (the stimulus). The sensory representations in DDM begin with activity in feature-tuned detectors (such as those in the middle temporal (MT) area with motion-direction preference). This sensory activity is sustained over time while the stimulus is presented and provides noisy evidence at each time point for how much leftward or rightward motion there is. This is called the momentary evidence. During evidence accumulation, a separate computation integrates the difference of the noisy responses that each feature-tuned detector emits at each moment in time. Thus, the accumulation will build toward a positive or negative boundary. Once this decision variable reaches either boundary, a decision is made. The DDM can thereby explain which choice was made and how long the decision took (reaction time). If the difference in momentary evidence between two detectors is large, this will drive accumulation toward a boundary faster, leading to shorter reaction times. The size of the difference in evidence strength between detectors determines the accumulation rate in the model. The accumulation starting point and placement of the boundaries are also model parameters that can be modified to capture decision biases or speed-accuracy trade-offs, for example.

Shadlen, 2000, 2007; Latimer et al., 2015). A minimal set of parameters for the DDM typically includes the starting point of the accumulator (e.g., zero if there is no bias), the rate of evidence accumulation (sometimes called drift rate), and the position of the decision thresholds (or the bounds). Additionally, when fitting response time data, a non-decision time parameter is often added to account for stimulus-independent sensory- and motor-related latencies in responding.

Sensory stage The sensory representation in the DDM is thought of as a sample of evidence for the presence of a particular stimulus feature at each moment in time. For instance, neurons such as those in the monkey middle temporal (MT) area that fire at a near-constant level while their preferred direction of motion is presented could be considered to generate constant momentary evidence for a particular motion direction. There is a clear relationship between sensory responses (momentary evidence) and accumulated evidence. For example, increasing the contrast of a stimulus or the coherence of motion will drive larger responses in sensory neurons (higher momentary evidence), which will cause an increase in the rate of evidence accumulation, thereby leading to faster threshold crossing and reaction times. This basic intuition explains why easy-to-perceive stimuli are both more accurately and more rapidly discriminated. Typically, in tasks with two alternatives, there is a representation of momentary evidence for one choice and a separate representation of momentary evidence for the other choice, and the difference between these signals feeds into the accumulator (see next section). The sensory signal is often modeled as successive independent draws from a Gaussian distribution with a mean centered on the strength of the momentary evidence (which could be a difference signal) and variance reflecting internal noise.

Decision stage The DDM accumulator can be thought of as a decision variable that, at some point, reaches a threshold, which marks the commitment to a decision. There is much debate, however, as to whether neurons in, for instance, the parietal cortex that have been argued to represent accumulated evidence are more sensory or motor in nature. In early experiments, neurons that behaved as evidence accumulators in the lateral intraparietal (LIP) area of the parietal cortex were chosen specifically because of their motor control functions in driving saccadic behaviors (Gold & Shadlen, 2007). That is, evidence for an accumulating decision variable has mostly been observed in neurons directly involved in the motor output that the animal must produce to indicate a specific percept. This may support an account on which the accumulation process is itself more motor in nature than sensory. However, a recent experiment dissociated more sensory from more motor LIP neurons by placing the saccade response targets in the opposite hemifield to the visual stimuli. Under these conditions, perceptual sensitivity decreased following inactivation of LIP neurons contralateral to the visual stimulus but ipsilateral to the saccade response targets, suggesting that the relevant neurons were more sensory (Zhou & Freedman, 2019).

It remains unclear, though, whether such sensory neurons have similar accumulation dynamics as the motor neurons and to what extent evidence accumulation is obligatory in any perceptual experience.

Although current evidence is not decisive about whether the accumulated evidence should be thought of as sensory, motor, or some intermediate representation, the threshold crossing is considered decisional in nature. Just as for SDT, we can distinguish between *decisional* and phenomenal boundaries. Decisional boundaries (the standard interpretation when fitting DDM models to behavior) can be flexibly adjusted to implement strategic control over one's decisions (e.g., when prioritizing response speed vs. accuracy). We do not expect decision strategy to change perception. So, we can reject a mapping of consciousness onto this decisional boundary crossing. Rather, here, we consider the possibility that there is a less flexible phenomenal boundary, the crossing of which would underlie the formation of a conscious percept.

Where is conscious perception in the DDM?   As with SDT, there are several viable options for how DDM components map onto phenomenology. With the addition of a temporal component to DDM, the options make notably distinct predictions about when a stimulus becomes conscious—i.e., how phenomenology evolves over time.

Options for mapping model components to phenomenology

Option 1: Phenomenal perception tracks the sensory representation (the momentary evidence) before it is fed into the accumulation process. This would imply that evidence accumulation and decision commitment are post-perceptual, and perhaps are only engaged when a choice about one's percept needs to be made. According to this option, then, perceptual experience represents stimuli as being a certain way (e.g., seeing dots moving upwards), but this is not enough to trigger an explicit decision that one is seeing the world in that way.

Option 2: Conscious perception tracks the evidence accumulation process. This would imply that conscious experience unfolds as the decision variable diffuses toward a boundary. This would also imply that perceptual experience of stimulus properties becomes stronger over time to the extent that evidence accumulation approaches one or another threshold.

Option 3: Conscious perception occurs once a decision threshold is reached. This would imply that momentary and accumulated evidence are unconscious and that phenomenal experience occurs only once a boundary is reached. It may also imply limited or no conscious access to the state of the decision variable until the threshold is reached. As with SDT, this option would need to posit a separate and more flexible decisional boundary such that decisions could be made in a perception-independent manner according to other task goals.

Evidence for mappings   Despite wide adoption of DDM as a model of perceptual decision making, as well as theoretical discussion of links between consciousness and accumulation processes (Shadlen & Kiani, 2011; Dehaene 2009; Dehaene et al., 2014), there have been surprisingly few empirical attempts to link particular aspects of the DDM to conscious perception of a stimulus. A notable exception is a recent experiment by Kang and colleagues (2017). In this experiment, a motion stimulus with one of two directions and variable coherence (across trials) was presented within an annular boundary surrounding fixation. At the center of fixation, observers also viewed a clock with a rotating hand. Observers were to report the direction of motion followed by the position of the clock hand at the moment they subjectively were aware of having made their choice. By fitting a DDM model using the reported moment of the decision as a substitute for reaction time, the authors could accurately predict observers' choice behavior across coherence levels. The accuracy of these predictions suggests that the DDM model captured the subjective reports of decision timing. Thus, the authors concluded that conscious awareness of having made a choice aligns with the threshold crossing postulated in the DDM, in line with option 3. This is not to say that there was no awareness of the stimulus prior to threshold crossing—observers were clearly seeing something (some moving dots). However, their awareness of the stimulus *as moving in a particular direction* may occur once evidence accumulation terminates at a boundary. Alternatively, this result may only show that observers are aware of when they committed to a decision, rather than indicating the moment when the conscious perception of directional motion occurred.

In humans, a signal thought to be similar to the parietal cortex evidence accumulation process in nonhuman primates is an electrophysiological signature known as the central parietal positivity (CPP; O'Connell, Dockree, & Kelly, 2012). Recently, Tagliabue and colleagues (2019) tracked the CPP during a contrast discrimination task. The contrast level of the stimulus varied across trials, and observers gave both objective responses about which contrast was presented and subjective responses about the visibility of the stimulus. Evidence accumulation, as indexed by the CPP, closely tracked subjective reports of stimulus awareness rather than the physical stimulus contrast. When trials were selected with equal subjective awareness levels, variation in contrast did not modulate the CPP. Notably, the CPP bears considerable resemblance—and may be identical (Twomey et al., 2015)— to the highly studied P300 electroencephalography component, which has been argued to be a neural marker of conscious perception (Dehaene & Changeux, 2011). The study by Tagliabue and colleagues (2019) is preliminary evidence that conscious perception of a simple contrast-defined stimulus tracks an evidence accumulation signal. More direct evidence that detection depends on accumulation processes consistent with DDM comes from single-neuron recordings from the posterior parietal cortex in a human participant performing a vibrotactile detection task. Neuronal firing rates ramped up for detected stimuli in a way that was related to choice and confidence behavior through a DDM model that assumed awareness occurs at threshold crossing (Pereira et al., 2020), as in option 3.

These findings raise several questions. Is evidence accumulation an index of awareness or only of reported awareness? For instance, using a no-report paradigm, Pereira and colleagues (2020) also found parietal cortex neurons that were sensitive to stimulus intensity, suggesting that part of the activity is conserved when an explicit decision does not need to be made. However, the temporal profile of this activity was notably different from the evidence accumulation signature (see figure 3 of Pereira et al., 2020), perhaps compatible with options 1 or 2. A further question is whether awareness unfolds contemporaneously with evidence accumulation or, as suggested by Kang and colleagues (2017), only once evidence accumulation reaches a threshold. If, as stated in option 3, conscious perception only occurs after boundary crossing, this leads to the counterintuitive prediction that as the decision variable might fluctuate above and below the threshold over time, the associated perceptual experience could appear and disappear in quick succession. Moreover, in many experiments, the momentary evidence signal is not simultaneously measured but rather is expected to be closely related to the accumulated evidence. Therefore, momentary sensory evidence, rather than accumulated evidence, could just as well reflect awareness, or both could be involved.

9.2.3 Probabilistic Population Codes
SDT and DDMs maintain only a point estimate related to the stimulus variable of interest. However, there are reasons to think that sensory representations

contain information about the uncertainty of stimulus variables as well. First, sensory signals are ambiguous. Retinal images, for example, are two-dimensional, whereas the world is three-dimensional. This loss of information means that retinal images are ambiguous with respect to world states. Retinal images therefore contain information that can be used to compute the probabilities of various world states without specifying a unique state. Second, sensory signals are noisy, or variable. Given repeated presentations of the same stimulus, the brain responds differently each time. The ambiguity and noise in sensory signals have motivated the idea that perception is an inference process (Helmholtz, 1856). That is, the perceptual system is trying to infer what things in the world most likely caused the image on the retina, given the uncertainty of the sensory signals (Knill & Richards, 1996; Kersten, Mamassian, & Yuille, 2004).

It would be advantageous, then, for the system to estimate and use its own uncertainty. We have empirical evidence that it does so. The incorporation of uncertainty into perceptual decisions has been found in many behavioral studies (Trommershäuser, Kording, & Landy, 2011; Ma & Jazayeri, 2014) and shows that the representations used for perceptual decision making are richer than point estimates. To illustrate how this might play out in an everyday activity, consider the following: when driving at night, you might be less certain about the distance between your car and the car in front of you than you would be during the day, and you might therefore keep further back as a result. A point estimate of the distance could be the same at night and during the day, but the greater uncertainty at night would lead to more cautious behavior. Behavioral studies have found both optimal and suboptimal perceptual decision making (Rahnev & Denison, 2018). However, probabilistic information can be incorporated into perceptual decisions even if the outcome is not strictly optimal (Ma, 2012).

PPC and sampling (section 9.2.4) are two widely used approaches for modeling sensory uncertainty. Both are formulated as neural models and describe how neurons could represent probability distributions over stimulus features.

Model characterization   In PPC (Jazayeri & Movshon, 2006; Ma et al., 2006), a population of neurons encodes a probability distribution over some sensory feature (Földiák, 1993; Sanger, 1996; Zemel, Dayan, & Pouget, 1998). For example, a population of orientation-tuned neurons would encode a probability distribution over stimulus orientation. PPC is a static model—it uses a population of neurons to encode a probability distribution at a single moment in time (or at steady state), and it has no time-varying internal processes.

Sensory stage   To describe the sensory stage of the PPC model, let's consider a population of orientation-tuned neurons. Each neuron has a tuning function that describes the mean response of the neuron when a stimulus of a particular orientation is presented (figure 9.4). However, the neuron's response is not perfectly predictable because of response variability. This variability reflects uncertainty in the stimulus representation. Response variability is also called noise. Biological neurons are often characterized as having Poisson noise—a particular noise distribution in which the response variance scales with the response mean. PPC assumes Poisson (or Poisson-like) noise, which turns out to have important theoretical consequences for population codes (Ma et al., 2006). The responses of a population of orientation-tuned neurons with Poisson noise can be mathematically combined using a Bayesian decoder to compute a probability distribution over the stimulus orientation (figure 9.4). Specifically, a log transform of the encoded probability distribution can be read out from the population using a weighted, linear combination of each neuron's response. To combine the responses in a Bayesian fashion, two pieces of information are required: (1) the orientation tuning functions of each neuron and (2) the noise distribution. Therefore, in PPC models, the response of a single neuron does not represent a probability. Rather, neural populations can encode probability distributions over stimulus features, where specific mathematical operations are required to calculate probabilities from the population activity.

Decision stage   Does the brain actually do the required calculations to compute probability distributions over sensory features? If so, this would be a job for the decision stage. Although PPC usually refers to the sensory encoding stage, there have been various ideas about how information could be read out from the sensory population code at a decision stage. The way such a readout is expected to work in the brain is that sensory neurons input to decision-related neurons, and the decision neurons integrate the sensory inputs. The specific way that the decision stage would integrate sensory inputs could vary, and here we consider different options.

First, and as introduced above, a log transform of the encoded probability distribution can be read out from the PPC using a linear operation. The

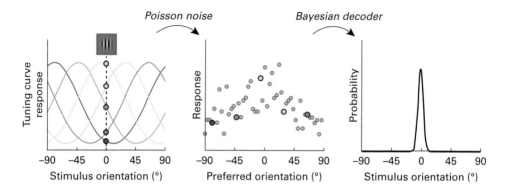

Figure 9.4

Probabilistic population codes. Left: A population of orientation-tuned neurons has tuning curves that tile orientation. Example tuning curves from the population are shown (curves), along with their mean responses to a stimulus with 0° (i.e., vertical) orientation (points). Middle: On a single trial, the response of each neuron (gray dots) is determined by its tuning curve and Poisson noise; the preferred orientation of each neuron is the orientation at which its tuning curve peaks. Right: A probability distribution can be read out from the single-trial population response using a Bayesian decoder, which combines the information from all neurons to calculate the likelihood of the population response given a stimulus of each orientation. See color plate 2.

linearity makes it plausible that a biological neural network could perform this computation, and specific proposals have been made for how it might do so (Jazayeri & Movshon, 2006; Ma et al., 2006; Beck, Ma, et al., 2008; Beck, Pouget, & Heller, 2012; Ma & Jazayeri, 2014). In this operation, each neuron's response is weighted by the log of its tuning function value at that stimulus value, and these weighted responses are then summed across the population to compute the log probability. This operation recovers the full probability distribution. Such information would enable optimal decisions in, for example, a two-choice discrimination task by simply comparing the probabilities associated with the two stimulus options. Confidence has been proposed to be based on the probability that the choice is correct (Meyniel, Sigman, & Mainen, 2015; Pouget, Drugowitsch, & Kepecs, 2016), though this idea has not been fully supported empirically (Adler & Ma, 2018). It could also be possible to use the PPC sensory representation to compute probability-dependent expected values or make probabilistic categorical decisions without reconstructing a full probability distribution over stimulus feature values.

Whether and under what circumstances downstream regions would read out a full probability distribution from sensory regions is unknown.

Second, key properties of the probability distribution encoded by the PPC can be extracted from the population activity without reading out the full distribution. For Gaussian distributions, the mean (and max) of the distribution is approximated by the stimulus value at which the population activity peaks, and the variance is inversely proportional to the height of this peak (Ma et al., 2006; figure 9.4). So, a simple downstream readout mechanism could estimate the stimulus value from the peak position and uncertainty from the peak amplitude or from the sum of the amplitudes across the population (Meyniel et al., 2015). Under such a readout, in a two-choice discrimination task, the stimulus option closer to the estimated value would be chosen, and confidence would be based on the uncertainty estimate.

In summary, PPC models show how probabilistic information could be encoded in sensory populations and read out from decision populations. Ongoing empirical work is actively testing whether real neural populations behave like the theorized PPC (Beck, Ma, et al., 2008; Fetsch et al., 2012; Bays, 2014; van Bergen et al., 2015; Hou et al., 2019; Walker et al., 2020) and whether humans have behavioral access to full probability distributions (Rahnev, 2017; Yeon & Rahnev, 2020).

Where is conscious perception in PPC?   In PPC models, the sensory representation is more complex than the sensory stage in SDT and DDM models. As a result, there are more possibilities for how the sensory representation itself could be realized in phenomenology (option 1). Alternatively, we can consider mappings between decision stage readouts and phenomenology (option 2).

Options for mapping model components to phenomenology
Option 1: Phenomenal experience maps to the sensory-stage population code. The content of conscious experience is specifically related to neural activity in the neural population that encodes the relevant probability distribution. Here, there are several alternatives for how phenomenal experience could relate to the population activity.
Option 1a: No transformation of the population code. We experience the population activity itself. This might look like an imprecise percept with graded strengths across a continuum of feature values, but those strengths would not be probabilities. If this is the case, then probabilistic information

is not directly available in phenomenal experience because a further transformation of the population activity is required to calculate the probability distribution.

Option 1b: Phenomenology reads out summary statistics of the population code. For example, we could experience a point estimate (such as the mean value represented by the population activity), a point estimate with some strength (such as the mean value plus the gain, or height of the peak), or a point estimate with some uncertainty estimate (Rahnev, 2017). If we experience just the mean, say, our conscious experience of some stimulus feature would consist of that single exact feature value.

Option 1c: Phenomenology consists of a probability distribution derived from the population code. We can divide this into two further sub-options. The first of these is that we experience the (log?) probability distribution. In this case, we could experience the distribution as probabilistic in the sense that our perceptual experience would simultaneously represent multiple alternatives with different likelihoods. This is the option that seems most compatible with the philosopher John Morrison's description of his "perceptual confidence" view (Morrison, 2016). In the second sub-option, alternatively, we could see the distribution but not see it as probabilistic (i.e., we could see all the possible values as simultaneously present with different strengths but not as alternative options about the state of the world). For example, if our vision is blurry, a dot may look smeared out in space, but we don't interpret the edge of the smear as a low probability dot location (Denison, 2017).

These options for translating the PPC into phenomenal experience are intended to span the range from minimal transformation to full probability distribution, but they are not exhaustive; there are many possible ways to transform PPC activity. It is also important to emphasize that in options 1b and 1c, summary statistics and probabilities are not explicitly represented in the sensory stage population itself.[6] Rather, the calculation of these quantities occurs (somehow) in the translation to phenomenal experience.

Option 2: Although a separate decision stage is not part of the PPC model, we can think of a second option wherein the information in the sensory-stage population code is read out by a decision stage, and only this explicit representation becomes conscious. Similar alternatives as described for option 1 would apply, but option 2 requires a decision stage in the brain to read out (i.e., explicitly compute) summary statistics, probability distributions, and so on from the sensory stage activity.

Evidence for mappings PPC models have rarely been related to conscious phenomenology. One exception is a study reporting that perceptual rivalry—in which perception of an ambiguous stimulus alternates between two perceptual interpretations—is consistent with Bayesian sampling from PPCs (Moreno-Bote, Knill, & Pouget, 2011; see also *Evidence for Mappings* in section 9.2.4). Lehky and Sejnowski (1999) have discussed phenomenal experience in relation to population coding models more generally, making the point that our subjective experience is not determined by physical variables as a physicist would describe them but rather by those physical variables transformed according to the ways in which our perceptual apparatus interacts with them. For example, our perception of color requires a transformation from physical variables involving wavelengths and reflectances into the psychological variable of hue.

Though clear links have not been made to phenomenal experience per se, testing whether PPC models can link behavioral performance to neural activity is an active area of research. One study decoded orientation uncertainty from fMRI responses in human visual cortex, and showed that this uncertainty correlated with both the variability and biases in perceptual decision behavior (van Bergen et al., 2015). Decoded likelihood functions from neural populations in the primate visual cortex likewise could predict trial-by-trial fluctuations in decision behavior (Walker et al., 2020). Therefore, the uncertainty information measured in PPC appears to be used in behavior. This fact means that the represented information is more than a single point estimate (see option 1b). Note, however, it does not mean that the full probability distributions are used, as required by option 1c (Rahnev, 2017). And it remains an open question whether the information used in behavior in these cases is also reflected in perceptual experience.

Some philosophers have argued that perceptual experience is probabilistic (Morrison, 2016; Munton, 2016; Morrison, 2017), in the sense that our conscious perceptual experience of a feature value includes the probability of that feature value (option 1c, sub-option 1). However, others have argued that experience is not probabilistic (all other options; Denison, 2017; Block, 2018; Beck, 2019; Cheng, 2018; Pohl, S., Perceptual representations are not probabilistic, unpublished). The extent to which probabilistic information is explicitly represented in the brain is relevant for this debate. Recall that at the sensory stage of PPC models, probability is encoded only implicitly; it requires a further step to be explicitly computed. That is, the explicit number corresponding to the probability of a given stimulus feature value can only be calculated by combining the neural responses in a specific way. There is no guarantee that such a calculation ever occurs; and if it does, it's not known whether it would have phenomenal content (Lehky & Sejnowski, 1999). A philosophical view that argues for probabilistic perception based on PPC models would need to specify how the computation of probability occurs—either somehow in the direct transformation of sensory representations to phenomenology (option 1) or explicitly in a separate decision stage (option 2).

9.2.4   Sampling

Whereas PPC models represent probability distributions through the activity of a large number of neurons at a single time, sampling models represent probability distributions through the activity of single neurons across time (Fiser et al., 2010). Therefore, unlike PPC models, sampling models are inherently dynamic.

Model characterization   Sampling models are strongly motivated by the idea of perception as probabilistic inference. In these models, neuronal response variability is a feature rather than a bug—it's a way to represent uncertainty over ambiguous world states. Sampling models propose that single neurons represent posterior probabilities of states of the world. They do so through a sequence of samples from the posterior probability distribution they represent (Hoyer & Hyvärinen, 2003).

Sensory stage   Each neuron corresponds to some stimulus property. A neuron's response represents a probability distribution over the possible values of that property. At an early sensory stage, these properties could be, for example, the intensity of an oriented patch[7] present in a stimulus at some location (Olshausen & Field, 1996; Hoyer & Hyvärinen, 2003; Haefner, Berkes, & Fiser, 2016; Orbán et al., 2016), and a given neuron's response might represent the intensity of a vertical orientation. Specifically, it would represent the probability of the vertical intensity of the stimulus given the image. The response of a neuron at a given time (e.g., its instantaneous firing rate)

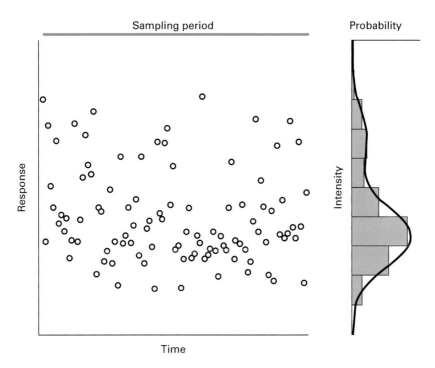

Figure 9.5
Sampling models. Left: The response (e.g., instantaneous firing rate) of a single neuron is plotted at successive time points. Right: The responses of this neuron are drawn from a probability distribution (black curve), such that the responses aggregated across the sampling period (gray histogram) approximate the probability distribution. If this neuron represents the intensity of a vertical stimulus, this series of responses indicates that it is most likely that the vertical intensity is low, but there is some probability that the intensity is higher.

is modeled as a sample from the posterior distribution over possible stimulus values. In our example, if there is high probability that the vertical intensity is low, the firing rate of the neuron will mostly be low. Samples collected over time will build up the probability distribution (figure 9.5). Given enough time, the probability distribution can be approximated with arbitrary precision. Recent theoretical work explores how sampling models could be implemented in biologically realistic neural networks (Buesing et al., 2011; Huang & Rao, 2014; Savin & Denève, 2014; Petrovici et al., 2016).

Decision stage In sampling models, there is not a sharp distinction between sensory and decision stages because all neurons represent posterior probabilities over world state variables. Decision making in sampling models therefore only requires reading the activity of neurons that explicitly represent the relevant variable for the task.

For example, in an orientation discrimination task, we imagine that there are neurons downstream of early sensory areas that correspond to decision variables: the presence of a counterclockwise stimulus and the presence of a clockwise stimulus (Haefner et al., 2016). These neurons would learn the posterior distributions over their decision variables from task training (Fiser et al., 2010). Then, the final decision computation would be simply to compare the counterclockwise response to the clockwise response over some length of time. This accumulation of evidence across time could resemble a drift diffusion process (Yang & Shadlen, 2007; Haefner et al., 2016).

Where is conscious perception in sampling?   For sampling models, as for PPC models, the complexity of the proposed representation makes multiple mappings possible. However, in sampling models, the complexity arises from a fundamentally different type of neural code, with its own mapping options.

Options for mapping model components to phenomenology   In sampling models, the responses of all neurons are samples from posterior distributions. So, the models imply that the content of conscious perception is based on these samples. There are several different options for how neural responses in sampling models could map to conscious perception.

Option 1: Our phenomenal experience is based on one sample: we experience the current sample. This option says that our conscious perception has no probabilistic content and only reflects the brain's current guess about the state of the world. This option implies that our perceptual experience changes with every new sample. Because neural activity is constantly changing and different parts of the brain receive information about the same stimulus at different times, it also requires a mechanism for samples across the brain to be synchronized in phenomenal experience: experience the current guess from all neurons in the brain *now*.

Option 2: Our phenomenal experience is based on multiple samples, which could be interpreted in one of two ways:

Option 2a: We experience a summary statistic based on samples across some time interval. For example, we may experience the mean value of the samples. In this case, our conscious perception would still be a point estimate. However, this option could also allow perception of uncertainty by extracting the variability across samples over the time interval.

Option 2b: We experience an approximated probability distribution based on samples across some time interval. We could either experience this distribution as probabilities, in which case, our phenomenal experience would be probabilistic, or we could see it as non-probabilistic—that is, we could simultaneously experience the different values of the distribution without experiencing these values as competing options.

Evidence for mappings There have been a few attempts to relate sampling models to conscious perception. Perhaps the most explored idea is that sampling models are well suited to predict the alternating perceptual interpretations observers see when viewing bistable images (Sundareswara & Schrater, 2008; Gershman & Tenenbaum, 2009; Moreno-Bote et al., 2011; Gershman, Vul, & Tenenbaum, 2012). In this case, the probability distribution over the world state is bimodal, and sampling models would sequentially sample from each of the modes of the distribution. Moreno-Bote and colleagues (2011) used sampling from PPCs to model perceptual alternations, combining the two probabilistic approaches. Recent theoretical work has described a unification of sampling and PPC models, showing that these probabilistic approaches are not mutually exclusive (Shivkumar et al., 2018). It has also been argued that perceptual reports during spatial and temporal selection tasks reflect probabilistic samples (Vul, Hanus, & Kanwisher, 2009) and that cognition more broadly reflects a Bayesian sampling process (Sanborn & Chater, 2016).

Testing perceptual sampling models in neural systems is in the early stages. Spontaneous activity in the visual cortex resembles evoked activity to natural images, which has been argued to be consistent with a prediction from sampling models. In the absence of stimulus-driven activity, neural responses will reflect the prior (in this case, the statistics of natural images; Berkes et al., 2011). Various statistical properties of responses in the visual cortex during visual stimulation and visual tasks are also consistent with sampling models (Haefner et al., 2016; Orbán et al., 2016; Bondy et al., 2018; Banyai et al., 2019).

When evaluating the relation between sampling models and phenomenology, it is important to keep in mind that even if phenomenal experience arises from a single sample or is a point-estimate summary statistic, perceptual decision behavior could still reflect uncertainty because (1) the current sample reflects the posterior probability distribution, which already incorporates the relative uncertainties of priors and likelihoods, and (2) decision readout mechanisms could still be based on multiple samples across time.

The hypothesis that phenomenal experience is based on multiple samples—a possibility offered only by the sampling model—raises several questions: Under this hypothesis, how many samples are required to experience something? Is our phenomenal experience always updated after a fixed number of samples? If not, how does the system know when to start a new round of sampling? Are successive percepts based on independent sets of samples, or a moving window of samples? Future empirical work could pursue these questions and perhaps, in the process, determine how well sampling models can explain and predict phenomenology.

9.3   Conclusions and Future Directions

9.3.1   Constrained, but not Unique, Mappings between Model Components and Phenomenal Experience

We considered four computational models—SDT, DDM, PPC, and sampling—which, taken together, have been broadly applied to perceptual decision making and perceptual coding in the brain. For each model, we find no unique mapping between model components and phenomenal experience. Rather, we find that each model is consistent with multiple possible mappings. Even SDT, the simplest model, affords different ways of relating internal responses, criteria, and their combination to phenomenal experience.

Importantly, the models cannot support any arbitrary mapping to phenomenal experience. Instead, each model offers a constrained set of possible mappings. The models we discussed here differ in three major ways in terms of the mappings they can support. First, the models differ in their uses of thresholds. SDT and DDM models use thresholds to define decision behavior, whereas PPC and sampling do not include thresholds and are more focused on the nature of the sensory representation. (This difference is perhaps related to the fact that SDT and DDM are specifically designed to model task-based decision behavior, whereas PPC and sampling are not.) In models with thresholds, different mappings predict that consciousness either is or is not tied to a threshold crossing. This distinction is relevant to ongoing debates about whether consciousness is all or nothing and whether an "ignition" process underlies conscious awareness of specific content (Dehaene, Sergent, & Changeux, 2003; Sergent & Dehaene, 2004; Fisch et al., 2009; Vul et al., 2009; Moutard, Dehaene, & Malach, 2015; Noy et al., 2015; van Vugt et al., 2018). Second, the models differ in how they represent uncertainty. PPC models could support a phenomenal experience of uncertainty instantaneously, whereas sampling and DDM models could only do so over time, and SDT models (with their single point estimate) could not do so at all. (NB: Whether we have a perceptual experience of uncertainty in addition to our cognitive sense of it is up for debate; see *Evidence for Mappings* in section 9.2.3.) Third, the models can support different kinds of predictions about the timing of phenomenal experience. In DDM, different mappings make distinct predictions about when consciousness occurs (before or after boundary crossing) and how gradual it is (tracking continuous momentary evidence, continuous evidence accumulation, or all or nothing once a boundary is reached). In sampling, different mappings can support phenomenal experience either for each sample or only after a period of sampling. The static models, SDT and PPC, don't afford predictions about timing.

The lack of unique mappings between model components and phenomenal experience is in some sense not surprising; these models were developed to predict perceptual decision behavior and brain activity rather than phenomenal experience. This gap raises two questions: What approaches can we take, in science and in philosophy, to develop tighter mappings between vision models and visual experience? And given the private nature of subjective experience, how far can we hope to get with such a project?

9.3.2   Directions for Science

We see considerable opportunity for advancing our understanding of the processes that underlie phenomenal experience using computational models. We can do this both by collecting more and different kinds of subjective reports in experiments and by developing our models to expand the kinds of reports they can predict. These two research directions work in tandem. Our goal should be vision models that predict a variety of objective and subjective perceptual reports.

There are several well-established psychophysical protocols for collecting subjective reports. Subjective reports can be classified as either reports about the observer's phenomenal experience of the stimulus or reports about the observer's objective decision. Subjective reports about phenomenal experience include visibility ratings, appearance judgments, and appearance matching. For example, an observer may be asked to report whether a test stimulus appears higher contrast than a stimulus with a standard contrast (Carrasco, Ling, & Read, 2004). This protocol allows estimation of the point of subjective equality: the contrast required for the observer to report that the test contrast is higher than the standard 50 percent of the time. Methods for measuring the perceived magnitude (e.g., brightness, loudness) of suprathreshold stimuli are called scaling methods.

Subjective reports about the observer's decision include confidence ratings, confidence comparisons (Barthelmé & Mamassian, 2009), and post-decision wagering (Persaud, McLeod, & Cowey, 2007). For example, an observer may be asked to perform an objective perceptual task and then report their confidence as the likelihood that their objective choice was correct. The topic of perceptual confidence has seen a surge of research interest in recent years (Mamassian, 2016; Pouget et al., 2016), including work modeling the computations underlying confidence reports (Kiani & Shadlen, 2009; Fetsch, Kiani, & Shadlen, 2014; Meyniel et al., 2015; Adler & Ma, 2018; Denison et al., 2018; Ott, Masset, & Kepecs, 2019). In contrast, appearance reports have received much less attention and little modeling. This is true, despite their critical role in the history of vision science. Perceptual matching tasks were used, for example, to establish the trichromatic theory of color vision and to demonstrate that perceived contrast is approximately constant across the visual field (Georgeson & Sullivan, 1975). Arguably, appearance reports allow more direct inferences about the content of phenomenal experience than metacognitive reports about the perceptual decision. We see particular potential in extending our current models to capture appearance comparison and similarity judgments. One promising current direction is maximum likelihood difference scaling—an approach for collecting and modeling suprathreshold appearance judgments (Maloney & Knoblauch, 2020).

9.3.3 Directions for Philosophy

We consider two directions for philosophy in this topic area. First, directly relevant to scientific practice, philosophers could consider the questions: (1) What should we try to explain and predict about phenomenal experience? (2) What can we hope to explain and predict about phenomenal experience? To make scientific progress on the nature of phenomenal experience, we must use the data of behavioral reports and neural measurements. In this way, inferring values on dimensions of phenomenal experience is a theoretical enterprise similar to others in science, such as inferring the masses of stars or the structures of atoms. Can we approach modeling phenomenal experience in the same way that we approach modeling these other entities? Or are there special considerations for modeling phenomenal experience? Are there any hard limits on what we can learn about phenomenal experience from behavioral and neural data, and if so, what are they? What is the role of neural data, in particular, in learning about phenomenal experience, and how can we best combine behavioral and neural data to make inferences about experience? The models we have considered are usually applied to perceptual decisions about a single visual feature. Is there a way to bridge to multiple simultaneous features or even visual phenomenology as a whole?

Second, as mentioned, our current models, if taken literally as models of phenomenal experience, would make very different predictions for what information about perceptual uncertainty could be available in phenomenal experience. The question of whether and how we may perceptually experience uncertainty has recently begun to attract interest in philosophy (see *Evidence for Mappings* in section 9.2.3). It is relevant for understanding not only the nature of experience (philosophy of mind; Morrison, 2016) but also whether perception must be probabilistic in order to justify probabilistic beliefs about the world (epistemology; Munton, 2016; Nanay, 2020). Currently, it is not even clear what the various options are for experiencing perceptual uncertainty in principle. We see this as an open direction in philosophy. The more such work in philosophy is able to relate to scientific models and data from perception science, the more productive we believe this line of work will be for both disciplines.

Acknowledgments

We would like to thank the following individuals who provided valuable feedback on earlier versions of this chapter: Ralf Haefner, Luke Huszar, Richard Lange, Brian Maniscalco, Matthias Michel, Jorge Morales, Dobromir Rahnev, Adriana Renero, Bas Rokers, Susanna Siegel, and Karen Tian. Thanks also to the attendees of the 2019 Summer Seminar in Neuroscience and Philosophy (SSNAP) at Duke University and the Philosophy of Mind Reading Group at New York University for helpful discussions.

Notes

1. Although the machinery of these models is dynamic, they are typically used to represent static stimuli or stable stimulus properties.

2. Although a single number could be used to represent a probability computed at a previous stage, here we use "probabilistic" to refer to the representational stage at which the probability is computed.

3. Philosophers distinguish between the core and total neural realizers of a conscious state. A total neural realizer is sufficient for a conscious state, but a core realizer is sufficient only given background conditions. The core realizer of the experience of red differs from the core realizer of the experience of green, but the background conditions that have to be added to either of these core realizers to obtain a sufficient condition of a conscious state may be the same (Shoemaker, 2007). Neural activity in the middle temporal (MT) cortical area correlates with perception of motion, but no one should think that if MT were removed from a brain and kept alive that activations in it would constitute motion experiences. Activity in MT is a core realizer of the experience of motion, but background conditions need to be added to it to obtain a total realizer.

A further important distinction is between background conditions that are constitutive of a total realizer of consciousness and background conditions that are contingently related to consciousness. Blood flow in the brain is a contingent background condition—you cannot have sustained neural activity without it—but there might be some other way of providing oxygen to neurons. Corticothalamic oscillations may be a constitutive background condition.

4. This prediction also assumes that stimulus-absent and stimulus-present reports faithfully reflect the observer's conscious experience.

5. One caveat of interpreting fMRI BOLD activity in a given brain region as a sensory response is that feedback from higher-order regions may contribute to it.

6. Different authors may use "explicit" differently. Pouget and colleagues (2016) call a representation explicit as long as it is linearly decodable. Here, we use "explicit" to mean that a neuron's response is proportional to the value of interest. One could also argue that representing the parameters of a probability distribution is sufficient for representing the full distribution, but here we intend that a full distribution is "explicitly" represented only when the probabilities associated with any given feature value map to some neuron's firing rate.

7. The patch could reflect multiple stimulus properties, including location, orientation, spatial frequency, color, disparity, and so on. In our example, we focus on orientation for simplicity.

References

Adler, W. T., & Ma, W. J. (2018). Comparing Bayesian and non-Bayesian accounts of human confidence reports. *PLoS Computational Biology, 14*, e1006572.

Azzopardi, P., & Cowey, A. (1997). Is blindsight like normal, near-threshold vision? *Proceedings of the National Academy of Sciences of the United States of America, 94*, 14190–14194.

Banyai, M., Lazar, A., Klein, L., Klon-Lipok, J., Stippinger, M., Singer, W., & Orban, G. (2019). Stimulus complexity shapes response correlations in primary visual cortex. *Proceedings of the National Academy of Sciences of the United States of America,* *116*, 2723–2732.

Barthelmé, S., & Mamassian, P. (2009). Evaluation of objective uncertainty in the visual system. *PLoS Computational Biology, 5,* e1000504.

Bays, P. M. (2014). Noise in neural populations accounts for errors in working memory. *Journal of Neuroscience, 34,* 3632–3645.

Beck, J. (2019). On perceptual confidence and "completely trusting your experience." *Analytic Philosophy, 61,* 174–188.

Beck, J. M., Ma, W. J., Kiani, R., Hanks, T., Churchland, A. K., Roitman, J., . . . Pouget, A. (2008). Probabilistic population codes for Bayesian decision making. *Neuron, 60,* 1142–1152.

Beck, J. M., Pouget, A., & Heller, K. A. (2012). Complex inference in neural circuits with probabilistic population codes and topic models. *Advances in Neural Information Processing Systems, 25,* 3059–3067.

Berkes, P., Orban, G., Lengyel, M., & Fiser, J. (2011). Spontaneous cortical activity reveals hallmarks of an optimal internal model of the environment. *Science, 331,* 83–87.

Block, N. (2011). Perceptual consciousness overflows cognitive access. *Trends in Cognitive Sciences, 15,* 567–575.

Block, N. (2018). If perception is probabilistic, why does it not seem probabilistic? *Philosophical Transactions of the Royal Society B: Biological Sciences, 373,* 20170341.

Bondy, A.G., Haefner, R.M., Cumming, B.G. (2018). Feedback determines the structure of correlated variability in primary visual cortex. *Nature Neuroscience, 21,* 598–606.

Brown, R., Lau, H., & LeDoux, J. E. (2019). Understanding the higher-order approach to consciousness. *Trends in Cognitive Sciences, 23,* 754–768.

Buesing, L., Bill, J., Nessler, B., & Maass, W. (2011). Neural dynamics as sampling: A model for stochastic computation in recurrent networks of spiking neurons. *PLoS Computational Biology, 7,* e1002211.

Carrasco, M., Ling, S., & Read, S. (2004). Attention alters appearance. *Nature Neuroscience, 7,* 308–313.

Chakravartty, A. (2017). Scientific realism. In E. N. Zalta (Ed.), *The Stanford encyclopedia of philosophy* (Summer 2017 ed.). Retrieved from https://plato.stanford.edu/archives/sum2017/entries/scientific-realism/.

Cheng, T. (2018). Post-perceptual confidence and supervaluative matching profile. *Inquiry,* 1–29.

Cowey, A., & Stoerig, P. (1991). The neurobiology of blindsight. *Trends in Neurosciences, 14*, 140–145.

Dehaene, S. (2009). Conscious and nonconscious processes: Distinct forms of evidence accumulation? *Séminaire Poincareé, XII*, 89–114.

Dehaene, S., & Changeux, J.-P. (2011). Experimental and theoretical approaches to conscious processing. *Neuron, 70*, 200–227.

Dehaene, S., Charles, L., King, J.-R., & Marti, S. (2014). Toward a computational theory of conscious processing. *Current Opinion in Neurobiology, 25*, 76–84.

Dehaene, S., Sergent, C., & Changeux, J.-P. (2003). A neuronal network model linking subjective reports and objective physiological data during conscious perception. *Proceedings of the National Academy of Sciences of the United States of America, 100*, 8520–8525.

Denison, R. N. (2017). Precision, not confidence, describes the uncertainty of perceptual experience: Comment on John Morrison's "Perceptual Confidence." *Analytic Philosophy, 58*, 58–70.

Denison, R. N., Adler, W. T., Carrasco, M., & Ma, W. J. (2018). Humans incorporate attention-dependent uncertainty into perceptual decisions and confidence. *Proceedings of the National Academy of Sciences of the United States of America, 115*, 11090–11095.

Fetsch, C. R., Kiani, R., & Shadlen, M. N. (2014). Predicting the accuracy of a decision: A neural mechanism of confidence. *Cold Spring Harbor Symposia on Quantitative Biology, 79*, 185–197.

Fetsch, C. R., Pouget, A., Deangelis, G. C., & Angelaki, D. E. (2012). Neural correlates of reliability-based cue weighting during multisensory integration. *Nature Neuroscience, 15*, 146–154.

Fisch, L., Privman, E., Ramot, M., Harel, M., Nir, Y., Kipervasser, S., . . . Malach, R. (2009). Neural "ignition": Enhanced activation linked to perceptual awareness in human ventral stream visual cortex. *Neuron, 64*, 562–574.

Fiser, J., Berkes, P., Orbán, G., & Lengyel, M. (2010). Statistically optimal perception and learning: From behavior to neural representations. *Trends in Cognitive Sciences, 14*, 119–130.

Földiák, P. (1993). The "ideal homunculus": Statistical inference from neural population responses. In F. H. Eeckman & J. M. Bower (Eds.), *Computation and neural systems* (pp. 55–60). New York: Springer.

Galvin, S. J., Podd, J. V., Drga, V., & Whitmore, J. (2003). Type 2 tasks in the theory of signal detectability: Discrimination between correct and incorrect decisions. *Psychonomic Bulletin and Review, 10*, 843–876.

Geisler, W. S. (2011). Contributions of ideal observer theory to vision research. *Vision Research, 51*, 771–781.

Georgeson, M. A., & Sullivan, G. D. (1975). Contrast constancy: Deblurring in human vision by spatial frequency channels. *Journal of Physiology, 252*, 627–656.

Gershman, S. J., & Tenenbaum, J. (2009). Perceptual multistability as Markov chain Monte Carlo inference. *Advances in Neural Information Processing Systems, 22*, 611–619.

Gershman, S. J., Vul, E., & Tenenbaum, J. B. (2012). Multistability and perceptual inference. *Neural Computation, 24*, 1–24.

Gold, J. I., & Shadlen, M. N. (2000). Representation of a perceptual decision in developing oculomotor commands. *Nature, 404*, 390–394.

Gold, J. I., & Shadlen, M. N. (2007). The neural basis of decision making. *Annual Review of Neuroscience, 30*, 535–574.

Haefner, R. M., Berkes, P., & Fiser, J. (2016). Perceptual decision-making as probabilistic inference by neural sampling. *Neuron, 90*, 649–660.

Haynes, J.-D., & Rees, G. (2005). Predicting the orientation of invisible stimuli from activity in human primary visual cortex. *Nature Neuroscience, 8*, 686–691.

Helmholtz, H. (1856). *Treatise on physiological optics*. London: Thoemmes Continuum.

Hou, H., Zheng, Q., Zhao, Y., Pouget, A., & Gu, Y. (2019). Neural correlates of optimal multisensory decision making under time-varying reliabilities with an invariant linear probabilistic population code. *Neuron, 104*, 1010–1021.e10.

Hoyer, P. O., & Hyvärinen, A. (2003). Interpreting neural response variability as Monte Carlo sampling of the posterior. *Advances in Neural Information Processing Systems, 15*, 293–300.

Huang, Y., & Rao, R. P. (2014). Neurons as Monte Carlo samplers: Bayesian inference and learning in spiking networks. *Advances in Neural Information Processing Systems, 27*, 1943–1951.

Iemi, L., Chaumon, M., Crouzet, S. M., & Busch, N. A. (2017). Spontaneous neural oscillations bias perception by modulating baseline excitability. *Journal of Neuroscience, 37*, 807–819.

Jazayeri, M., & Movshon, J. A. (2006). Optimal representation of sensory information by neural populations. *Nature Neuroscience, 9*, 690–696.

Kang, Y. H. R., Petzschner, F. H., Wolpert, D. M., & Shadlen, M. N. (2017). Piercing of consciousness as a threshold-crossing operation. *Current Biology, 27*, 2285–2295.

Kanai, R., Walsh, V., & Tseng, C.-H. (2010). Subjective discriminability of invisibility: A framework for distinguishing perceptual and attentional failures of awareness. *Consciousness and Cognition, 19*, 1045–1057.

Kersten, D., Mamassian, P., & Yuille, A. (2004). Object perception as Bayesian inference. *Annual Reviews in Psychology, 55,* 271–304.

Kiani, R., & Shadlen, M. N. (2009). Representation of confidence associated with a decision by neurons in the parietal cortex. *Science, 324,* 759–764.

Knill, D. C., & Richards, W. (Eds.). (1996). *Perception as Bayesian inference.* New York: Cambridge University Press.

Ko, Y., & Lau, H. (2012). A detection theoretic explanation of blindsight suggests a link between conscious perception and metacognition. *Philosophical Transactions of the Royal Society B: Biological Sciences, 367,* 1401–1411.

Koeing, D., & Hofer, H. (2011). The absolute threshold of cone vision. *Journal of Vision, 11,* 21.

Kupers, E. R., Carrasco, M., & Winawer, J. (2019). Modeling visual performance differences "around" the visual field: A computational observer approach. *PLoS Computational Biology, 15,* e1007063.

Latimer, K. W., Yates, J. L., Meister, M. L. R., Huk, A. C., & Pillow, J. W. (2015). Single-trial spike trains in parietal cortex reveal discrete steps during decision-making. *Science, 349,* 184–187.

Lau, H. (2019). Consciousness, metacognition, and perceptual reality monitoring. Retrieved from https://psyarxiv.com/ckbyf/.

Lee, S. H., Blake, R., & Heeger, D. J. (2005). Traveling waves of activity in primary visual cortex during binocular rivalry. *Nature Neuroscience, 8,* 22–23.

Lehky, S. R., & Sejnowski, T. J. (1999). Seeing white: Qualia in the context of decoding population codes. *Neural Computation, 11,* 1261–1280.

Ma, W. J. (2012). Organizing probabilistic models of perception. *Trends in Cognitive Sciences, 16,* 511–518.

Ma, W. J., Beck, J. M., Latham, P. E., & Pouget, A. (2006). Bayesian inference with probabilistic population codes. *Nature Neuroscience, 9,* 1432–1438.

Ma, W. J., & Jazayeri, M. (2014). Neural coding of uncertainty and probability. *Annual Review of Neuroscience, 37,* 205–220.

Maloney, L. T., & Knoblauch, K. (2020). Measuring and modeling visual appearance. *Annual Review of Vision Science, 6,* 13.1–13.19.

Mamassian, P. (2016). Visual Confidence. *Annual Review of Vision Science, 2,* 459–481.

Marr, D. (1982). *Vision: A computational investigation into the human representation and processing of visual information.* San Francisco, CA: W. H. Freeman.

Mazor, M., Friston, K. J., & Fleming, S. M. (2020). Distinct neural contributions to metacognition for detecting, but not discriminating visual stimuli. *eLife, 9,* e53900.

Meuwese, J. D. I., van Loon, A. M., Lamme, V. A. F., & Fahrenfort, J. J. (2014). The subjective experience of object recognition: Comparing metacognition for object detection and object categorization. *Attention, Perception, and Psychophysics, 76,* 1057–1068.

Meyniel, F., Sigman, M., & Mainen, Z. F. (2015). Confidence as Bayesian probability: From neural origins to behavior. *Neuron, 88,* 78–92.

Moreno-Bote, R., Knill, D. C., & Pouget, A. (2011). Bayesian sampling in visual perception. *Proceedings of the National Academy of Sciences of the United States of America, 108,* 12491–12496.

Morrison, J. (2016). Perceptual confidence. *Analytic Philosophy, 57,* 15–48.

Morrison, J. (2017). Perceptual confidence and categorization. *Analytic Philosophy, 58,* 71–85.

Moutard, C., Dehaene, S., & Malach, R. (2015). Spontaneous fluctuations and non-linear ignitions: Two dynamic faces of cortical recurrent loops. *Neuron, 88,* 194–206.

Munton, J. (2016). Visual confidences and direct perceptual justification. *Philosophical Topics, 44,* 301–326.

Nanay, B. (2020). Perceiving indeterminately. *Thought: A Journal of Philosophy, 9*(3), 160–166.

Noy, N., Bickel, S., Zion-Golumbic, E., Harel, M., Golan, T., Davidesco, I., . . . Malach, R. (2015). Ignition's glow: Ultra-fast spread of global cortical activity accompanying local "ignitions" in visual cortex during conscious visual perception. *Consciousness Cognition, 35,* 206–224.

O'Connell, R. G., Dockree, P. M., & Kelly, S. P. (2012). A supramodal accumulation-to-bound signal that determines perceptual decisions in humans. *Nature Neuroscience, 15,* 1729–1735.

Olshausen, B. A., & Field, D. J. (1996). Emergence of simple-cell receptive field properties by learning a sparse code for natural images. *Nature, 381,* 607–609.

Orbán, G., Berkes, P., Fiser, J., & Lengyel, M. (2016). Neural variability and sampling-based probabilistic representations in the visual cortex. *Neuron, 92,* 530–543.

Ott, T., Masset, P., & Kepecs, A. (2019). The neurobiology of confidence: From beliefs to neurons. *Cold Spring Harbor Symposia on Quantitative Biology, 83,* 038794.

Overgaard, M., Fehl, K., Mouridsen, K., Bergholt, B., & Cleeremans, A. (2008). Seeing without seeing? Degraded conscious vision in a blindsight patient. *PLoS One, 3,* e3028.

Pereira, M., Mégevand, P., Tan, M. X., Chang, W., Wang, S., Rezai, A., . . . Faivre, N. (2020). Evidence accumulation determines conscious access. *bioRxiv.* Advance online publication. doi:10.1101/2020.07.10.196659.

Persaud, N., McLeod, P., & Cowey, A. (2007). Post-decision wagering objectively measures awareness. *Nature Neuroscience, 10,* 257–261.

Petrovici, M. A., Bill, J., Bytschok, I., Schemmel, J., & Meier, K. (2016). Stochastic inference with spiking neurons in the high-conductance state. *Physical Review E, 94,* 042312.

Polonsky, A., Blake, R., Braun, J., & Heeger, D. J. (2000). Neuronal activity in human primary visual cortex correlates with perception during binocular rivalry. *Nature Neuroscience, 3,* 1153–1159.

Pouget, A., Drugowitsch, J., & Kepecs, A. (2016). Confidence and certainty: Distinct probabilistic quantities for different goals. *Nature Neuroscience, 19,* 366–374.

Rahnev, D. (2017). The case against full probability distributions in perceptual decision making. *bioRxiv.* Advance online publication. doi:10.1101/108944.

Rahnev, D., & Denison, R. N. (2018). Suboptimality in perceptual decision making. *Behavioral and Brain Sciences, 41,* 1–107.

Ratcliff, R., Smith, P. L., Brown, S. D., & McKoon, G. (2016). Diffusion decision model: Current issues and history. *Cognitive Science, 20,* 260–281.

Ress, D., & Heeger, D. J. (2003). Neuronal correlates of perception in early visual cortex. *Nature Neuroscience, 6,* 414–420.

Sanborn, A. N., & Chater, N. (2016). Bayesian brains without probabilities. *Trends in Cognitive Science, 20,* 883–893.

Sanger, T. D. (1996). Probability density estimation for the interpretation of neural population codes. *Journal of Neurophysiology, 76,* 2790–2793.

Savin, C., & Denève, S. (2014). Spatio-temporal representations of uncertainty in spiking neural networks. *Advances in Neural Information Processing Systems, 27,* 2024–2032.

Sergent, C., & Dehaene, S. (2004). Is consciousness a gradual phenomenon? Evidence for an all-or-none bifurcation during the attentional blink. *Psychological Science, 15,* 720–728.

Shadlen, M. N., & Kiani, R. (2011). Consciousness as a decision to engage. In S. Dehaene & Y. Christen (Eds.), *Characterizing consciousness: From cognition to the clinic?* (Vol. 31, pp. 27–46). Berlin: Springer.

Shivkumar, S., Lange, R. D., Chattoraj, A., & Haefner, R. M. (2018). A probabilistic population code based on neural samples. Retrieved from https://arxiv.org/abs/1811.09739.

Shoemaker, S. (2007). *Physical realization.* Oxford: Oxford University Press.

Sundareswara, R., & Schrater, P. R. (2008). Perceptual multistability predicted by search model for Bayesian decisions. *Journal of Vision, 8,* 12.11–19.

Tagliabue, C. F., Veniero, D., Benwell, C. S. Y., Cecere, R., Savazzi, S., & Thut, G. (2019). The EEG signature of sensory evidence accumulation during decision formation closely tracks subjective perceptual experience. *Scientific Reports, 9*, 4949.

Trommershäuser, J., Kording, K., & Landy, M. S., (Eds.). (2011). *Sensory cue integration.* Oxford: Oxford University Press.

Twomey, D. M., Murphy, P. R., Kelly, S. P., & O'Connell, R. G. (2015). The classic P300 encodes a build-to-threshold decision variable. *European Journal of Neuroscience, 42*, 1636–1643.

van Bergen, R. S., Ma, W. J., Pratte, M. S., & Jehee, J. F. M. (2015). Sensory uncertainty decoded from visual cortex predicts behavior. *Nature Neuroscience, 18*, 1728–1730.

van Vugt, B., Dagnino, B., Vartak, D., Safaai, H., Panzeri, S., Dehaene, S., & Roelfsema, P. R. (2018). The threshold for conscious report: Signal loss and response bias in visual and frontal cortex. *Science, 23*, eaar7186.

Vul, E., Hanus, D., & Kanwisher, N. (2009). Attention as inference: Selection is probabilistic; responses are all-or-none samples. *Journal of Experimental Psychology: General, 138*, 546–560.

Walker, E. Y., Cotton, R. J., Ma, W. J., & Tolias, A. S. (2020). A neural basis of probabilistic computation in visual cortex. *Nature Neuroscience, 23*, 122–129.

Weiskrantz, L. (1996). Blindsight revisited. *Current Opinion Neurobiology, 6*, 215–220.

Wixted, J. T. (2019). The forgotten history of signal detection theory. *Journal of Experimental Psychology: Learning, Memory, and Cognition, 46*, 201–233.

Wunderlich, K., Schneider, K. A., & Kastner, S. (2005). Neural correlates of binocular rivalry in the human lateral geniculate nucleus. *Nature Neuroscience, 8*, 1595–1602.

Yang, T., & Shadlen, M. N. (2007). Probabilistic reasoning by neurons. *Nature, 447*, 1075–1080.

Yeon, J., & Rahnev, D. (2020). The suboptimality of perceptual decision making with multiple alternatives. *Nature Communications, 11*, 3857.

Zemel, R. S., Dayan, P., & Pouget, A. (1998). Probabilistic interpretation of population codes. *Neural Computation, 10*, 403–430.

Zhou, Y., & Freedman, D. J. (2019). Posterior parietal cortex plays a causal role in perceptual and categorical decisions. *Science, 365*, 180–185.

Color plate 1

A

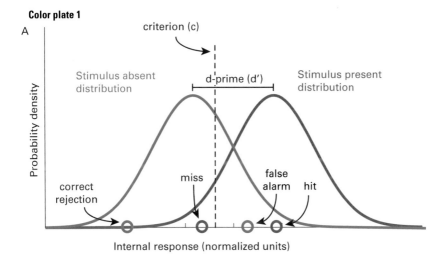

criterion (c)

Stimulus absent
distribution

d-prime (d′)

Stimulus present
distribution

Probability density

correct
rejection

miss

false
alarm

hit

Internal response (normalized units)

B

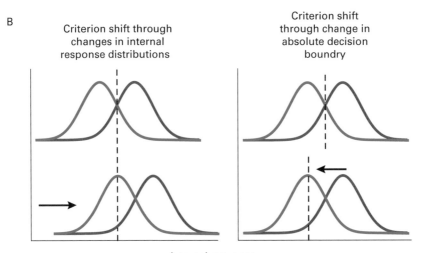

Criterion shift through
changes in internal
response distributions

Criterion shift
through change in
absolute decision
boundry

Internal response

Color plate 1 (previous page)

Signal detection theory (SDT) illustrated for a detection task. (A) The sensory stage of SDT is a single internal response, generated on every trial (open circles on the x-axis represent four example trials). When no stimulus is presented, the internal response is drawn from a Gaussian probability distribution (stimulus-absent distribution). When the stimulus is presented, the internal response is drawn from another Gaussian distribution with a higher mean (stimulus-present distribution). When the internal responses are normalized by the standard deviation of the Gaussian, the discriminability of the two distributions (d') is the difference in distribution means. The criterion represents the decision stage and is independent of d'. It defines the magnitude of internal response needed to report that the stimulus was present. When the stimulus is absent, an internal response below the criterion (report absent) gives a correct rejection, and an internal response above the criterion (report present) gives a false alarm (open circles). When the stimulus is present, an internal response below the criterion gives a miss, and an internal response above gives a hit (open circles). (B) The SDT criterion computed from behavior is a relative measure: it indicates how much evidence is needed to make a decision relative to the intersection point of the two distributions. However, when relating behavior to neural activity, we may be interested in the absolute decision boundary—for example, what neural response magnitude is required to report that the stimulus was present. This example illustrates a potential difficulty in inferring an absolute decision boundary from the SDT criterion in a detection task. The criterion estimated from behavior could change due to either a shift in the internal response distributions with no change in absolute decision boundary (left) or a shift in the absolute decision boundary with no change in internal response distributions (right).

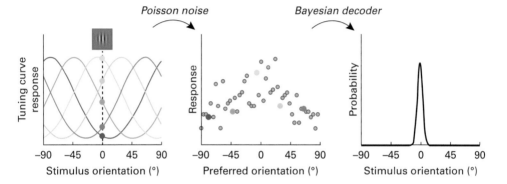

Color plate 2

Probabilistic population codes. Left: A population of orientation-tuned neurons has tuning curves that tile orientation. Example tuning curves from the population are shown (curves), along with their mean responses to a stimulus with 0° (i.e., vertical) orientation (points). Middle: On a single trial, the response of each neuron (gray dots) is determined by its tuning curve and Poisson noise; the preferred orientation of each neuron is the orientation at which its tuning curve peaks. Right: A probability distribution can be read out from the single-trial population response using a Bayesian decoder, which combines the information from all neurons to calculate the likelihood of the population response given a stimulus of each orientation.

My breakfast this morning

My typical breakfast

Typical breakfast foods in my culture

Meal eaten in the morning

Autobiographical   *Intersubjective*

Color plate 3

Hypothesized placement of examples of different kinds of memories in terms of their level of autobiographicality versus intersubjectivity/sharedness. (Images are open-source with free license.)

10   The Neural Substrates of Conscious Perception without Performance Confounds

Jorge Morales, Brian Odegaard, and Brian Maniscalco

10.1   Introduction

Imagine you are driving with your friend Austin at night on a poorly lit road. Both of you have 20/20 vision, are used to driving at night, and are attentively looking at the road to ensure there are no surprises. Suddenly, Austin yelps, "Watch out!"—there is a large branch in the middle of the lane. You avoid it just in time, but only thanks to your friend's warning: you had not noticed the branch before you were alerted to it. How can this be? How could, under similar perceptual circumstances, Austin experience the obstacle while you completely miss it? One plausible explanation is that your friend consciously saw the branch while you did not. At the crucial moment, your visual experience of the road did not include any debris—you were unaware of it.

This example illustrates important aspects of how philosophers and neuroscientists think about consciousness, which is commonly characterized as "what it is like" to be in a particular mental state (Nagel, 1974). For example, there is something that it is like to see the branch, while presumably there is nothing it is like to be a camera that records the visual properties of the road. This notion of consciousness can be extended beyond visual experiences to include other sensory modalities (e.g., auditory or olfactory), feelings and emotions, pains, and perhaps even the conscious experience of having thoughts and desires.

This subjective feeling of what it is like to be conscious of a particular content (e.g., the branch on the road) is referred to as phenomenal consciousness. In contrast, access consciousness describes the functional aspects of being consciously aware of contents as they become accessible to cognitive systems such as working memory, reasoning, categorization, planning, decision making, and, more generally, rational control of action (Block, 1995, 2005, 2007).

Were you *phenomenally* aware of the branch but failed to *access* a conscious representation of it, such that you could not use your motor control mechanisms to steer the wheel? Perhaps your conscious experience was of a rich and detailed branch, but it was not accessible by your categorization and decision-making systems—which are supposed to have a more limited capacity and, at least in principle, are distinct and independent from your phenomenal consciousness (Block, 2007, 2011; Lamme, 2010). Alternatively, perhaps your phenomenal experience of the road lacked any branch altogether: there was no phenomenally conscious branch that your cognitive mechanisms failed to access (as would occur if, for example, it were impossible to have truly unconscious experiences; Phillips, 2016, 2018b, 2020). Of course, it might well be the case that this sensible conceptual distinction does not reflect how things are split up in the mind and brain. Perhaps there is no phenomenal consciousness without access consciousness (Cohen & Dennett, 2011; Cohen et al., 2012; Dehaene, Changeux, et al., 2006), or perhaps access consciousness capacity is not limited with respect to phenomenal consciousness (Phillips, 2011; Gross & Flombaum, 2017; Quilty-Dunn, 2019). And even if these two types of consciousness are distinct in principle, it could be impossible to know what phenomenal experiences you are in if you cannot access them (Kouider, de Gardelle, et al., 2010; Kouider, Sackur, & de Gardelle, 2012).

Orthogonal to the phenomenal and access distinction, different things could be meant when we talk about consciousness (Rosenthal, 1993). It could mean transitive consciousness, also known as content-consciousness, namely, when one is conscious of a particular content[1] (e.g., being conscious of the branch); state-consciousness, namely, when a mental state itself is conscious (e.g., the conscious experience of seeing the branch in contrast to perceptually processing the branch albeit unconsciously); and creature-consciousness, namely, the overall conscious state of someone as an individual (e.g., someone awake compared to someone asleep, anaesthetized, or in a coma; Bayne, Hohwy, & Owen, 2016).

10.1.1 The Scientific Study of Consciousness

Can we know, from a scientific point of view, what explains the difference in conscious contents between you and your friend? Theoretical and practical concerns may cause one to question the possibility of a scientific study of consciousness. From a purely theoretical standpoint, many philosophers and scientists share the intuition that studying access consciousness in general, and perceptual processing in particular, is "easy." That is, while understanding how we perceive the environment is challenging, understanding access consciousness and perceptual processing does not seem to pose a distinct theoretical challenge compared to other psychological and neural phenomena we study: perceptual and decision-making mechanisms compute information, and that is something that, at least in principle, we know how to study. In contrast, understanding phenomenal consciousness is sometimes considered to be "hard" (Chalmers, 1996). The idea is that even if we found what the neural correlates of conscious experiences are, these would still fail to explain why those biophysical processes give rise to those subjective experiences. This so-called "hard problem of consciousness" has garnered much attention in the last twenty-five years. However, not everyone shares the intuition that we should be troubled by the alleged irreducibility of consciousness (Bickle, 2008; Godfrey-Smith, 2008). The metaphysical assumptions of the problem can be rejected, as they involve a notion of deductive explanation that is too stringent (Taylor, 2015). Furthermore, phenomenal consciousness is supported in some way or another by brain activity. So, regardless of one's metaphysical inclinations, understanding the neural substrates of consciousness should be within the purview of scientific research.

To study consciousness scientifically, researchers aim to create conditions that probe the thresholds of awareness, where stimuli yield graded levels of awareness that range from complete unconsciousness to clear, full-blown awareness. In other words, scientists create conditions where subjects respond correctly to the stimuli they are presented with in a majority of the trials (but not in all of them), and only in a portion of those trials do subjects consciously see the stimuli; in the rest, they report having guessed and not being aware of the stimuli. These conditions may be achieved by, for example, presenting a mask right before or after the stimulus (forward/backward masking); presenting distinct images to each eye, effectively yielding one of them invisible (binocular rivalry and continuous flash suppression); degrading the contrast or the presentation duration of the stimulus; using constant stimulation that, however, can be perceived in different ways (bistable figures); or disrupting visual processing with transcranial magnetic stimulation (TMS). Thus, these conditions allow scientists to contrast subjects' experiencing something against experiencing nothing (Fleming, 2020; Michel, 2021), that is, *detection* (e.g., a branch vs. nothing); or they can contrast experiencing *this* compared to experiencing *that*, that is, *discrimination* (e.g., a branch vs. a snake). Importantly, these contrasts can be characterized in an all-or-nothing fashion, or they can take into account relative levels of awareness too. For example, you could be either aware or unaware of the branch, but you could also be less aware of the branch than your friend is, or you could be more aware now than you were before your friend yelped.

When searching for the neural substrates of consciousness, scientists look for the minimally jointly sufficient neural events required for having a conscious experience (Chalmers, 2000). To find these substrates, they compare the neural activity of subjects when they are (more) aware of stimuli against neural activity when they are not (or less) aware of them. When subtracting neural activity of the less conscious states from the more conscious ones, the remaining activity should reveal the unique neural processes that support consciousness. Besides this kind of subtraction, scientists can also compare patterns of activity or connectivity profiles across conditions. Ideally, to guarantee that the neural substrates of consciousness—and nothing but the neural substrates of consciousness—are isolated, the only difference between these two contrast conditions should be phenomenal consciousness. For instance, the story at the beginning of the chapter would not be so surprising if you did not have 20/20 vision, if you were not paying attention, or if your friend had much more experience driving at night than you. Translating this scenario to the lab, this means that we need to ensure that the perceptual, attentional, and cognitive demands of a task, as well as the subjects' performance in it, are matched when subjects are aware and unaware. Then, and only then, we can expect to learn what the neural substrates of consciousness are.

Nevertheless, in practice, it is quite challenging to eliminate confounds and irrelevant differences between conscious and unconscious conditions. In particular, there is an often-neglected confound that is, however, crucial to eliminate from neuroimaging studies: task performance.

10.1.2   Task Performance: A Confound in Neuroimaging Studies

Task performance is the objective effectiveness with which subjects achieve an experiment's goal. On the road, with respect to the goal of detecting debris, your friend is objectively more effective at this task than you, making their task performance better than yours. In the lab, consider a task that consists in identifying the shape of a stimulus that is presented on the screen on multiple trials. A straightforward way of measuring someone's task performance is by computing the percentage of correct responses they provide across all the trials. Task performance is an important reflection of subjects' capacity to process sensory information (which is required for succeeding at the task at hand). However, when task performance differs across conscious and unconscious conditions, behavioral, perceptual, and cognitive profiles can be expected to differ as well. Most of the time, performance is higher in conscious trials than in unconscious trials. On the road, when your friend is conscious of the branch, they are also more likely to detect its presence, to discern its location, to identify it as a branch and not a snake, and so on. However, if variations in awareness are closely correlated with variations in task performance, a direct comparison of the neural activity of conscious versus unconscious trials is not as straightforward as it seems. In particular, a simple subtraction of the neural activity recorded during less conscious trials from activity recorded during more conscious trials may reveal differences in the neural substrates of *perception* in general (and perhaps other capacities as well), rather than, or in addition to, the neural substrates of *conscious* perception in particular. Consequently, matching performance is crucial in neuroimaging studies that compare neural activity across awareness conditions.

In the following sections, we discuss the benefits and challenges of matching performance in consciousness research. In section 10.2, we discuss the difference between subjective and objective measures of consciousness and how they dissociate, and we argue that consciousness research needs to focus on subjective measures while keeping objective performance constant. Then, in section 10.3, we elaborate on the logic of considering task performance a confound in neuroimaging studies of consciousness. In section 10.4, we discuss signal detection theory (SDT) accounts of how performance-matched differences in awareness can occur, and show how these accounts can inform the design of stimuli specifically created to match performance and still obtain differences in awareness. In section 10.5, we discuss potential technical and theoretical issues that stem from matching performance across conditions of awareness. Finally, in section 10.6, we discuss future directions in consciousness research, and introduce the notion of "triangulation" for designing comprehensive experimental sets that can better reveal the neural substrates of consciousness.

A note on the terminology and scope of this chapter: While we acknowledge the long-standing debate about separating reportability and phenomenal consciousness (Block, 1995, 2007; Cohen & Dennett, 2011; Phillips, 2018a), unless otherwise specified, by "consciousness," here we will refer to phenomenal consciousness of visual contents as revealed by subjective reports in detection and discrimination tasks. The context should make clear whether we are discussing cases of all-or-nothing consciousness or cases of relative levels of awareness.

10.2 Subjective and Objective Measures of Consciousness

To analyze neural data, experimenters need to know when subjects are conscious of the stimuli they are presented with and when they are not. A straightforward way to achieve this is by asking subjects to report their subjective state, for example "I saw the branch" or "I did not see a branch." For obvious reasons, this kind of *subjective* measure is widely used. However, subjective measures have been criticized in both philosophy and neuroscience. From a behavioral standpoint, critics argue that introspective reports of consciousness are prone to mistakes, biases, and response criterion effects (Irvine, 2012; Phillips, 2016; Schwitzgebel, 2011; Spener, 2019). Subjects could report more or less frequently that they saw a stimulus due to their response strategies and not due to reliable introspective judgments of their conscious experiences. By using *objective* measures that assess subjects' perceptual sensitivity, that is, their ability to detect and discriminate stimuli independently of whether they take themselves to have consciously seen them, experimenters could bypass the problem of the response criterion and the fallibility of introspection. From a neuroscientific perspective, an additional concern is that by eliciting subjective reports of consciousness, we risk capturing the neural correlates of *the report* of consciousness instead of *consciousness itself* (Tsuchiya et al., 2015). To address this potential issue, critics have suggested using no-report paradigms where subjects' conscious status can be inferred by some indirect means other than direct subjective reports.

In this section, we discuss—and reject—the use of objective measures. Instead, we argue that objective and subjective measures can come apart: a subject may report being subjectively unaware of a stimulus, and yet their behavior demonstrates that they are objectively able to detect or discriminate it (and vice versa). In the next section, we address the neuroscientific objections against subjective reports and argue that task performance is a confound in neuroimaging studies of consciousness.

10.2.1 Objective Measures

To assess the objective performance of a subject during a visual task, one simple method is to compute the percentage of their correct responses. But percentage correct estimates do not disentangle perceptual sensitivity from response bias. A more sophisticated method is estimating subjects' d', which is a measure of perceptual sensitivity that stems from SDT (Green & Swets, 1966; Macmillan & Creelman, 2005). Importantly, with d', one can estimate subjects' objective perceptual sensitivity (i.e., their perceptual signal-to-noise ratio; e.g., their ability to discern whether a line is tilted left or right) *independently* from their response bias (e.g., their overall propensity for reporting "left tilt" or "right tilt"). According to proponents of objective measures of consciousness, subjects' awareness of a stimulus can be equated with their perceptual sensitivity. Thus, if subjects do not perform a perceptual task above chance levels (i.e., $d'=0$), one could assume that they did not see the stimuli consciously (Holender, 1986; Kouider et al., 2007). And a case for the opposite direction could be made too: if a subject demonstrates $d'>0$, one should assume, at least prima facie, that they saw the stimuli consciously (Phillips, 2020).

Unfortunately, the use of objective measures ignores a fundamental aspect of consciousness—in fact, it ignores what makes it an interesting phenomenon in the first place: its subjective character. In normal scenarios, perceptual sensitivity *may* track consciousness. For example, objectively discriminating branches from a clear road might coincide with the subjective report of experiencing a branch and the subjective report of experiencing no debris, respectively. However, as we show below, objective and subjective measures can dissociate: one can perceptually discriminate stimuli without awareness, and one can enjoy conscious experiences without any perceptual sensitivity. During illusions or hallucinations, conscious experiences do not entail perceptual discrimination above chance—during a hallucination there is nothing to discriminate! Alternatively, above-chance discrimination does not entail consciousness. For instance, artificial systems can make successful discriminations of visual stimuli, but with the

292   J. Morales, B. Odegaard, and B. Maniscalco

current state of technology, it is unlikely they are conscious (Dehaene, Lau, & Kouider, 2017). Moreover, blindsight patients deny being conscious of perfectly visible stimuli presented in a blind region of their visual field, and yet they are able to detect or discriminate these otherwise invisible stimuli significantly above chance (but see Phillips, 2020, for a recent challenge to this traditional interpretation of blindsight). If we made d' the measure of awareness, we would need to reject patients' subjective reports. Rather than ignoring subjective reports, we should value them as an important window to awareness, which is distinct and dissociable from objective performance.

10.2.2 Subjective Measures

Subjective reports can be obtained using a wide variety of procedures, such as reports of awareness (e.g., "seen" vs. "not seen" or "seen" vs. "guess," as in, e.g., Lau & Passingham, 2006), reports on the visibility of the stimulus (e.g., from "clearly visible" to "not visible," as in, e.g., Sergent & Dehaene, 2004), the method of adjustment or comparative judgments between two stimuli, which allows estimation of the point of subjective equality (e.g., "this stimulus is more visible than this other one," as in, e.g., Knotts, Lau, & Peters, 2018), reports of awareness using the Perceptual Awareness Scale (PAS; where 0 = no awareness, 1 = brief glimpse, 2 = almost clear awareness, 3 = clear awareness; Ramsøy & Overgaard, 2004), confidence ratings (e.g., 1 = not confident, 2 = barely confident, 3 = somewhat confident, 4 = very confident, as in, e.g., Maniscalco & Lau, 2012), or post-decision wagering (e.g., high vs. low wager of points or money, as in, e.g., Persaud, McLeod, & Cowey, 2007).

Although there are important differences among these subjective methods, they all aim to probe the qualities of subjects' conscious experiences. The first four methods require subjects to introspect and report on the nature of their experiences. Even though confidence ratings are more indirect, they are very commonly used in consciousness research. When asked to provide confidence ratings, subjects are asked about their *subjective* impression regarding their *objective* performance in the task. Despite being less direct, confidence ratings can provide similar insights into a subject's conscious experience as those given by direct introspective reports, while also potentially offering some advantages (Morales & Lau, In Press; but see Rosenthal, 2019). Empirically, confidence ratings often correlate with reports of subjective awareness (Michel, 2019; Peters & Lau, 2015; Sandberg et al., 2010). This empirical correlation reflects the fact that one's confidence in a visual task is largely shaped by one's phenomenology. If one sees clearly what is on the screen, in general, one should be more confident that one responded correctly about the stimulus presence/identity. Alternatively, if one is not clearly aware of the stimulus, one should be less confident in the correctness of their response—it should feel more like guessing (cf. Rausch & Zehetleitner, 2016). One potential advantage of confidence ratings is that it might be easier for subjects to understand what is being asked of them when providing confidence ratings than when they are asked to introspect about the nature of their subjective experience. A second advantage is that confidence ratings are more interpretable than awareness reports for assessing subjects' metacognitive capacity, which itself can potentially offer a meaningful window into subjective conscious states.

Metacognition is the capacity to monitor and evaluate one's own cognitive processes (Flavell, 1979; Fleming, Dolan, & Frith, 2012; Proust, 2013). Confidence ratings can be viewed as metacognitive judgments about the likelihood that a given response in a task is correct. As a consequence, it is possible to compute "objective" measures of metacognitive performance from subjective confidence ratings by quantifying how well confidence correlates with accuracy. In particular, SDT analyses can provide a measure of metacognitive sensitivity free from response bias analogous to d', termed "meta-d'" (Maniscalco & Lau, 2012). This "objective" measure thus offers the tantalizing potential for having the best of both worlds when studying awareness: taking subjective report seriously (like subjective measures), while sidestepping thorny issues of response bias (like objective measures; Kunimoto, Miller, & Pashler, 2001). However, it is possible for blindsight patients to have above-chance metacognitive performance in their blind field (Persaud et al., 2011), and conceptually it is possible to have chance-level metacognition about phenomenological experiences (e.g., due to hallucination; Shaver, Maniscalco, & Lau, 2008), suggesting that the presence or absence of metacognitive sensitivity cannot be taken as a hard and fast indicator of the presence or absence of phenomenology (Maniscalco & Lau, 2012). Nonetheless, measures of metacognitive sensitivity may have heuristic value in assessing levels of stimulus awareness, as presumably one's metacognitive sensitivity would tend to dwindle with reductions in phenomenological stimulus awareness. For instance, Persaud and colleagues (2011) showed that although their blindsight patient had above-chance metacognitive performance in the blind field, this was still lower than metacognitive performance in the normally sighted field, despite the fact that visual task performance in the two fields was matched by appropriately adjusting stimulus contrast.

10.2.3 Objective and Subjective Measures Can Dissociate

The idea that subjective and objective measures of consciousness can dissociate, and that their dissociation represents a unique opportunity to isolate the neural basis of conscious awareness, is not new. More than twenty-five years ago, Weiskrantz, Barbur, and Sahraie (1995) suggested that blindsight patients offer a stunning demonstration of how subjectivity and objectivity differ (Lau, 2008). Blindsight occurs when patients have damage to the primary visual cortex (V1). These patients can perform many perceptual tasks at above-chance levels and yet report no phenomenological experience associated with this ability. In some patients, performance in the blind part of the visual field can be as high as that of the unimpaired field, and phenomenological experience can be found in one but not the other. Thus, blindsight patients provide a critical proof of principle in demonstrating how subjective and objective measures can dissociate within a single individual.

It is beyond the scope of this chapter to review the literature exhaustively. Objective and subjective measures can dissociate in healthy and atypical populations in many ways (Lau, 2008; Maniscalco, Castaneda, et al., 2020; Phillips, 2020), but the examples of matched performance/different awareness findings discussed below constitute one salient subset of such dissociations. Importantly, the dissociability of objective and subjective measures entails not only that objective measures may be unreliable indicators of consciousness, but also that differences in objective performance associated with differences in awareness can pose as confounds that must be controlled for in isolating the cognitive and neural properties of consciousness.

10.3 The Importance of Matching Task Performance

An important challenge faced when trying to isolate the neural bases of consciousness is the need to distinguish, on one hand, the neural substrates of consciousness proper and, on the other hand, the pre and post processing that enables and follows conscious experiences, respectively (Aru et al., 2012). Equally important is to distinguish the processing that occurs concurrently with conscious processes, but that is ultimately irrelevant for

supporting them. As indicated above, the proper neural substrates of consciousness are only those that are jointly minimally sufficient for sustaining a conscious experience with a given content (Chalmers, 2000; see also
Shoemaker, 1981). There is, however, a multitude of pre, concurrent, and
post processes that are not necessary or sufficient for sustaining conscious
experiences. Some of these might be necessary for perceptually processing
the stimulus (albeit unconsciously). Perhaps they are even necessary for
giving rise to the neural events that are in fact the basis of consciousness
without themselves being a neural correlate of consciousness. Crucially,
these irrelevant processes need to be eliminated or matched across conscious and unconscious conditions. In this section, we first offer a general
characterization of these processes that are unrelated to consciousness, and
then we focus on performance as the most salient example.

10.3.1   Preconditions, Concurrent Processing, and Post-Processing Effects

Consider comparing the neural activity of someone with their eyes open
and then closed. They are more likely to see a stimulus consciously when
their eyes are open than when they are closed. However, comparing their
neural states in these two conditions would hardly reveal the neural correlates of consciousness: so many other things are different! This extreme case
illustrates what happens in more subtle scenarios where there are differences
in pre, concurrent, and post processing. For instance, consider the general
excitability of neuronal populations. Oscillating pre-stimulus brain activity
can reliably predict whether a subsequent stimulus is perceived (Benwell
et al., 2017; Mathewson et al., 2009; Samaha, Iemi, & Postle, 2017). When
contrasting conscious and unconscious trials, these differences in neural
activity are likely to be reflected in neuroimaging data (specifically, in the
phase of pre-stimulus alpha oscillations obtained by electroencephalography). However, these enabling pre-stimulus oscillations are not the neural
substrate of consciousness.

Consider now post processing. Consciously experiencing a stimulus is likely
to have ripple effects in subsequent neural processing that are either lacking or
reduced during unconscious perception (Dennett, 1991, 2018). Some of these
might be cognitive consequences that are not associated with consciousness
at all (Block, 2019; Phillips & Morales, 2020). For example, sustained maintenance of information in working memory, access to long-term memory, verbal

reports, or intentional behavior are examples of post-perceptual processing that could be more markedly revealed in neural activity during conscious trials compared to unconscious trials. This post-processing neural activity, however, is not the neural substrate of consciousness proper, as it only happens *after* consciousness has already started taking place.

This concern has led some researchers to argue that we need to eliminate any kind of cognitive difference between conscious and unconscious conditions with what has been recently labeled "no-cognition" paradigms (Block, 2019). But cognition cannot be perfectly matched (Phillips & Morales, 2020), so some researchers argue we should at least try to eliminate subjective reports from consciousness research (Tsuchiya et al., 2015). The worry about asking for subjective reports is that requiring them might contaminate results with post-processing neural activity associated with access and report itself but not consciousness. However, we can assuage these specific worries about subjective reports (and cognition more generally) by highlighting an important constraint: processing unrelated to consciousness is most problematic when it is not matched across conditions. As long as subjects have similar cognitive and reporting requirements across conscious and unconscious trials, subjective reports need not be a confound (Michel & Morales, 2020).

Concurrent processing of the stimulus (e.g., perceptual processing independent from consciousness such as distinguishing signal from noise, feature extraction, categorization, etc.), which is fundamental for performing the task successfully, takes place alongside processes supporting consciousness. But those perceptual processes are not part of the neural basis of consciousness, since, presumably, these are perceptual processes that are also present during unconscious perception.

One might wonder whether there is any neural activity left. One important lesson from thinking about the importance of matching background conditions and cognitive processes across conditions of awareness is that the neural activity that supports consciousness may indeed be quite subtle. For instance, it might only be detectable with highly sensitive neuroimaging methods such as single-cell recording, sophisticated statistical methods such as multivariate (rather than univariate) analyses, and in localized rather than brain-wide activity (Morales & Lau, 2020). So, when pre, concurrent, and post processes are not matched across conditions, experimenters risk conflating them with the neural substrates of consciousness proper. Unfortunately, while these differences might be conceptually clear, in practice, it can be challenging to distill all these types of neural activity (Giles, Lau, & Odegaard, 2016). Part of the difficulty is that there is no clear temporal differentiation between relevant and irrelevant types of neural activity for consciousness. Activity related to preconditions could continue after stimulus presentation when neural activity related to consciousness begins. Similarly, the consequences of conscious awareness could begin to manifest while subjects are still aware of the stimulus, effectively creating temporally overlapping neural activity pertaining to distinct processes. Naturally, concurrent processes are especially hard to disentangle from consciousness-related processes. Moreover, nothing we know about neurobiology rules out a priori that pre, concurrent, and post processing recruit at least a subset of the same neuronal populations recruited by consciousness processes.

An effective way to eliminate, or at least reduce, these confounds is to match the testing conditions across conscious and unconscious trials. As long as the preconditions, concurrent processes, and post-processing effects of consciousness are sufficiently similar across conscious and unconscious trials, one may not need to worry about distilling them from the neural data pertaining to consciousness proper. Presumably, perceptual and cognitive processes that are similar across conscious and unconscious trials will be subserved by similar neural activity, and such neural activity will be canceled out when computing the difference in neural activity between conscious and unconscious trials. Some of the dimensions along which tasks are often matched include type, duration and strength of stimulation, response demands (e.g., sensorimotor and cognitive requirements for report), and cognitive demands (e.g., attention, working memory load, task difficulty, cognitive control, etc.). However, an important, yet often neglected, dimension that experimenters should aim to match across conscious and unconscious trials is task performance.

10.3.2 Performance Matching Is Key

Matching subjects' performance in conscious and unconscious trials ensures that concurrent perceptual signal processing is comparable. This is important both in itself and because it helps matching other types of processing. For instance, similarity in perceptual processing increases the odds that pre- and post-processing neural activity is comparable. If one wants to find the neural basis of consciousness proper and distinguish it from the objective capacity to perceive a stimulus, performance matching is required. But it is also important because it correlates with other cognitive capacities. Whereas task performance can be straightforwardly computed (e.g., percentage of correct trials or d'), it is hard to objectively quantify cognitive processes such as cognitive effort, working memory load (beyond number of items to be reported), and so on. But matching these cognitive demands is important. By making sure that task performance is the same across conditions, we increase the chances that cognitive effort, working memory load, and other cognitive demands are similar as well.

While matching performance is highly desirable in theory, it is hard to achieve in practice, and it is in all likelihood impossible to achieve without creating differences somewhere else (see section 10.6). To make someone unaware of an otherwise visible stimulus, some change in the testing conditions needs to take place (Kim & Blake, 2005). These changes can be applied to the stimulus (e.g., decreasing stimulus strength or duration, adding a mask or changing the mask's duration), to the task (e.g., increasing task difficulty), or to participants themselves (e.g., distracting participants' attention or altering their brain states directly via TMS).

It is important to emphasize that the goal of performance matching is to match perceptual signal processing—in other words, perceivers' capacity to process the perceptual signal triggered in their visual system such that it can disentangle signal from noise and eventually create a perceptual representation of the stimulus. To illustrate this point, consider the following case. Imagine an experiment where subjects detect stimuli correctly more frequently when they are conscious of them than when they are not—that is, an experiment where performance, and hence perceptual signal processing, is not matched across conscious and unconscious trials. To fix this, one could try to match for performance artificially a posteriori by only analyzing the neural data of correct trials, leaving out incorrect trials. This way, performance in the selected trials would be, by necessity, matched at 100% in both cases. But this artificial correction would not match perceptual signal processing capacity and its supporting brain states across different awareness conditions. For instance, some correct detections of the stimulus in unaware trials could occur due to lucky guesses rather than to perceptual processing. One could attempt more sophisticated corrections to guesses in unaware trials by taking into account subjects' guessing rate (Lamy, Salti, & Bar-Haim, 2009). But this approach is insufficient for matching the underlying perceptual capacity and the corresponding neural activity that drives correct trials in aware and unaware conditions (Morales, Chiang, & Lau, 2015). Thus, artificially matching performance by post hoc selection of subsets of trials should be avoided because such post hoc procedures fail to achieve the intended effect whereby conditions exhibiting similar task performance do so by virtue of having similar underlying perceptual processing. By contrast, if experimental manipulations yield matched overall performance across all trials in two experimental conditions, we can have higher confidence that the matched performance indicates similarity in the underlying perceptual processes responsible for yielding that level of performance.

10.3.3 Performance Matching Reveals Neural Correlates of Consciousness in the Prefrontal Cortex, in Agreement with Higher-Order Theories

One seminal demonstration of performance matching and its importance for revealing the neural correlates of awareness comes in a metacontrast masking study by Lau and Passingham (2006). In their behavioral experiment, subjects were presented with a brief visual target and were required to discriminate its identity (either diamond or square) and indicate whether they consciously saw the target. Critically, a metacontrast mask was presented with varying stimulus onset asynchrony (SOA) after the visual target. Behavioral results showed that two distinct SOAs yielded similar levels of performance on the discrimination task, but different levels of awareness (the percentage of trials subjects reported seeing the stimulus). Functional magnetic resonance imaging revealed that while activations in many cortical areas distinguished performance levels in general (i.e., correct vs. incorrect trials), only dorsolateral prefrontal cortex (dlPFC) activity reflected differences in two SOA conditions with matched performance and different awareness.

Maniscalco and Lau (2016) replicated the behavioral effect and conducted a model comparison analysis to test the ability of various candidate theories to capture the data. They found that the data were best captured by models embodying principles of higher-order theories of consciousness (Brown, Lau, & LeDoux, 2019; Lau & Rosenthal, 2011), in which task performance is determined by first-order processing and conscious awareness is determined by subsequent higher-order processing that evaluates first-order processing. Lau and Passingham's finding that performance-matched differences in awareness are associated with activity in dlPFC but not sensory cortices can be well accommodated by higher-order theory. This is so given the broad observation that various forms of first-order processing tend to occur in posterior sensory cortices, whereas higher-order processing is more localized to the PFC (Brown et al., 2019).

The special role of the PFC in supporting subjective awareness, independently of objective task performance, is supported by a number of other studies. Disruption of dlPFC function by TMS (Rounis et al., 2010; Ruby, Maniscalco, & Peters, 2018) or concurrent task demands (Maniscalco & Lau, 2015) selectively impairs metacognitive sensitivity but not objective performance in perceptual tasks. Patients with anterior PFC lesions exhibit selective impairment of metacognitive sensitivity on a perceptual task relative to temporal lobe patients and healthy controls, even when task performance is matched across groups (Fleming et al., 2014). In a blindsight patient, frontoparietal areas in the brain are more activated for stimulus perception in the healthy visual field than in the blind visual field, even when task performance across the fields is equated (Persaud et al., 2011). Metacognitive sensitivity and task performance dissociate over time as one continuously performs a demanding task without rest, and this dissociation can be accounted for by individual differences in gray matter volume in the anterior PFC (Maniscalco, McCurdy, et al., 2017). Higher pre-stimulus activity in the dorsal attention network is associated with lower confidence ratings but not altered task accuracy (Rahnev, Bahdo, et al., 2012). Further examples of matched performance with different awareness are discussed in the next section.

10.4   Understanding and Designing Matched Performance/Different Awareness Stimuli with Signal Detection Theory

SDT is a simple yet powerful framework for understanding how observers make perceptual decisions in the presence of noisy sensory evidence (Green & Swets, 1966; Macmillan & Creelman, 2005). In the paradigmatic case, an observer must decide whether a noisy perceptual sample e was generated by stimulus class S1 (e.g., "stimulus absent" or "grating tilting left") or S2 (e.g., "stimulus present" or "grating tilting right"), where e can be represented by a real number (which could correspond to magnitudes such as strength of sensory evidence or neural firing rate). SDT assumes that across repeated presentations, the same stimulus can yield different values of e, for example due to random noise in the sensory environment or internal neural processing, such that across many presentations, the distribution of *e* is Gaussian. The dimension along which these values of *e* are distributed is called the decision axis.

Different stimulus classes generate distributions of *e* with different means. For instance, in a stimulus detection task, *e* might represent the strength of the sensory evidence in favor of saying the stimulus was present rather than absent on a given trial, and thus the typical value of *e* would tend to be higher for stimulus-present trials than for stimulus-absent trials. Figure 10.1A depicts a scenario like this, where, for example, the distribution of sensory evidence occurring on stimulus-absent versus stimulus-present trials would correspond to the left (lower mean) and right (higher mean) Gaussian distributions, respectively. The greater the distance between the means of these distributions relative to their standard deviation, the less the distributions overlap, and so the easier it is for the observer to discriminate S1 from S2. This signal-to-noise ratio for the two distributions is the SDT measure of task performance, d'.

Because the Gaussian distributions always have some degree of overlap, in principle it is always possible for a given value of *e* to have been generated by S1 or S2. Thus, the observer is faced with an inherent ambiguity in deciding how to classify *e*. According to SDT, the observer classifies *e* as S2 if it exceeds a response criterion *c* (solid vertical line in figure 10.1A) and as S1 otherwise. Whereas d' reflects aspects of stimulus properties and perceptual processing that are typically beyond the observer's control, *c* reflects a response strategy determined by the observer's preferences and goals. The observer may strategically set their response criterion so as to accomplish a certain objective, for example maximizing the probability of responding correctly or maximizing rewards contingent on trial outcomes. For instance, in the common case where S1 and S2 occur with equal frequency across trials and have the same standard deviation, the observer can maximize their probability of responding correctly by placing their criterion where the two distributions intersect, that is, where *e* is equally likely to have been generated by S1 or S2. Ratings of awareness or confidence can also be characterized as resulting from a criterion-setting process (dashed lines in figure 10.1A), as discussed further below.

A general principle that has been employed to both explain and generate matched performance with different awareness data is that task performance depends on the signal-to-noise ratio of the Gaussian distributions (d'), whereas awareness often depends more so on absolute levels of

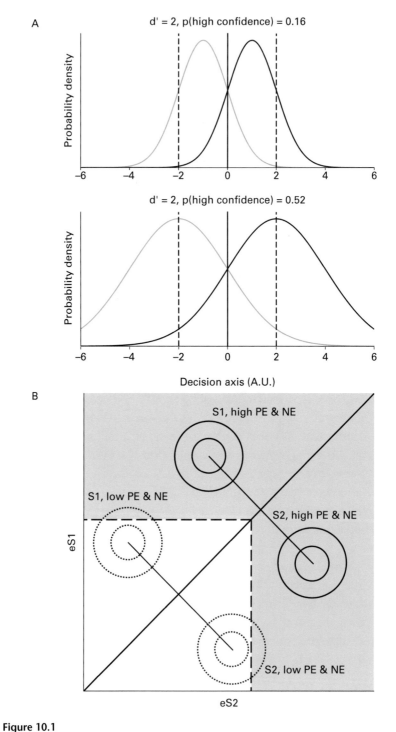

Figure 10.1
Explaining matched performance/different awareness with one- and two-dimensional signal detection theory. (A) In the one-dimensional case, consider a scenario where a subject needs to decide whether a given evidence sample on the decision axis

perceptual evidence (i.e., the magnitude of e). For instance, consider an SDT model of an experiment where an observer must discriminate stimulus classes S1 and S2 in two experimental conditions, A and B (figure 10.1A). Suppose that in condition A (figure 10.1A, top panel), the S1 and S2 distributions have means at the decision axis values −1 and +1, respectively, and standard deviations of 1. The subject responds "S2" if the evidence value e on the current trial exceeds 0, and endorses classification responses with high confidence if $e<-2$ or $e>2$ (corresponding to strong evidence for S1 or S2, respectively). Now suppose that in condition B (figure 10.1A, bottom panel), the S1 and S2 distributions have means of −2 and +2, and standard deviations of 2, but that the subject's decision rules for classifying and rating confidence remain the same. Conditions A and B then have identical task performance due to having identical signal-to-noise ratio. In both cases, the means of the evidence distributions are two standard deviations apart (i.e., $d'=2$), meaning it is equally difficult to infer whether a given perceptual sample originated from S1 or S2. However, confidence is higher in condition B, since in this case the absolute levels of perceptual evidence

Figure 10.1 (continued)

was drawn from class S1 (gray distribution) or S2 (black) and rate confidence. The discrimination judgment depends on whether the given sample drawn is above or below a decision criterion, which in this example is set at 0 (solid line). If the sample is below 0, the subject selects S1, and if the sample is above 0, the subject selects S2. The rating of low or high confidence depends on where the sample falls with respect to the confidence criteria (dashed lines). In this example, samples greater than 2 or less than −2 yield high confidence ratings, while samples within this range yield low confidence. It follows that in the bottom panel, average confidence is higher due to higher evidence variance, in spite of task performance (signal-to-noise ratio, d') being the same as in the top panel. (B) In the two-dimensional case, the two axes represent evidence for each stimulus class (eS1 and eS2). Circles represent bivariate normal distributions, and samples drawn from these distributions thereby contain evidence for both eS1 and eS2. Thus, the means of these distributions represent various positive evidence (PE)/negative evidence (NE) levels. Discriminating whether the stimulus is S1 or S2 involves evaluating whether the sample falls above or below the solid diagonal line. Confidence, however, involves evaluating the magnitude of the response-congruent evidence, which is shown by the confidence criteria (dashed lines) separating the white and gray regions. In this example, the high PE/NE stimuli (solid circles) have higher confidence than the low PE/NE stimuli (dotted circles) due to having more probability mass exceeding the confidence criteria, in spite of task performance (distance between the means of the distributions divided by standard deviation, d') being the same.

are more extreme and therefore more frequently exceed the criteria for high confidence. In this way, provided that the subject uses the same decision strategy across conditions,[2] higher absolute levels of evidence will cause higher confidence, even for matched signal-to-noise ratios.

This model has successfully explained performance-matched differences in awareness due to attentional manipulations (Rahnev, Maniscalco, et al., 2011), as well as simultaneous decreases in task performance and increases in confidence caused by TMS (Rahnev, Maniscalco, et al., 2012). It has also generated novel matched performance/different awareness findings by informing the experimental design of stimuli. For instance, experiments 1B and 2B of Koizumi, Maniscalco, and Lau (2015) used random dot motion stimuli in which a subset of dots moved coherently left or right, and the rest moved randomly. Across conditions, the fraction of coherently moving dots (i.e., signal-to-noise ratio) was the same, but the overall number of dots (i.e., absolute levels of perceptual evidence) differed. As expected, confidence was higher for stimuli with higher dot density, even though task performance was the same. Maniscalco, Castaneda, and colleagues (2020) replicated and expanded upon this result by constructing entire type 2 psychometric curves of confidence versus d' across a full spectrum of d' values (ranging from near chance to near ceiling) for several levels of dot density. They found that density logarithmically modulates confidence across a wide range of task performance. Importantly, the effect was stronger when density levels were randomly interleaved across trials rather than organized into predictable blocks, lending further support to the notion that randomly interleaving experimental conditions help ensure that subjects use a consistent decision strategy across conditions, thus allowing the matched-performance, different-awareness effect to emerge. An approach similar to that of Koizumi and colleagues (2015) and Maniscalco, Castaneda, and colleagues (2020), but manipulating the signal-to-noise ratio and absolute evidence levels of noisy gratings rather than dot motion, was employed by Samaha and colleagues (2016) and Samaha, Switzky, and Postle (2019). Across experimental conditions, the ratio of grating contrast to noise contrast was identical, but overall contrast of the composite stimulus differed, yielding higher confidence in the higher contrast stimuli despite equivalent performance.

A similar principle and accompanying method of stimulus construction comes from findings that confidence follows a response-congruent evidence rule. That is, confidence depends heavily on evidence congruent with the perceptual decision while downweighting or ignoring evidence that contradicts the perceptual decision (Zylberberg, Barttfeld, & Sigman, 2012; Peters et al., 2017). Exploiting this finding, experiments 1A and 2A of Koizumi and colleagues (2015) used stimuli with different levels of positive evidence (PE) and negative evidence (NE), where PE is evidence supporting the correct perceptual decision and NE is evidence supporting the incorrect decision. Specifically, they used oriented gratings embedded in noise, where a higher-contrast grating (PE) tilted left or right was superimposed with a lower-contrast grating (NE) tilted in the opposite direction, and the correct tilt response corresponded to the higher-contrast grating. By manipulating the contrasts of PE, NE, and noise, they created conditions where performance was similar but PE and NE levels differed. Crucially, since confidence depends on response-congruent evidence, confidence was higher in the conditions with higher PE and NE levels.

An illustration of the logic of capitalizing on the response-congruent evidence rule to create matched performance/different awareness stimuli by manipulating PE and NE levels is shown in figure 10.1B. Following Maniscalco, Peters, and Lau (2016), we use a two-dimensional SDT representation in which the two axes, eS1 and eS2, correspond to evidence for the two stimulus classes S1 and S2. Generalizing from the one-dimensional case (figure 10.1A), we assume that each stimulus class generates a bivariate normal distribution, such that the perceptual evidence elicited by a stimulus on a given trial is a random draw of an (eS1, eS2) pair from the corresponding stimulus distribution. Circles in the plot represent contours of the distributions as three-dimensional hills seen from above, similar to a topographic map. The mean of the distributions corresponds to PE and NE levels. For instance, an S1 stimulus with high PE and intermediate NE will have a high mean value along the eS1 dimension and an intermediate mean value along the eS2 dimension. Given evidence (eS1, eS2) on a given trial, the subject responds "S2" if eS2>eS1 (region of the plot below the solid diagonal line eS1=eS2) and "S1" otherwise. Crucially, the subject rates confidence by comparing the magnitude of response-congruent evidence to a criterion value (corresponding to the dashed horizontal and vertical lines), yielding high confidence for evidence pairs located in the shaded region of the plot.

In figure 10.1B, we show stimulus distributions for two experimental conditions: one with low PE and NE, and one with high PE and NE. Task performance (d') is determined by the distance between the distributions along the line connecting their means divided by their common standard deviation. Thus, the low and high PE/NE conditions shown here have matched levels of d'. However, a greater proportion of the high PE/NE distributions lies within the shaded region of the plot than the low PE/NE distributions, thus yielding higher confidence. Note that this arrangement depends on the response-congruent evidence rule in order to yield differences in confidence: if confidence depended on the magnitude of the difference in evidence eS2–eS1, then the dashed confidence criterion lines would be 45° (parallel to the solid perceptual decision criterion), and the proportion of the distributions lying in the shaded (high confidence) regions would be equivalent for the high and low PE/NE stimuli.

Notably, the PE/NE method of creating stimuli yielding matched performance and different awareness has the advantage that it allows for overall stimulus energy (e.g., contrast or dot density) to be matched across conditions, since increases in PE and NE energy can be offset by decreases in noise energy (Koizumi et al., 2015). By contrast, the signal/noise method requires there to be higher overall stimulus energy in the condition with higher awareness, thus posing an undesirable confound. On the other hand, PE/NE manipulations potentially induce response conflict in a way that signal/noise manipulations do not (by virtue of PE and NE priming opposing perceptual decisions/responses), which can also be undesirable.

Note that the models discussed in this section are not meant to be exhaustive explanations for all cases. It is possible that other kinds of computational processes can also produce matched performance/different awareness data, such as the higher-order model of Maniscalco and Lau (2016) mentioned previously. Nonetheless, the methods discussed in this section are powerful insofar as they not only provide potential post hoc explanations, but actually enable us to *design* stimuli that yield matched performance and different awareness using well-understood computational principles.

10.5   Theoretical Caveats and Nuances

We can summarize the logic of performance matching as follows: to isolate subjective awareness of a stimulus precisely from confounding factors, we should conduct experiments that satisfy the following criteria:

1. **Dissociable processing identified:** We have some notion of which sensory and perceptual processing of the stimulus is dissociable from awareness and thus needs to be controlled for when experimentally isolating awareness.
2. **Dissociable processes matched:** We empirically confirm that the dissociable processing identified in (1) is matched across experimental conditions by demonstrating equal performance on a task that probes such processing.
3. **Awareness differs:** Average subjective awareness of the stimulus differs across conditions.

Here, we highlight some nuances and potential difficulties in each of these criteria that should inform the way we conduct and interpret performance-matching studies and the study of subjective awareness more broadly. In brief, the nuances explored for each criterion are:

1. **Uncertainty about dissociable processing:** There is some uncertainty about which perceptual processing is dissociable from awareness and which is not.
2. **Multidimensionality of dissociable processing:** There are potentially many dimensions of stimulus processing that are dissociable from awareness other than the task probed in the experiment.
3. **Absolute versus relative levels of awareness:** Interpreting a difference in awareness requires considering not just the relative difference in reported awareness across conditions, but also the absolute level of awareness within each condition.

10.5.1 Uncertainty about Dissociable Processing

In order to argue that we should control for some aspect of perceptual processing P when studying awareness, we must have some prior reason for thinking that P is dissociable from awareness to begin with. For instance, based on data from blindsight patients, we have strong reason to believe that forced-choice discrimination of simple stimulus features can proceed without awareness (Weiskrantz, 1986; but see Phillips, 2020). There is also evidence for above-chance forced-choice stimulus discrimination without awareness in healthy observers (e.g., Kouider, Dehaene, et al., 2007; Merikle, Smilek, & Eastwood, 2001; Snodgrass, Bernat, & Shevrin, 2004), although such findings are more contentious (Eriksen, 1960; Hannula, Simons, & Cohen, 2005; Lloyd, Abraha-myan, & Harris, 2013). And of course, as reviewed above, there is ample evi-dence that awareness can differ across conditions with matched forced-choice discrimination performance.

Stances on what aspects of stimulus processing are dissociable from awareness versus which are inseparable from or deeply intertwined with it are influenced not just by evidence but also by theory. For instance, some theoretical frameworks—such as higher-order theories (Brown et al., 2019; Lau & Rosenthal, 2011; Rosenthal, 2005) and some interpretations or imple-mentations of SDT (Maniscalco & Lau, 2012)—lend themselves naturally to viewing task performance and subjective reports of awareness as strongly dissociable, whereas other frameworks posit a tighter relationship in which cleanly separating task performance and awareness might not always be so straightforward. For instance, in global workspace theory (e.g., Baars, 1997; Dehaene, 2014), a content becomes conscious by virtue of entering a global workspace, but also enjoys enhanced processing by virtue of being in the workspace, such that the enhanced processing of the content may not be completely separable from awareness of the content per se.

Importantly, these theoretical orientations affect not just predictions about what sorts of stimulus processing should be dissociable from aware-ness, but also interpretation of extant demonstrations of such dissociations. For instance, for a higher-order theorist, matched performance dissocia-tions are straightforward demonstrations of the theoretically expected sepa-rability of task performance and awareness. By contrast, a global workspace theorist might hold that even though feature discrimination and stimulus awareness are partially dissociable, awareness of a stimulus nonetheless plays some direct participatory role in the full-blown kind of feature dis-crimination present in conditions of full stimulus awareness. (An instance of such a view is the model of Del Cul et al., 2009.) For such a theorist, matching discrimination performance when studying awareness might eliminate too much, removing the confounds of non-conscious contribu-tions to feature discrimination while also masking the contributions of con-sciousness itself, leaving only some minimal difference in awareness that happens to be insufficient to manifest as a difference in task performance.

More generally, to whatever extent awareness directly participates in some aspect of perceptual processing P, that aspect of awareness must necessarily be masked by experimental procedures that match P across conditions. Demonstrating that awareness can differ to some extent when *P* is matched does not necessarily entail that awareness plays no part in *P* whatsoever. Rather, it only conclusively demonstrates that in some conditions, it is possible for the observed difference in awareness to fail to manifest as a difference in *P*.

There is thus a kind of circularity that poses some difficulty for different theoretical camps to agree on basic aspects of methodology in consciousness science: our theories should be constrained by empirical results, but the interpretation of those results and how they should refine our theories is itself theory dependent. Continued advances in empirical findings and theoretical developments will presumably lead to increasing convergence on both theory and methodology, but achieving such convergence is nontrivial in the face of these issues.

10.5.2   Multidimensionality of Dissociable Processing

To this point, we have focused the discussion on matching task performance for the task being probed in the experimental design. For instance, if the task requires the subject to discriminate left versus right grating tilt and then report awareness, we would recommend studying awareness by comparing two experimental conditions where tilt discrimination performance is equal and yet average subjective report differs. However, it is of course the case that the subject performs many other perceptual operations ("tasks") that are not directly probed by such a design, for example detecting the presence of the grating, identifying the detected stimulus *as* an oriented grating, discerning the exact degree of its tilt (as opposed to making a binary left/right classification), and so on.

We can therefore draw a distinction between *probed* task performance (performance on the task explicitly measured in the experiment, e.g., tilt discrimination) and *latent* task performance (performance on perceptual "tasks" that were not explicitly probed in the experiment but could have been, e.g., stimulus detection, object identification, etc.). The question then becomes whether matching probed task performance is sufficient for matching latent task performance. Presumably, as general quality of stimulus processing improves (e.g., due to stronger stimulus drive, improved attention, etc.), different dimensions of perceptual processing (detection, feature discrimination, identification, etc.) will all improve as well. The existence of such a correlation in perceptual performance across different dimensions of stimulus processing helps address concerns about possible confounds in latent task performance when probed task performance is matched. Yet, there is no general guarantee that matching performance on the probed task entails matching performance on all latent tasks. For instance, it can be readily demonstrated with an SDT model that identical levels of performance for discriminating between stimuli A and B are compatible with different levels of performance for detecting A and B. For instance, by increasing the means and variances of the evidence distributions appropriately, d' for the discrimination task can remain unchanged (as in figure 10.1A), but such increases in mean and variance will yield altered detection performance with respect to a stimulus-absent noise distribution with fixed mean and variance.

Of course, it is impossible in practice to probe all relevant kinds of perceptual processing in a single experiment, and so a pragmatic approach is just to match performance on a representative task (such as feature discrimination) and assume that this does an acceptable job of matching latent task performance. However, it is worth keeping in mind that the same logic that would lead us to worry about matching probed task performance should also lead us to worry about matching latent task performance. If latent task performance is not matched, then between-condition differences in behavior or neural activity could potentially be attributed to the difference in a "latent task" rather than the difference in awareness per se. Additionally, it is possible that in some situations we might have reasons to believe that some aspect of latent performance is *not* matched, despite matched performance on the probed task, and such situations would require special care (e.g., caution in interpreting the results, or designing a new study that properly controls for the latent task performance in question).

10.5.3   Absolute Levels of Awareness

Not all differences in awareness are created equal. For instance, imagine an experiment where subjects use the PAS (Ramsøy & Overgaard, 2004), a standardized scale for rating visual awareness with four levels: 0 = no awareness, 1 = brief glimpse, 2 = almost clear awareness, and 3 = clear awareness. A performance-matched difference in PAS levels of 0 (no reported awareness whatsoever) and 1 (the first hints of entry of the stimulus into awareness) would then indicate something very different from a difference in PAS levels of 2 (almost clear awareness) and 3 (clear awareness). In turn, this would have consequences for interpreting the performance-matched difference in awareness in terms of cognitive functions or neural mechanisms. An experiment achieving a performance-matched difference of a PAS rating of 0 versus 1 would allow inferences about what cognitive functions and neural mechanisms correspond to the transition of a stimulus representation from complete unconsciousness to the first faint entries into conscious awareness.[3] By contrast, an experiment achieving a performance-matched difference of a PAS rating of 2 versus 3 would not allow inferences about the functions and mechanisms of a representation's being conscious as such, but rather would be limited to inferences about the cognitive functions and neural mechanisms corresponding to increases in the relative intensity or clarity of contents that are already conscious (Michel, 2019; Morales, 2021).

Studies on awareness are often centrally interested in the cognitive functions and neural mechanisms of a stimulus representation being conscious as such. In principle, the ideal way to approach this research question from a performance-matching perspective would be to achieve performance matching for a "completely unconscious" condition[4] (i.e., PAS=0) and a "somewhat conscious" condition (PAS>0). In practice, performance-matching studies to this point have typically compared conditions in which subjects report an intermediate level of stimulus awareness in both conditions (Koizumi et al., 2015; Lau & Passingham, 2006; Maniscalco & Lau, 2016; Samaha et al., 2016; Maniscalco, Castaneda, et al., 2020), making them ideally suited to investigating the relative degree of intensity or clarity of contents of awareness rather than awareness per se (Fazekas, Nemeth, et al., 2020).[5] Furthermore, it has been difficult to unambiguously demonstrate above-chance performance without awareness in healthy subjects (Eriksen, 1960; Hannula et al., 2005; Lloyd et al., 2013; Peters & Lau, 2015). Thus, using the performance-matching framework to study the cognitive functions and neural mechanisms of consciousness as such, as opposed to the functions and mechanisms of changes in intensity or clarity of contents that are already conscious, faces significant practical hurdles still in need of addressing in future work.

Another way in which the absolute levels of awareness in performance-matched conditions matter is in interpreting the potential role of awareness in supporting further cognitive functions. For instance, Koizumi and colleagues (2015) used specially designed grating stimuli to yield performance-matched differences in confidence for discriminating grating tilt. Confidence was rated on a scale of 1 to 4. Across two levels of d' for tilt discrimination, mean confidence for the low and high confidence stimuli was about (low = 2, high = 2.3) and (low = 2.3, high = 2.5), respectively. Koizumi and colleagues then used the tilt of the performance-matched stimuli as cues in go/no-go and task set preparation tasks to probe the role of performance-matched differences in confidence on cognitive control. They found that higher confidence did not confer an advantage in either cognitive control task. As with any null effect, this finding needs to be interpreted with caution. Failure to find an effect could be due to a true absence of an effect (figure 10.2A) or failure to detect a true but weak effect (figure 10.2B).

In addition to these possibilities, it is also possible that failure to find an effect reflects a ceiling effect (figure 10.2C) or floor effect (figure 10.2D) attributable to the absolute levels of confidence probed in this study. For instance, it is possible that increases in performance-matched confidence do increase cognitive control, but that this effect is most pronounced at lower levels of confidence (confidence <2) and is already saturated for the levels of confidence probed in Koizumi and colleagues (confidence >2; figure 10.2C). Alternatively, it could be that performance-matched increases in confidence only manifest as increases in cognitive control for higher absolute levels of confidence than were probed in Koizumi and colleagues (figure 10.2D).

10.6 Future Directions: Triangulating on Consciousness

We have argued that task performance is a serious yet underappreciated confound in the neuroscientific study of consciousness. Yet, even if we can find conditions yielding different levels of stimulus awareness while task performance is matched—and even if we satisfactorily address the caveats and nuances discussed in the previous section—the unfortunate fact remains that *some* confound in the comparison between the "more conscious" and "less conscious" conditions must be present. Namely, there must be some difference between the conditions that causes awareness to differ, whether it is a difference in stimulus properties, attention, brain stimulation, or some other factor.

In practice, stimulus confounds are the type of confound most likely to be salient for performance matching studies. The matched signal-to-noise ratio/different variance method for designing performance-matched

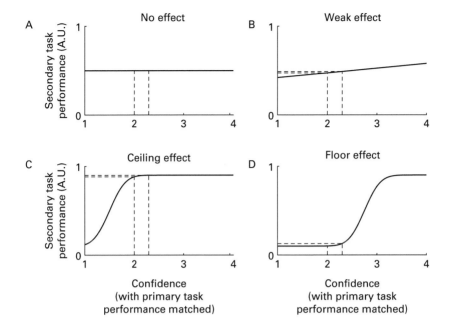

Figure 10.2
Possible ways in which performance-matched differences in awareness could fail to yield differences on a secondary task. Plots show confidence on the x-axes, with the idealized assumption that these levels of confidence are all achieved at a constant level of primary task performance. Plotted on the y-axes are performance on a secondary task (such as the cognitive control task in Koizumi et al., 2015) using the matched performance/different awareness stimuli (e.g., using grating tilt to inform task set preparation for a separate, upcoming task). Vertical lines indicate two levels of performance-matched confidence probed in a hypothetical experiment, and horizontal lines show the corresponding difference in the secondary task under different scenarios. (A) No effect: performance-matched confidence does not affect the secondary task. (B) Weak effect: the influence of performance-matched confidence is small and therefore difficult to detect in an experiment. (C) Ceiling effect: performance-matched confidence does influence the secondary task, but the effect is saturated at the levels of confidence probed in the experiment. (D) Floor effect: performance-matched confidence does influence the secondary task, but the effect is stronger at higher levels of confidence than those probed in the experiment.

stimuli discussed previously (figure 10.1A) requires energy in the signal, noise, and overall stimulus to be larger in the "more conscious" condition. The positive evidence/negative evidence method (figure 10.1B) allows for overall stimulus energy to be matched, but only if energy in stimulus noise in the "conscious" condition is reduced to compensate for the increases in the energy of positive and negative evidence necessary to yield higher levels of awareness (Koizumi et al., 2015). These stimulus confounds are more severe than is typically encountered in more traditional consciousness experiments, where stimulus confounds are frequently minimal (e.g., differences in the temporal gap between stimulus and mask on the order of tens of milliseconds, as in Dehaene, Naccache, et al., 2001) or non-existent (e.g., a fixed stimulus repeatedly presented at threshold contrast so that it is sometimes consciously experienced and other times not, as in Baria, Maniscalco, & He, 2017).

Yet, of course, these studies invoking minimal or no stimulus confound typically suffer from drastic performance confounds. (A notable exception here is the metacontrast masking paradigm employed in Lau & Passingham, 2006, and Maniscalco & Lau, 2016, which can achieve performance matching with a difference in stimulus-mask onset asynchrony in the order of tens of milliseconds.) One can argue that it is preferable to have stimulus confounds than performance confounds, as the latter presumably affect brain dynamics in a more global and complex way. However, significant stimulus confounds are clearly also undesirable.

Indeed, if any method of generating a difference in consciousness across experimental conditions must be contaminated with some confounding factor or other, it would seem that there may be no single experimental design that could reveal the "pure," uncontaminated neural substrates of consciousness. However, a possible way forward is to *triangulate* on these substrates by combining the results from multiple experimental designs with disjoint sets of confounds into one overarching analysis, rather than counting on any one given design being the silver bullet. A simple illustration of the idea is as follows: if experimental design A matches for stimulus properties but suffers from performance confounds, and design B matches for performance but suffers from stimulus confounds, then perhaps analysis of the combined data could reveal the common subset of neural activity that correlates with consciousness in both experiments. In the idealized case, this common subset of neural activity would be confound free, since the confounds in design A that co-vary with consciousness are completely absent in design B and vice versa. In other words, such an analysis approach could potentially reveal the "pure" neural basis of consciousness. In practice, the triangulation approach faces significant challenges, not the least of which is the possibility that the neural substrate of consciousness might interact with different confounding factors in distinct and nonlinear ways, thus complicating the distillation of the "pure" substrate of consciousness across experiments. Nonetheless, we regard the general premise of the triangulation approach as promising and worthy of development in future work.

10.7   Conclusion

We have presented theoretical considerations for why it is crucial to control for task-performance confounds in the neuroscientific study of consciousness. The feasibility and value of this approach is demonstrated by a growing body of literature in which performance-matched differences in awareness have been successfully isolated and computationally modeled. However, the performance-matching approach comes with a number of caveats and nuances that require careful consideration. A promising way forward may be to combine performance-matching approaches with other complementary approaches so as to triangulate on the "pure" confound-free neural substrate of consciousness.

Acknowledgments

For their discussion and comments on a previous version of this chapter, we thank Megan Peters, Jason Samaha, Rachel Denison, Anna Leshinskaya, Enoch Lambert, Maria Khoudary, Ian Phillips, J. D. Knotts, Matthias Michel, Hakwan Lau, the editors Felipe De Brigard and Walter Sinnott-Armstrong, and participants at the 2019 Summer Seminars in Neuroscience and Philosophy (SSNAP).

Notes

1. "Content" in philosophy is used as a technical term that refers to what a mental state is about. For example, believing that it is raining and hoping that it is raining are psychologically different mental states with the same content. In perception, the content of a conscious state is what the perceptual state is about or, more generally,

what is conveyed to the subject by her perceptual experience. In the example from the introduction, the contents of your friend's conscious experience included a branch, whereas the contents of your conscious experience did not. Despite its surface simplicity, there is much controversy about what the admissible contents of perception are and how to best individuate them. For an overview, see Siegel (2016).

2. Note that experimental designs in which conditions are randomly interleaved across trials can help ensure that decision strategy is constant across conditions, since human subjects have difficulty dynamically adjusting response criteria from trial to trial, even when it would be optimal to do so (Gorea & Sagi, 2000). See also the discussion below of the results by Maniscalco, Castaneda, and colleagues (2020).

3. For simplicity, here we bracket legitimate concerns about response biases and how to measure them that complicate taking such reports at face value.

4. Again, bracketing response bias concerns.

5. The concern that performance-matching studies compare conditions at intermediate levels of stimulus awareness could be somewhat alleviated if it could be demonstrated that the difference in average awareness is driven strongly by different frequencies of "completely unconscious" or PAS = 0 trials.

References

Aru, J., Bachmann, T., Singer, W., & Melloni, L. (2012). Distilling the neural correlates of consciousness. *Neuroscience and Biobehavioral Reviews, 36*(2), 737–746.

Baars, B. J. (1997). Global workspace theory: A rigorous scientific theory of consciousness. *Journal of Consciousness Studies, 4*(4), 292–309.

Baria, A. T., Maniscalco, B., & He, B. J. (2017). Initial-state-dependent, robust, transient neural dynamics encode conscious visual perception. *PLoS Computational Biology, 13*(11), e1005806.

Bayne, T., Hohwy, J., & Owen, A. M. (2016). Are there levels of consciousness? *Trends in Cognitive Sciences, 20*(6), 405–413.

Benwell, C. S. Y., Tagliabue, C. F., Veniero, D., Cecere, R., Savazzi, S., & Thut, G. (2017). Prestimulus EEG power predicts conscious awareness but not objective visual performance. *eNeuro, 4*(6).

Bickle, J. (2008). Real reduction in real neuroscience: Metascience, not philosophy of science (and certainly not metaphysics!). In J. Hohwy & J. Kallestrup (Eds.), *Being reduced: New essays on reduction, explanation, and causation* (pp. 34–51). Oxford: Oxford University Press.

Block, N. (1995). On a confusion about a function of consciousness. *Behavioral and Brain Sciences, 18*(2), 227–247.

Block, N. (2005). Two neural correlates of consciousness. *Trends in Cognitive Sciences, 9*(2), 46–52.

Block, N. (2007). Consciousness, accessibility, and the mesh between psychology and neuroscience. *Behavioral and Brain Sciences, 30*(5–6), 481–548.

Block, N. (2011). Perceptual consciousness overflows cognitive access. *Trends in Cognitive Sciences, 15*(12), 567–575.

Block, N. (2019). What is wrong with the no-report paradigm and how to fix it. *Trends in Cognitive Sciences, 23*(12), 1003–1013.

Brown, R., Lau, H., & LeDoux, J. E. (2019). Understanding the higher-order approach to consciousness. *Trends in Cognitive Sciences, 23*(9), 754–768.

Chalmers, D. (1996). *The conscious mind: In search of a fundamental theory*. New York: Oxford University Press.

Chalmers, D. (2000). What is a neural correlate of consciousness? In T. Metzinger (Ed.), *Neural correlates of consciousness* (pp. 17–39). Cambridge, MA: MIT Press.

Cohen, M. A., Cavanagh, P., Chun, M. M., & Nakayama, K. (2012). The attentional requirements of consciousness. *Trends in Cognitive Sciences, 16*(8), 411–417.

Cohen, M. A., & Dennett, D. C. (2011). Consciousness cannot be separated from function. Notes*Trends in Cognitive Sciences, 15*(8), 358–364.

Dehaene, S. (2014). *Consciousness and the brain*. New York: Viking Penguin.

Dehaene, S., Changeux, J.-P., Naccache, L., Sackur, J., & Sergent, C. (2006). Conscious, preconscious, and subliminal processing: A testable taxonomy. *Trends in Cognitive Sciences, 10*(5), 204–211.

Dehaene, S., Lau, H., & Kouider, S. (2017). What is consciousness, and could machines have it? *Science, 358*(6362), 486–492.

Dehaene, S., Naccache, L., Cohen, L., Bihan, D. L., Mangin, J.-F., Poline, J.-B, & Rivière, D. (2001). Cerebral mechanisms of word masking and unconscious repetition priming. *Nature Neuroscience, 4*(7), 752–758.

Del Cul, A., Dehaene, S., Reyes, P., Bravo, E., & Slachevsky, A. (2009). Causal role of prefrontal cortex in the threshold for access to consciousness. *Brain: A Journal of Neurology, 132*(Pt 9), 2531–2540.

Dennett, D. C. (1991). *Consciousness explained*. Boston, MA: Little, Brown and Company.

Dennett, D. C. (2018). Facing up to the hard question of consciousness. *Philosophical Transactions of the Royal Society B: Biological Sciences, 373*(1755).

Eriksen, C. (1960). Discrimination and learning without awareness: A methodological survey and evaluation. *Psychological Review, 67*(5), 279–300.

Fazekas, P., Nemeth, G., & Overgaard, M. (2020). Perceptual representations and the vividness of stimulus-triggered and stimulus-independent experiences. *Perspectives on Psychological Science, 15*(5), 1200-1213.

Flavell, J. H. (1979). Metacognition and cognitive monitoring. *American Psychologist, 34*(10), 906–911.

Fleming, S. M. (2020). Awareness as inference in a higher-order state space. *Neuroscience of Consciousness, 6*(1), 1–9.

Fleming, S. M., Dolan, R. J., & Frith, C. D. (2012). Metacognition: Computation, biology and function. *Philosophical Transactions of the Royal Society B: Biological Sciences, 367*(1594), 1280–1286.

Fleming, S. M., Jihye, R., Golfinos, J. F., & Blackmon, K. E. (2014). Domain-specific impairment in metacognitive accuracy following anterior prefrontal lesions. *Brain, 137*(10), 2811–2822.

Giles, N., Lau, H., & Odegaard, B. (2016). What type of awareness does binocular rivalry assess? *Trends in Cognitive Sciences, 20*(10), 719–720.

Godfrey-Smith, P. (2008). Reduction in real life. In J. Hohwy & J. Kallestrup (Eds.), *Being reduced: New essays on reduction, explanation, and causation* (pp. 52–74). Oxford: Oxford University Press.

Gorea, A., & Sagi, D. (2000). Failure to handle more than one internal representation in visual detection tasks. *Proceedings of the National Academy of Sciences of the United States of America, 97*(22), 12380–12384.

Green, D. M., & Swets, J. A. (1966). *Signal detection theory and psychophysics.* New York: John Wiley.

Gross, S., & Flombaum, J. I. (2017). Does perceptual consciousness overflow cognitive access? The challenge from probabilistic, hierarchical processes. *Mind and Language, 32*(3), 358–391.

Hannula, D. E., Simons, D. J., & Cohen, N. J. (2005). Imaging implicit perception: Promise and pitfalls. *Nature Reviews Neuroscience, 6*(3), 247–255.

Holender, D. (1986). Semantic activation without conscious identification in dichotic listening, parafoveal vision, and visual masking. *Behavioral and Brain Sciences, 9*, 1–66.

Irvine, E. (2012). *Consciousness as a scientific concept: A philosophy of science perspective.* Dordrecht, Netherlands: Springer Science & Business Media.

Kim, C.-Y., & Blake, R. (2005). Psychophysical magic: Rendering the visible "invisible." *Trends in Cognitive Sciences, 9*(8), 381–388.

Knotts, J. D., Lau, H., & Peters, M. A. K. (2018). Continuous flash suppression and monocular pattern masking impact subjective awareness similarly. *Attention, Perception, and Psychophysics, 80*(8), 1974–1987.

Koizumi, A., Maniscalco, B., & Lau, H. (2015). Does perceptual confidence facilitate cognitive control? *Attention, Perception, and Psychophysics, 77*(4), 1295–1306.

Kouider, S., de Gardelle, V., Sackur, J., & Dupoux, E. (2010). How rich is consciousness? The partial awareness hypothesis. *Trends in Cognitive Sciences, 14*(7), 301–307.

Kouider, S., Dehaene, S., Jobert, A., & Le Bihan, D. (2007). Cerebral bases of subliminal and supraliminal priming during reading. *Cerebral Cortex, 17*(9).

Kouider, S., Sackur, J., & de Gardelle, V. (2012). Do we still need phenomenal consciousness? Comment on block. *Trends in Cognitive Sciences, 16*(3), 140–141.

Kunimoto, C., Miller, J., & Pashler, H. (2001). Confidence and accuracy of near-threshold discrimination responses. *Consciousness and Cognition, 10*(3), 294–340.

Lamme, V. (2010). How neuroscience will change our view on consciousness. *Cognitive Neuroscience, 1*(3), 204–220.

Lamy, D., Salti, M., & Bar-Haim, Y. (2009). Neural correlates of subjective awareness and unconscious processing: An ERP study. *Journal of Cognitive Neuroscience, 21*(7), 1435–1446.

Lau, H. (2008). Are we studying consciousness yet? In L. Weisenkrantz & M. Davies (Eds.), *Frontiers of consciousness: Chichele lectures* (pp. 245–258). Oxford: Oxford University Press.

Lau, H., & Passingham, R. E. (2006). Relative blindsight in normal observers and the neural correlate of visual consciousness. *Proceedings of the National Academy of Sciences of the United States of America, 103*(49), 18763–18768.

Lau, H., & Rosenthal, D. (2011). Empirical support for higher-order theories of conscious awareness. *Trends in Cognitive Sciences, 15*(8), 365–373.

Lloyd, D. A., Abrahamyan, A., & Harris, J. A. (2013). Brain-stimulation induced blindsight: Unconscious vision or response bias? *PLoS One, 8*(12), e82828.

Macmillan, N. A., & Creelman, C. D. (2005). *Detection theory.* (2nd ed.). Mahwah, NJ: Lawrence Erlbaum Associates.

Maniscalco, B., Castaneda, O. G., Odegaard, B., Morales, J., Rajananda, S., & Peters, M. A. K. (2020). The metaperceptual function: Exploring dissociations between confidence and task performance with type 2 psychometric curves. Retrieved from https://psyarxiv.com/5qrjn/

Maniscalco, B., & Lau, H. (2012). A signal detection theoretic approach for estimating metacognitive sensitivity from confidence ratings. *Consciousness and Cognition, 21*, 422–430.

Maniscalco, B., & Lau, H. (2015). Manipulation of working memory contents selectively impairs metacognitive sensitivity in a concurrent visual discrimination task. *Neuroscience of Consciousness, 2015*(1), 1–13.

Maniscalco, B., & Lau, H. (2016). The signal processing architecture underlying subjective reports of sensory awareness. *Neuroscience of Consciousness, 2016*(1), 292.

Maniscalco, B., McCurdy, L. Y., Odegaard, B., & Lau, H. (2017). Limited cognitive resources explain a trade-off between perceptual and metacognitive vigilance. *Journal of Neuroscience, 37*(5), 1213–1224.

Maniscalco, B., Peters, M. A. K., & Lau, H. (2016). Heuristic use of perceptual evidence leads to dissociation between performance and metacognitive sensitivity. *Attention, Perception, and Psychophysics, 78*(3), 923–937.

Mathewson, K. E., Gratton, G., Fabiani, M., Beck, D. M., & Ro, T. (2009). To see or not to see: Prestimulus alpha phase predicts visual awareness. *Journal of Neuroscience, 29*(9), 2725–2732.

Merikle, P. M., Smilek, D., & Eastwood, J. D. (2001). Perception without awareness: Perspectives from cognitive psychology. *Cognition, 79*(1–2), 115–134.

Michel, M. (2019). The mismeasure of consciousness: A problem of coordination for the perceptual awareness scale. *Philosophy of Science, 86*(5), 1239–1249.

Michel, M. (2021) Calibration in consciousness science. *Erkenntnis*. Advance online publication.

Michel, M., & Morales, J. (2020). Minority reports: Consciousness and the prefrontal cortex. *Mind and Language, 35*(4), 493–513.

Morales, J. (2021) Introspection is signal detection. *The British Journal for the Philosophy of Science*. Advance online publication.

Morales, J., Chiang, J., & Lau, H. (2015). Controlling for performance capacity confounds in neuroimaging studies of conscious awareness. *Neuroscience of Consciousness, 1*(1).

Morales, J., & Lau, H. (2020). The neural correlates of consciousness. In U. Kriegel (Ed.), *Oxford handbook of the philosophy of consciousness* (pp. 233–260). Oxford: Oxford University Press.

Morales, J., & Lau, H. (In Press). Confidence tracks consciousness. In J. Weisberg (Ed.), *Qualitative consciousness: Themes from the philosophy of David Rosenthal.* Cambridge: Cambridge University Press.

Nagel, T. (1974). What is it like to be a bat? *Philosophical Review, 83*, 435–450.

Persaud, N., Davidson, M., Maniscalco, B., Mobbs, D., Passingham, R. E., Cowey, A., & Lau, H. (2011). Awareness-related activity in prefrontal and parietal cortices in blindsight reflects more than superior visual performance. *NeuroImage, 58*(2), 605–611.

Persaud, N., McLeod, P., & Cowey, A. (2007). Post-decision wagering objectively measures awareness. *Nature Neuroscience, 10*(2), 257–261.

Peters, M. A. K., & Lau, H. (2015). Human observers have optimal introspective access to perceptual processes even for visually masked stimuli. *eLife, 4.*

Peters, M. A. K., Thesen, T., Ko, Y. D., Maniscalco, B., Carlson, C., Davidson, M., . . . Lau, H. (2017). Perceptual confidence neglects decision-incongruent evidence in the brain. *Nature Human Behaviour, 1,* 0139.

Phillips, I. (2011). Perception and iconic memory: What Sperling doesn't show. *Mind and Language, 26*(4), 381–411.

Phillips, I. (2016). Consciousness and criterion: On Block's case for unconscious seeing. *Philosophy and Phenomenological Research, 93*(2), 419–451.

Phillips, I. (2018a). The methodological puzzle of phenomenal consciousness. *Philosophical Transactions of the Royal Society B: Biological Sciences, 373*(1755), 20170347.

Phillips, I. (2018b). Unconscious perception reconsidered. *Analytic Philosophy, 59*(4), 471–514.

Phillips, I. (2020). Blindsight is qualitatively degraded conscious vision. *Psychological Review.* Advance online publication. doi:10.1037/rev0000254.

Phillips, I., & Morales, J. (2020). The fundamental problem with no-cognition paradigms. *Trends in Cognitive Sciences, 24*(3), 165–167.

Proust, J. (2013). *The philosophy of metacognition.* Oxford: Oxford University Press.

Quilty-Dunn, J. (2019). Is iconic memory iconic? *Philosophy and Phenomenological Research, 59*(175), 171–123.

Rahnev, D. A., Bahdo, L., de Lange, F. P., & Lau, H. (2012). Prestimulus hemodynamic activity in dorsal attention network is negatively associated with decision confidence in visual perception. *Journal of Neurophysiology, 108*(5), 1529–1536.

Rahnev, D. A., Maniscalco, B., Graves, T., Huang, E., de Lange, F. P., & Lau, H. (2011). Attention induces conservative subjective biases in visual perception. *Nature Neuroscience, 14*(12), 1513–1515.

Rahnev, D. A., Maniscalco, B., Luber, B., Lau, H., & Lisanby, S. H. (2012). Direct injection of noise to the visual cortex decreases accuracy but increases decision confidence. *Journal of Neurophysiology, 107*(6), 1556–1563.

Ramsøy, T. Z., & Overgaard, M. (2004). Introspection and subliminal perception. *Phenomenology and the Cognitive Sciences, 3,* 1–23.

Rausch, M., & Zehetleitner, M. (2016). Visibility is not equivalent to confidence in a low contrast orientation discrimination task. *Frontiers in Psychology, 7,* 591.

Rosenthal, D. (1993). State consciousness and transitive consciousness. *Consciousness and Cognition, 2,* 355–363.

Rosenthal, D. (2005). *Consciousness and mind.* Oxford: Oxford University Press.

Rosenthal, D. (2019). Consciousness and confidence. *Neuropsychologia, 128,* 255–265.

Rounis, E., Maniscalco, B., Rothwell, J. C., Passingham, R. E., & Lau, H. (2010). Theta-burst transcranial magnetic stimulation to the prefrontal cortex impairs metacognitive visual awareness. *Cognitive Neuroscience, 1*(3), 165–175.

Ruby, E., Maniscalco, B., & Peters, M. A. K. (2018). On a "failed" attempt to manipulate visual metacognition with transcranial magnetic stimulation to prefrontal cortex. *Consciousness and Cognition, 62*(July), 34–41.

Samaha, J., Barrett, J. J., Sheldon, A. D., LaRocque, J. J., & Postle, B. R. (2016). Dissociating perceptual confidence from discrimination accuracy reveals no influence of metacognitive awareness on working memory. *Frontiers in Psychology, 7,* 851.

Samaha, J., Iemi, L., & Postle, B. R. (2017). Prestimulus alpha-band power biases visual discrimination confidence, but not accuracy. *Consciousness and Cognition, 54,* 47–55.

Samaha, J., Switzky, M., & Postle, B. R. (2019). Confidence boosts serial dependence in orientation estimation. *Journal of Vision, 19*(4), 25.

Sandberg, K., Timmermans, B., Overgaard, M., & Cleeremans, A. (2010). Measuring consciousness: Is one measure better than the other? *Consciousness and Cognition, 19*(4), 1069–1078.

Schwitzgebel, E. (2011). *Perplexities of consciousness.* Cambridge, MA: MIT Press.

Sergent, C., & Dehaene, S. (2004). Is consciousness a gradual phenomenon? Evidence for an all-or-none bifurcation during the attentional blink. *Psychological Science, 15*(11), 720–728.

Shaver, E., Maniscalco, B., & Lau, H. (2008). Awareness as confidence. *Anthropology and Philosophy, 9*(1–2), 58–65.

Shoemaker, S. S. (1981). Some varieties of functionalism. *Philosophical Topics, 12*(1), 93–119.

Siegel, S. (2016). The contents of perception. In E. N. Zalta (Ed.), *The Stanford encyclopedia of philosophy* (Winter 2016 ed.). Retrieved from https://plato.stanford.edu/archives/win2016/entries/perception-contents/.

Snodgrass, M., Bernat, E., & Shevrin, H. (2004). Unconscious perception at the objective detection threshold exists. *Perception and Psychophysics, 66*(5), 888–895.

Spener, M. (2019). Introspecting in the twentieth century. In A. Kind (Ed.), *Philosophy of mind in the twentieth and twenty-first centuries* (pp. 148–174). London: Routledge.

Taylor, E. (2015). Explanation and the explanatory gap. *Acta Analytica, 31*(1), 77–88.

Tsuchiya, N., Wilke, M., Frässle, S., & Lamme, V. A. F. (2015). No-report paradigms: Extracting the true neural correlates of consciousness. *Trends in Cognitive Sciences, 19*(12), 757–770.

Weiskrantz, L. (1986). *Blindsight: A case study and implications.* Oxford: Oxford University Press.

Weiskrantz, L., Barbur, J. L., & Sahraie, A. (1995). Parameters affecting conscious versus unconscious visual discrimination with damage to the visual cortex (V1). *Proceedings of the National Academy of Sciences of the United States of America, 92*(13), 6122–6126.

Zylberberg, A., Barttfeld, P., & Sigman M. (2012). The construction of confidence in a perceptual decision. *Frontiers in Integrative Neuroscience, 6*(September), 1–10.

11   Memory Structure and Cognitive Maps

Sarah Robins, Sara Aronowitz, and Arjen Stolk

11.1   Introduction

Over the course of any given day, we are exposed to vast amounts of information, and yet our memory systems are capable of encoding and later retrieving this information. This would be difficult, if not impossible, unless the stored information were structured—that is, organized across various dimensions such as space, time, and semantic content. The use of structure to facilitate effective retrieval can be thought of as a general mnemonic activity, both in terms of the sub-personal processes that organize the memory system and in terms of the personal-level strategies that we can use to intentionally facilitate recall of particular pieces of information (Aronowitz, 2018). Cognitive scientists interested in memory have thus long been focused on investigations of memory structure. How do we organize information and experiences so as to make subsequent retrieval possible?

A common way to conceptualize memory structures in the cognitive sciences is as a cognitive map. Cognitive maps, in the most literal sense, are mental representations that are structured in a way that reflects the features of real space and which aid in navigation. Grounding the structure of memory systems in this basic and general ability that is conserved across a wide range of species has obvious appeal. Cognitive maps thus offer hope of theoretical and interspecies unity, as well as the opportunity to learn more about the structure of human memory by investigating the neural systems and behavior of model organisms such as rats and mice, where more extensive and precise interventions are available.

Cognitive maps also present a puzzle. The appeal to these maps begins literally: as an account of how spatial information is represented. Their intended use, however, is more ambitious. Cognitive maps are meant to scale up and provide the basis for our more sophisticated memory capacities (e.g., Bellmund et al., 2018). Our memory systems, as well as those of animals, surely represent a variety of nonspatial information, and at least in humans, some of this information is richly conceptual and linguistic. The extension is not meant to be metaphorical, but the sense in which these richer mental structures are supposed to remain map-like is rarely made explicit. How precisely is this process of scaling up meant to go? How do cognitive maps represent nonspatial information? There are a range of ways that generalization and abstraction could occur, each of which comes with a unique set of empirical consequences and a distinct view of mental representation and memory structure. Each, too, comes with a set of particular concerns and challenges. Our aim in this chapter is not to defend any particular view, but instead to provide a framework for exploring the available options.

This project is important for the neuroscience of memory because clarifying what cognitive maps represent and why has consequences for the methodology of identifying cognitive maps, the relationship between kinds of information in memory, and the relationship between memory and other forms of cognition. From a philosophical perspective, thinking carefully about cognitive maps is a window into understanding the nature of mental representation in memory and cognition more broadly. It would be an understatement to say that the nature of perceptual representations has attracted serious philosophical interest—and yet, the corresponding question in the philosophy of memory remains understudied. We also hope that this chapter can shed light on a debate about map-like forms of representation more generally (e.g., Camp, 2007, 2018; Rescorla, 2009).

A few caveats: the aim of this chapter is to understand what cognitive maps are and how they are incorporated into memory research. As such, we will not start by defining a cognitive map. Instead, we'll consider empirical work that appeals to this concept, taking note of definitions given by others along the way, attempting to derive a working definition that fits at least the majority of this research. We do not intend our review of this empirical work to be exhaustive. When determining what to include, our primary focus is on the views of cognitive maps that have been developed into accounts of memory structure. We recognize, but do not discuss, the extensive literature on cognitive maps as competing models of spatial navigation and representation in animal cognition (see Bermudez, 1998, and Rescorla, 2017, for reviews).

We begin, in section 11.2, with a survey of two traditions. The first of these traditions is the foundational work on cognitive maps that assumes these maps represent information in a spatial structure. The second is a review of alternative, nonspatial representational structures. From the former, we identify a set of themes widely shared by proponents of cognitive maps. From the latter, we extract general lessons for accounts of cognitive structure. With this background, in section 11.3, we turn to several cutting-edge projects that are engaged in the task of scaling up cognitive maps so as to accommodate nonspatial information. These projects each do so in interestingly different ways. Some kinds of nonspatial information may also be represented in a map-like form because they are organized along dimensions that are substantially analogous to spatial information. In other cases, nonspatial information is represented as an abstraction from spatial information. And still other cognitive maps embed nonspatial information in a spatial structure. Putting these cases alongside one another reveals the variety of options available for building cognitive maps and the distinctive limitations of each. We conclude by reflecting on where these results take us in terms of understanding the place of cognitive maps in memory.

11.2   Foundational Work on Cognitive Structures

11.2.1   Cognitive Maps as Spatial Structures

Thinking of cognitive structures in terms of cognitive maps has a long history in psychology and neuroscience. The view began as an explanation of maze-running abilities in rats and, over time, has developed and changed as it has been used to capture a range of activities, from semantic knowledge structures to the navigational expertise of London taxi drivers (Collins & Loftus, 1975; Maguire, Frackowiak, & Frith, 1997). Throughout, theorists have aimed to make connections between these abilities in experimental animals and humans, but they have offered subtly different accounts of why these maps exist, what's essential to their structure, and how the extension from basic neural structure to broader human competences is characterized.

Tolman (1948) is often identified as originating the idea of a cognitive map in this literature.[1] Tolman's account of the cognitive map emerged from his work on maze running and spatial learning in rats—the dominant method and experimental framework in early twentieth-century psychology. For Tolman, cognitive maps were part of an argument that explaining the navigational abilities of rats required more cognitive, representational structure than was allowed for by the stimulus–response approach, which was dominant at the time. Specifically, Tolman documented rats' ability to learn shortcuts in mazes—an ability inexplicable in terms of the animal's learned association with particular places in the maze as individual stimuli. Tolman further observed that rats were capable of latent or non-reinforced learning. That is, rats that were simply allowed to explore mazes while fully fed—not receiving nor wanting any reinforcement for their exploration— were able to learn routes through the maze. In order to explain this behavior, Tolman (1948) argued, the rat needed to be credited with the possession of a cognitive map that provided a "field map of the environment" (p. 192). Although Tolman's evidence was based on the maze-running behavior of rats, the cognitive maps that he posited to explain this behavior were intended to apply to a much wider range of cognitive creatures. Indeed, his 1948 paper was titled "Cognitive Maps in Rats and Men." The paper even concludes with a few pages of speculation on how particular features of human personality and social organization may be explicable within this framework.

Tolman's (1948) initial proposal was solidified into a theory of neural structure with the publication of O'Keefe and Nadel's *Hippocampus as a Cognitive Map* (1978). For O'Keefe and Nadel, cognitive maps were not simply a general framework for thinking about mental structure; the cognitive map was posed as a theory of hippocampal function. It was the first such proposal, and a highly systematic one, which helps to explain both the initial excitement about the idea and its lasting influence. There are further differences between Tolman's use of the term and use in O'Keefe and Nadel's framework, the latter of which serves as the basis of "cognitive map theory" as it is now understood. First, O'Keefe and Nadel take the notion of a map far more literally than Tolman. The claim is not that the information processing of the hippocampus can be understood as map-like or spatial-ish, but rather that these cognitive maps are inherently spatial. In putting forward their theory, O'Keefe and Nadel make continued, explicit appeal to the Kantian idea of spatial structures as an organizing feature of cognition. These spatial maps are considered innate structures endemic to all cognitive creatures. Second, the extension of these maps to humans is not a metaphorical abstraction from the idea of a spatial map, but instead is characterized as an expansion of the kind of inputs that the spatial system can incorporate and process.

This is best illustrated by their account of cognitive maps in humans: "the left hippocampus in humans functions in *semantic mapping*, while the right hippocampus retains the spatial mapping function seen in infra-humans. On this view, species differences in hippocampal function reflect changes in the inputs to the mapping system, rather than major changes in its mode of operation" (O'Keefe & Nadel, 1978, p. 3).

The centerpiece of cognitive map theory is the discovery of place cells (O'Keefe & Dostrovsky, 1971): neurons in the hippocampus that fire preferentially—that is, exhibiting a burst of action potentials—in response to a specific location in the organism's environment. When a rat navigates a maze, for example, some place cells fire at the beginning of the maze, others at the first fork, still others at the second fork, and so on. These place cells are organized topographically, so that their collective firing pattern reflects the rat's route. After the maze has been run, the pattern is rehearsed, establishing a "map" that allows the rat to navigate this environment more easily the next time it is encountered.

The discovery of grid cells further enriches our understanding of the maps created by the hippocampal system (Hafting et al., 2005). Grid cells are found in the medial entorhinal cortex and, in contrast to hippocampal place cells, fire at multiple regularly spaced locations in the environment. Seen over the rat's trajectory, the spatial firing patterns of these cells provide a grid-like representation of the organism's environment. Other cells select for additional elements of the map—for example, cells that track objects, landmarks, and other agents (Høydal et al., 2019); head direction cells that fire selectively based on the way the organism's head is oriented relative to its route (Taube, Muller, & Rank, 1990); and those that encode information about the distance to borders and edges (Solstad et al., 2008).

In our brief survey of this work, we want to highlight two important features of this literature as we see it. First, even though work on cognitive maps and memory structure has been done mostly with rodents and has focused on low-level neural structure, the intent has always been to make claims about the role of such maps in cognitive creatures more general. That is, the aim was not simply to move away from overly simplistic stimulus–response models of nonhuman animal learning, but rather to think about the cognitive structures available to these nonhuman animals in terms of a framework that would encompass cognitive processes and cognitive creatures more generally. How the cross-species and beyond-spatial generalizations of the framework are envisioned differs across particular accounts and interpretations of cognitive maps.

Second, cognitive map theory remains influential and controversial. The framework continues to serve as a serious guide to inquiry into neural structure, especially with regards to the hippocampus (Redish, 1999; Bellmund et al., 2018). The view also serves as a steady target for alternative conceptions of neural structure and cognitive processing. Many of these criticisms involve claims and/or evidence that the information represented in these "maps" is nonspatial, including, for instance, findings of hippocampal cells encoding temporal context (Eichenbaum, 2014). That such interpretations of the content of cognitive maps are available is certain. Whether this sense of cognitive maps is an extension of the original framework or an objection to it is more contentious. Answering this question depends on asking, first, how the notion of "map" should be understood. We address this in section 11.3.

11.2.2 Nonspatial Cognitive Structures

Cognitive psychology has, since its beginnings, been interested in how humans and other cognitive creatures organize their vast amounts of knowledge so as to support efficient and effective search and retrieval. Although some of these cognitive structures are referred to as "maps," in such cases the term is being stretched to nonstandard or metaphorical use. In this section, we'll survey some of the foundational work on structures in cognitive science that do not seem to be map-like and do not primarily encode spatial content.

We'll start with emergent conceptual structures (figure 11.1A and B). These are ways of organizing and relating information that emerge from amassing overlapping conceptual content and acquire their particular structures from patterns in the accumulated information. For example, this type of memory structure is often thought to support language comprehension and production. Adele Goldberg (2019) presents a view of language intended to explain the differences between the following kinds of minimal pair (p. 57):

1. I'll cry myself to sleep.

2. I'll cry myself asleep.

The former sentence is perfectly felicitous, whereas native speakers judge the latter to be odd. Perhaps sentence 2 is odd because it is novel or unusual, but as Goldberg notes, we are perfectly happy with unusual sentences like this (p. 76):

3. She'd smiled herself an upgrade.

Goldberg explains what is special about sentence 2 by appealing to the role of long-term memory organization. On her account, we encode much of the language we hear in a high-dimensional conceptual space that is structured by syntactic form, meaning (in context), phonetic features, and so on. Since there are systematic relationships between many of these features, over time, clusters emerge. Bits of language that are encountered more frequently are selectively strengthened, whereas the connection between a word or phrase and its initial context is weakened if it then fails to occur in similar contexts. We have only noisy, implicit access to this space. Thus, the problem with sentence 2 is that it is close to a stronger competitor, sentence 1. Conversely, sentence 3, while unusual, does not trigger us to recall a more common alternative formulation, and so we judge it to be felicitous.

This example helps us extract several key features of an emergent conceptual structure. Over time, in the case of language, certain features of language stand out as principal components. For instance, in English, sentences with the form of the double-object construction—for example, she (x) passed him (y) something (z)—almost always have the meaning that x causes y to receive z. This is a regularity that relates sentential form to semantic content, and crucially Goldberg (2019) argues that this regularity arises in memory without any need to learn an explicit rule or start off with innate knowledge. Instead, the emergent conceptual structure allows for all kinds of regularities between a wide variety of features to be learned over sufficient exposure. Such a structure must therefore be (1) highly dimensional in order to catch the relevant features, (b) associative in order to relate these features flexibly without a prior model, and (c) content addressable in order to utilize stored information to produce responses efficiently.

A second kind of nonspatial structure is a graphical model, a family of models within which we will focus on Bayesian networks (figure 11.1C). In the case we just considered, the regularity between the double-object construction and the type of causal-agential content was represented as a clustering or association. We neither represent the syntactic structure as dependent on the semantic content nor vice versa; the association is not specific enough to represent anything more than organization in terms of similarity. But in a Bayesian network, relationships between features are represented as a set of conditional (in)dependencies. For example, I might encode information about academic lectures as in figure 11.1C.

In this model, the nodes are random variables, and the edges represent dependencies (sets of local probability models). This allows us to assume that a node is independent of any other node, conditional on its parents: for instance, conditional on "conference quality," "speaker style" is independent of "talk content". Notice that in the above graph, an edge connects "conference quality" to "talk content," but it seems unlikely that "conference quality" is a cause of "talk content." A narrower class of these models, causal Bayesian networks, interprets dependence and independence

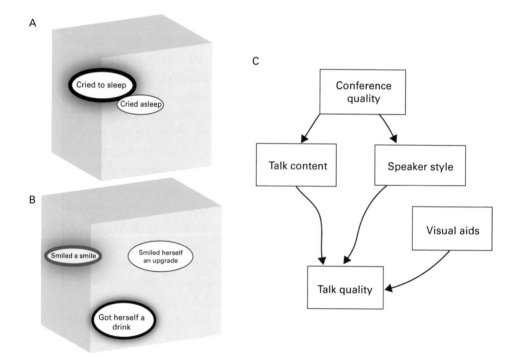

Figure 11.1
Conceptual structures: A and B represent highly dimensional spaces, as described by Goldberg (2019), which combine semantic information with other dimensions such as syntax and phonetics with dark lines representing phrase frequency. We find "cried asleep" to be more awkward than "smiled herself an upgrade" because of its nearness to a common alternative. (C) A graphical model, with nodes representing event features and edges representing probabilistic dependencies.

causally. Consequently, laws of causality can be applied to the graph structures, such as transitivity, asymmetry, and nonreflexivity. As is, this graphical representation is completely equivalent to an enumeration of the local probability models. However, when treated as a representation structure, the graphical representation can have properties not shared by the set of local models. For instance, we might search the graphical representation with an algorithm designed specifically for search in a graph, which would produce different behavior than search over other forms of representing the same information (e.g., Dechter & Mateescu, 2007). Whether it is true that conceptual knowledge is in fact *represented with*, in addition to *representable by*, graphical models, this hypothesis provides an interesting model of nonspatial mental structures.

Glymour (2001) analyzes a wide swath of human cognition in terms of causal Bayesian networks. These representations have three functions on his account: control, prediction, and discovery. To varying degrees, these functions could be fulfilled just as well, no matter the format of the probabilistic and causal information. However, the graphical format is significant as soon as the thinker employing the models is not perfect. Graphical representations figure directly in heuristics and inductive biases, such as a preference for an explanation that appeals to fewer causes (Lombrozo, 2007). Graphical representations allow simple access to points of potential intervention (Gopnik et al., 2004). As we noted above, we can define distinctive algorithms for search over graphical representations, and both noise and lesions to the model will operate differently, depending on representational format.

Thus, our second class of nonspatial models, causal Bayesian networks, is used to represent all kinds of causal knowledge. These representations function to identify interventions, predict new outcomes, and enumerate new possible theories. Bayesian networks are well suited to performing these functions because they organize information according to principles (i.e., the principles of causation) that (1) apply to the entire domain and (2) align with our interests in manipulating the environment.

Emergent conceptual structures and causal Bayesian networks are both structures that have been posited as operant in memory. Neither of these structures is in any notable way spatial or especially suited for spatial information. Both of these structures are functional: they are thought to have certain features that map onto computational advantages for the thinkers who employ them. Emergent conceptual structures are more flexible than causal Bayesian networks, since the latter can only represent causal relationships, whereas the former can represent non-causal associations. Correspondingly, causal Bayesian networks can express a more complex set of relationships within the causal domain, differentiating cause and effect, and identifying potentially complex causal patterns. Emergent conceptual structures represent many features in exactly one relationship: similarity.

Considering these two structures leaves us with a few key takeaways. First, even for models of abstract, domain-bridging, and perhaps distinctively human knowledge, cognitive structures are still thought to be tailored to particular functions. Second, there seems to be a trade-off between the generality of a representation (i.e., the kinds of features it could in principle represent) and its inferential power (i.e., the conclusions that can be derived from the connections among representational subunits). When data and processing limitations are held fixed, we could either utilize a structure with more flexible (and hence weak) connections or with less flexible (but more inferentially generative) connecting links. This idea is fairly intuitive, following from a more general connection between flexibility and informativeness. In the case of emergent conceptual structures, we saw an advantage of flexibility at work: Goldberg's (2019) model allows speakers not just to track semantic or syntactic patterns separately, but also to combine all the information we have about a string of language and thereby to learn patterns of association that crosscut traditional linguistic categories. Causal Bayesian networks displayed one of the advantages of inferential power: by representing causal structures in graphs, we made the task of determining points of intervention vastly easier. These cases offer helpful comparisons for considering how to manage these trade-offs in characterizing the functions of spatial structures that serve as the basis for cognitive maps.

11.3   Cognitive Maps and Nonspatial Information

The foregoing sections divided recent and historical work on memory structures into two categories: spatial (or spatially grounded) cognitive maps and nonspatial cognitive structures. In this section, we'll look at how the line between them can be blurred, such that cognitive maps might be used to encode less obviously spatial information. Specifically, we ask how minimal the spatial format can be while still leaving us with a meaningful notion of a cognitive map as a particular kind of functional memory structure.

To do so, we require a more extensive understanding of the basic notion of a map from which the idea of a cognitive map is generated. There is no canonical account of cartographic representation available, but we can provide a sketch by building off of a set of features proposed by Rescorla (2017). For Rescorla, maps (1) represent geometric aspects of physical space,[2] (2) have veridicality conditions, (3) have geometric structure, and (4) are veridical only if they replicate salient geometric aspects of the region being represented. Cognitive maps are, then, maps in a strict sense when they consist of mental representations with these properties.

However, the definition proposed by Rescorla will not capture crucial elements of the spatial maps we've already discussed, since his definition focuses solely on synchronic, intrinsic features.[3] The kind of cognitive maps we've surveyed are also used in navigation, and interpreted and updated accordingly.[4] This addition is crucial: a representation that has all the right internal properties but is never used in navigation is not really a map—and likewise with one that does not even potentially keep step with changes of information about the environment. Combining this functional role with Rescorla's conditions also lets us derive a fairly distinctive feature of maps, both cognitive and otherwise: we often update a piece of a map, such as a representation of the rooms on my floor, without even assessing a possible re-mapping of global relations, such as the distance between my room and Samarkand. We'll call this feature "locality."

When extending the notion of cognitive maps to nonspatial information, we relax the definition to capture a more general (or perhaps analogical) sense of map. Most directly, point 1 will always be false because the information being represented is not spatial. This will require, in turn, changes to how the veridicality conditions in point 4 are understood and what navigation might mean.

In this section, we consider three ways of extending the cognitive map. Each involves a distinct way of modifying the first condition on cartographic representations—that is, what is being represented: (1) encoding content that is nonspatial but in some sense isomorphic to spatial content; (2) encoding content that is an abstraction over first-order spatial information; and (3) embedding nonspatial information within a spatial context.

Before we begin, there is a caveat. Our interest in this section is in representations of nonspatial information that are in some sense utilizing the map-like representations traditionally associated with space (for a description of these parameters, see O'Keefe, 1991). This neither entails nor follows from a relationship between the neural-level realizers of spatial and nonspatial representations. Howard Eichenbaum and associated scholars have long tried to challenge the cognitive map picture by pointing to nonspatial uses of neural resources thought to be part of cognitive maps. For example, by showing that cells in the hippocampus represent events within their temporal context and not just their spatial context (see also MacDonald et al., 2011; Aronov, Nevers, & Tank, 2017; Wood, Dudchenko, & Eichenbaum, 1999). Thus, this line of research is not a case of extending the cognitive map to encompass nonspatial information, so long as the claim is about a shared neural substrate rather than a shared representational structure.

11.3.1   Spatial-Isomorphic Information

Spatial-isomorphic content is a kind of content that is structured according to dimensions that functionally correspond to spatial dimensions. By functional correspondence, we mean that the regularities, limitations, and inference patterns that we commonly apply to spatial dimensions will for the most part apply to these nonspatial dimensions. For example, (Euclidian) spatial distance is symmetric: if my office is ten feet from the coffee machine, then the coffee machine is ten feet from my office.[5] Spatial-isomorphic content, since its dimensions functionally correspond to spatial dimensions, will tend to have a "distance-like" measure that is symmetric in the same way. It seems reasonable that, were we to have a dedicated cognitive mapping system for dealing with spatial content, this system might also be used for dealing with spatial-isomorphic content.

Constantinescu, O'Reilly, and Behrens (2016) offer some preliminary evidence that some of the core processes for representing spatial maps can be used for spatial-isomorphic content. Unlike related work on spatial representations of nonspatial information (e.g., Tavares et al., 2015), the authors went beyond neurally co-locating spatial and nonspatial activity. Instead, they focused on a signature of spatial representation: coding of a space into a hexagonal lattice, such that the rate of cell firing corresponds to the orientation of movement relative to the orientation of the lattice. Because strongest firing occurs at 60° increments in orientation, the 360°

of phase space are divided into six identical regions, giving rise to the lattice's hexagonal symmetry.[6] The authors looked for this hexagonal symmetry as a mark of what are sometimes called human grid cells. Unlike the grid cells discussed in section 11.2.1, these neurons are not thought to be restricted to regions of the medial temporal lobe but instead are thought to occur throughout (some of) the brain regions that also form the default mode network, including the ventromedial prefrontal and posterior cingulate cortex. Still, previous work has associated these more distributed cells with spatial representations (e.g., Doeller, Barry, & Burgess, 2010). Rather than a strictly spatial task, Constantinescu and colleagues (2016) taught participants a pattern of association between the appearance of a bird and a set of symbolic cues. The bird figure varied according to neck height and leg lengths, which allowed for a representation of possible bird figures in a two-dimensional space structured by these two features. The bird-cue relationships were chosen so that each cue picked out a single region of this "bird space." The authors indeed found hexagonally symmetric responses (measured in fMRI) in a variety of default mode brain regions that seemed to correspond to hexagonal, grid-like representations of "bird space."

The bird space used in this study was spatial-isomorphic, since it was structured according to two dimensions (neck height and leg length) that could be used to carve up a feature space with several space-like functional dimensions: it was a two-dimensional Euclidean space, with distance and orientation operating just as they would in a real space. Intuitively, the bird space is space-like in that it articulates a "conceptual space," but also space-like in that neck height and leg length are themselves literally spatial dimensions. However, the design of this study allows Constantinescu and colleagues (2016) to differentiate between these two spatial aspects of the stimulus: because the bird space and the bird's features in regular space are two distinct spaces, moving through each would produce different patterns of symmetrical activation. Since the stimuli were carefully chosen to avoid passing through phase space in the same way, the observed symmetries in fMRI signal should not reflect the bird's position in visual space.

The use of hexagonal coding itself, if the authors are correct, suggests a second kind of isomorphism. Hexagonal coding is thought to be optimal for spatial representation in particular. Mathis, Stemmler, and Herz (2015), for example, present an optimal model that ranks hexagonal coding highest for spatial resolution in the two-dimensional plane. In ordinary space, we don't normally privilege one dimension over another—that is, the north–south axis is not in general more informative than the east–west axis. This allows us to value spatial resolution uniformly across the plane. But we do typically privilege those two axes over the up–down axis in navigation. These two features must be assumed in order to show the hexagonal lattice is optimal in the spatial domain. Neither feature needs to obtain in conceptual space. For instance, resolution in the bird neck-height dimension may be more valuable than information in the bird leg-length dimension. Were this to be true, the hexagonal symmetries observed by Constantinescu and colleagues (2016) would reflect a suboptimal representation. And so we can conclude that the use of a hexagonal symmetry code either reflects (a) a genuine isomorphism between the conceptual space and real space, or (b) a representational choice that favors spatial-isomorphism over customization to the optimal division of conceptual space.

Another kind of spatial isomorphism centers on temporal rather than conceptual structure. Researchers commonly motivate the division of a temporal sequence into parts by analogy with the division of a spatial layout into parts. For instance, Zacks and Swallow (2007) write:

> For quite a while, psychologists have known that in order to recognize or understand an object people often segment it into its spatial parts (e.g., Biederman, 1987). A new body of research has shown that just as segmenting in *space* is important for understanding objects, segmenting in *time* is important for understanding events (p. 83).

This literature on event segmentation asks how and why we draw boundaries between events. While Zacks and Swallow take the process of segmentation to be somewhat automatic, DuBrow and colleagues (2017) present contrasting evidence suggesting that segmentation can be active, abrupt, and driven by top-down goals.

Is this use of space merely a helpful metaphor, or is event structure genuinely spatial-isomorphic? One genuine isomorphism comes from the local structure of both representations—that is, a ubiquitous feature of cognitive maps is their locality. While I have a clear idea of how things in my apartment complex are oriented, and a good idea of how things in the Philadelphia Museum of Art are oriented, I do not necessarily have a joint map that neatly connects the two. Kuipers (1982, 2007) views this as a key starting assumption of cognitive maps even in machines: breaking a map into smaller, local maps allows the agent to remain noncommittal about global connections. This locality of representation seems to hold for temporal segmentation as well. Upon hearing a story, I might build a temporal "map" of the events of my friend's adventure last week without forming any particular representation of how the details of the events she is describing fit into a temporal sequence of my own schedule last week. Locality naturally arises from the use of schemas. Baldassano, Hasson, and Norman (2018) found that temporal boundaries in event schemas across different kinds of events had a common neural signature, provided they shared an abstract schematic structure—that is, schematic representations impose a local structure relative to the device of the schema itself (e.g., from when you enter a restaurant to when you pay the check). Anchoring event segmentation in local (temporal) structure, then, creates an abstract isomorphism with spatial maps, which are anchored in local (spatial) structures.

We could point to a long tradition locating isomorphisms between space and time, tracing at least back to Kant (1781/1787, A33/B49–50). The strength of this tradition, however, is a double-edged sword. The abundance of spatial language used in our everyday talk about time makes it hard to conceive genuinely of the capacity to represent space and the capacity to represent time as distinct. The question of isomorphism between space and time may, from this perspective, be ill formed if the two capacities are more than accidentally linked to one another.

In summary, one way to extend the core notion of a cognitive map to nonspatial information is to treat the nonspatial information as spatial-isomorphic. These expansions are most efficient in cases where the nonspatial domain has significant regularities that mirror regularities that compose our representations of space, such as a symmetrical distance measure, roughly equal value assigned to discriminability among the dimensions on a two-dimensional plane, and a representation of related "spaces" that can be composed locally and independently.

11.3.2   Abstractions over Spatial Information

Another way to extend cognitive map theory—integrating work on neural-level spatial maps and cognitive-level structure—is to explore ways in which the neural systems that support cognitive maps can process and represent abstractions from spatial information. Here, we consider two kinds of spatial abstraction: (1) a structure where the abstraction itself is still isomorphic to the lower-order representation of space, and (2) abstractions over space that are no longer spatial-isomorphic.

Michael Hasselmo (2011) has used cognitive map theory—with its place, grid, and head direction cells—to build a map-based account of episodic memory. In keeping with key themes of cognitive map theory, Hasselmo's theory is derived largely from work with rats but is meant to provide an account of episodic memory that can scale to humans. His book-length articulation of the view is entitled *How We Remember*, and the "we" here encompasses all mammals with similar hippocampal structure. Critical to Hasselmo's particular version of cognitive map theory is the idea that the hippocampus and surrounding structures are a phase-coding mechanism, where the map-making activity of place and grid cells is integrated into maps of the environment at multiple spatial and temporal scales—that is, the hippocampus produces a series of cognitive maps, in Rescorla's (2009) loose sense, representing the environment in more or less detail, as a function of the scale imposed by the cells from which they are activated. Together, these maps represent a particular event or experience, serving as the content of an episodic memory. To support the idea of multiple maps, Hasselmo incorporates neurons from the entire hippocampus into his model rather than focusing primarily on the dorsal portions of the hippocampus, as is common in much of the literature.

Hasselmo argues that neurons across the hippocampus share the mapping function. The differences between dorsal and ventral neurons are a matter of the size of their receptive fields not their general function. As one moves across the hippocampus, from dorsal to ventral, the receptive field size of the neurons increases. This increase in receptive field size results in a comparable increase in the scalar proportion of the map. Maps featuring place cells with the smallest receptive fields represent the organism's immediate surroundings in detail, whereas larger maps are produced by place cells with larger receptive fields, situating the experience within its (increasingly broad) spatial and temporal context. Importantly, the broadest "maps" may remain spatial in only the loosest or most metaphorical sense, situating the event within a social, conceptual, or experiential context.

The result is a mechanism that produces representations rich enough to support episodic remembering. The existence of multiple maps allows for a single episode to be recorded at several "scales of experience" (Hasselmo, 2008), capturing the episode as occurring not only at a particular place and time but as associated with various objects and events. For example, consider my episodic memory of walking from my campus office to the university library to return a book this morning. On Hasselmo's view, the representation of this episode is a conjoined set of maps of the event at different scales of experience. We can think of the smallest-scale map of the event in terms of traditional cognitive map approaches—as an allocentric map of the campus, along with my route from my office to the library. But other maps associated with this episode will represent this event at different spatial, temporal, and contextual scales. The more abstract spatial maps may represent campus with relation to the part of town, city, state or continent in which I live. More abstract temporal maps will represent my route through this map as part of my schedule for the day, or schedule for the week, or activities characteristic of this time in the academic year. Further contextual maps will also be available, where the items represented in the map situate the landmarks along the route on different contextual scales—for example this trip to the library as a stage in a particular research project, trees along this route at this time of the year, campus construction at this time, and so on.

Hasselmo's model proposes that the cognitive map system can process increasingly abstract characterizations of space and time that can then serve as the content for more elaborate and higher-order episodic memories. However, his hierarchical picture would seem to preserve some degree of structural similarity between levels. On the other hand, Behrens and colleagues (2018) also provide an account of abstraction from first-order spatial information in the form of eigenvectors corresponding to transformations between first-order spatial (and nonspatial) environments. Unlike a hierarchical, nested representation of experience, an eigenvector is an abstraction that does not share a structure with its first-order counterparts. Eigenvectors fall into a broader class discussed by Behrens and colleagues (2018), including inductive biases and factorizations. These are all features applying to a set of environments or state spaces that aid in learning but seem to require additional representational resources.

The authors argue for a common set of abstractive capacities operating over both spatial and nonspatial representation, which would utilize higher-order features to drive first-order prediction and planning. Presumably, whatever representational resources would be needed to supplement first-order maps with these higher-order features must be integrated tightly with the first-order maps themselves. Behrens and colleagues (2018) provide a few suggestions, but how this integration might work is still very much an open question.

11.3.3 Embedding Nonspatial Information in a Spatial Format

A third way to expand spatial representations would be to keep the spatial structure intact and embed nonspatial information within that structure. The most prominent example of memory success, the method of loci (MoL), can be understood as organizing information in this way. In the MoL, subjects typically memorize a list of unstructured items such as random words or phone numbers by imagining these items in a structured environment such as a familiar childhood walk through the neighborhood. The MoL thus appears to be a way of using a useful feature of spatial memory to store nonspatial content. The explicit process in the MoL involves two stages: (1) a strategy for encoding items through visualization, and (2) a strategy for retrieving items through a parallel visualization. For example, I could encode a list by imagining a walk through my childhood home, and then later recall the items by imagining walking through the home again and picking up the items.

Questions about how and why mnemonic structure works have received very little attention from memory theorists and scientists. The scant evidence that exists is, however, suggestive and intriguing, and invites us to ask more detailed questions about memory structure. Both the testimony of memory champions and some preliminary studies of expert performance reveal that success in the use of mnemonics does not require any particular level of intelligence or distinct cognitive skills. Instead, the key to success using mnemonic techniques is simply practice (Ericsson, 2003; Wilding & Valentine, 1997). These behavioral reports are complemented by neuroimaging studies indicating that those who use these techniques regularly differ from controls in functional but not structural brain features (Maguire et al., 2003; Raz et al., 2009). A recent and intriguing paper showed that after only six weeks of training, cognitively matched novices exhibited the same functional changes seen in memory champions (Dresler et al., 2017). Similarly, Yoon, Ericsson, and Donatelli (2018) have just shown that a person trained to increase their digit span to more than 100 digits thirty years ago has retained many of the associated skills, despite undertaking no training in the meantime.

In the majority of these cases, the items to be memorized do not have an interesting nonspatial structure themselves (e.g., digit span, presentation order of a shuffled deck of cards). However, looking more closely at the history of this mnemonic technique reveals that it has also been used for semantically structured information. In *De Oratore*, Cicero recommended the technique for memorizing speeches, lyric poems, and the like. The

Classicist Minchin (2001, p. x) argued that Homer used this method to compose and perform the *Iliad* and *Odyssey*. In the Middle Ages, it was common for monks to use this technique to memorize the Bible's 150 Psalms, as well as long passages from other scholarly texts (Carruthers, 2008). This is continued in some forms of contemporary use, as when students use such techniques to help them remember conceptual blocks of information (Kerr & Neisser, 1983; Roediger, 1980; Wang & Thomas, 2000). The ability to achieve additional mnemonic gains by translating information that is already semantically or conceptually structured into a spatial format suggests that there is something about spatial structure in particular that is useful for memory retrieval.

We have surveyed three ways in which spatial maps might be "scaled up" to accommodate nonspatial information. First, spatial structures might be repurposed to represent nonspatial content with a suitably isomorphic structure: for instance, a conceptual "space" can be modeled as a two-dimensional Euclidean plane. Second, spatial structures might be used to represent abstractions over spatial information, whether in the form of higher-order but still spatially structured representations, or with summary statistics that are to some degree represented jointly with traditional maps. Third, nonspatial information might be embedded in a spatial map, overriding or supplementing nonspatial structure with an exogenous spatial structure.

Taking stock of these results in light of the two traditions discussed in section 11.2, we can draw two key conclusions. First, following on from Tolman's original essay, the idea of seeing how far the basic components of the cognitive map can be stretched to fit nonspatial content is still very much an open research question. Second, the trade-off between flexibility and inferential power that we observed in the case of conceptual representations characterizes the question of expanding the cognitive map as well. The more we stretch the notion, the more we reduce inferential power in favor of flexibility.

11.4   Why Extend the Cognitive Map?

The work that we have surveyed in this chapter testifies to a persistent interest in understanding the edges of applicability of the concept of a cognitive map. Given the difficulties of extending the cognitive map concept to nonspatial information, why are researchers pursuing this program? What benefits come from thinking about the structure of mental representations within this framework?

One reason to extend the cognitive map concept is in service of the broader project of tying cognitive functions to evolutionary history. Neuroscientists have long argued for functional and anatomical homologies across a broad range of mammalian species (e.g., Clark & Squire, 2013, but see also Zhao, 2018). As a theory of hippocampal function, cognitive maps offer the potential for an explanation of navigational and cognitive systems across all species in which these basic anatomical structures are preserved. A cognitive map, in the core spatial context, links a somewhat general capacity for long-term spatial memory to a more obviously adaptive capacity for spatial navigation. Likewise, a model on which some kinds of conceptual memory utilize a cognitive map provides a potential basis for understanding the evolution of conceptual memory. Viewed from this perspective, the cognitive map project embodies an explanatory aim: not just to have accurate models of mental faculties, but also to find models that explain how a creature with our lineage could have developed these faculties. Under this aim, the three types of extension we discussed have divergent implications. Spatial-isomorphic extension is a way of reusing structures that were tailored for a particular spatial purpose for new tasks that happen to have similar structure. Extension by abstraction, on the other hand, is a richer ability applied to the very same core domain. In comparison with these two, embedding nonspatial content in space appears more like a culturally developed trick that makes clever use of evolutionarily adapted structures. In short, there is a story to tell for each way the cognitive map could have become extended, but there are important differences in how this story goes in each case.

Another reason to persist with cognitive maps may be a general commitment to parsimony across cognitive systems. An enriched and extended notion of cognitive maps could provide the representational basis for a broad range of cognitive processes, providing a competitor to the logically structured language of thought, as developed primarily by Fodor (1975, 1987) and Pylyshyn (1984). An example of such a project has been explicitly advanced for spatial maps by Rescorla (2009). In fact, one might think of the work on mental models done by Johnson-Laird (1983) as a version of this kind of proposal, though his mental models are not as distinctively spatial.[7] The more forms of memory, and cognition more broadly, that can be understood from within this basic framework, the more streamlined and unified our understanding of cognitive systems becomes. This idea of parsimony as part of the explanatory aim in understanding the mind is notably different from a concern with parsimony within the mind as a matter of representational efficiency, though both might motivate an interest in conserving structures across cognitive domains.

As we noted in section 11.2.2, however, parsimony has consequences for the trade-off between flexibility and informativeness—that is, the more cognitive subsystems employ the same structures, the more flexibly (and thereby less informative) the structures must be. Again, much depends on which form of generalizing to the nonspatial we adopt. Under spatial-isomorphic extension, there is a genuine form of parsimony in applying similar structures to isomorphic domains. But the utility of this approach will be limited to the number of domains that share these structures. There is little reason to hope or expect that *all* domains will be spatial-isomorphic in the requisite ways. The same argument might be harder to leverage for the other two kinds of extension. The discussion of the models by Hasselmo and by Behrens and colleagues revealed that abstractions of spatial information might be spatial-isomorphic, but they need not be. Embedding nonspatial information in maps likewise would not be predicted by the idea that related functions should in general utilize related structures. Instead, to motivate embeddings of this kind, we'd need a much stronger notion of parsimony—for example, that an explanation with fewer structures should be preferred even when it explains across distinct functions. This stronger form of parsimony would of course undermine the previous rationale under which we prefer a one-to-one fit of structure to function, since it encodes a preference for a one-to-many fit.

In the cognitive science tradition, memory has long served as one of the best candidates for inter-level integration. It is often, literally, the textbook model of how the various cognitive sciences can all contribute to a shared understanding of a particular cognitive process (e.g., Bermudez, 2014). Similarly, philosophers of neuroscience use memory as a case study in mechanistic levels of explanation (Craver, 2007). The appeal of cognitive map theory can be understood as a commitment to this aspirational model. The origins of cognitive map theory reflect this. O'Keefe and Nadel were inspired by the relatively simultaneous discovery of place cells, long-term potentiation, and other cellular and molecular processes in the hippocampus and the observation of the patient H. M.'s loss of episodic memory in response to hippocampal damage during neurosurgery. The promise of inter-level integration relies on a shared understanding of the cognitive process at each of the cascading levels in the model. Each version of cognitive map theory's extension offers a different account of the system whose levels of understanding and inquiry are being integrated. On the spatial-isomorphic interpretation of the extension, the system is one for representing regular and symmetric structures, spatial or otherwise. On the abstraction approach, spatial navigation is the core cognitive function or ability of this inter-level system. On the embedding approach, it is declarative remembering that is central to the inter-level project. In short, inter-level integration in cognitive science requires an understanding of the cognitive ability/system/function of interest, from which the process of decomposition and unification of distinct methodological approaches can begin. Such integration may be possible on each of these three forms of cognitive map theory, but they will each offer a distinct account of the cognitive ability being integrated.

These three reasons *why* one might be interested in expanding the cognitive map to nonspatial domains thus push us in different directions when it comes to *how* to extend the map. But beyond that, they also reveal a puzzle inherent in locating the cognitive map in memory. That is, to what extent is this information structure a proper part of memory? Given that alternative, nonspatial cognitive structures—such as those explored in section 11.2.2— can also be understood as accounts of memory structure, the need to defend the reliance on cognitive maps, however extended, becomes more critical.

11.5   Conclusion

What is a cognitive map? This concept is pervasive but hard to define comprehensively. At its core, a cognitive map represents the environment by taking advantage of the structure of space. As a metaphor, it offers a way of unifying our understanding of the representational structures endemic to cognition, giving a sense of shared function across a range of species and abilities. Insofar as cognitive maps remain an active research program, more attention should be devoted to the conceptual space between the literal and metaphorical versions.

The concept of a cognitive map is deeply intertwined with contemporary research on memory. In this chapter, we've surveyed a series of recent attempts to extend the concept to cover representations that are not obviously spatial.

These projects, seen in the light of historical developments in the theory of spatial memory and cognitive structure, reveal both the ambitions and limitations of research into general-purpose representations.

Notes

1. O'Keefe and Nadel (1978), in their subsequent work on cognitive maps in the hippocampus (discussed later in this chapter), find historical inspiration in the work of Gulliver (1908).

2. Dabaghian, Brandt, and Frank (2014) argue that hippocampal maps represent topological (i.e., ordinal) features of space rather than geometric properties such as absolute distances and angles. We suspect this difference with Rescorla is at least partly terminological. Thus, we take point 1 to be satisfied by the model proposed by Dabaghian and colleagues.

3. We can see several potential ways to derive updating behaviors from Rescorla's conditions. However, the same cannot be done for navigation, since it is clearly possible for a creature which does not behave at all, let alone navigate, to have a cognitive map on his definition.

4. Camp (2007) focuses more on these dynamic factors—on her view, a map is a representational system with a semi-compositional structure that determines what we can infer, how maps can be assembled, and how updating works.

5. Interestingly, path representations are not always symmetric—see Kuipers (1982) for a theoretical computer science perspective on how this asymmetry interacts with the "map" metaphor.

6. However, recent work in rodents (e.g., Stensola et al., 2015) has found a variety of cases where hexagonal symmetry in grid cells is distorted.

7. Thanks to Felipe De Brigard for this suggestion.

References

Aronov, D., Nevers, R., & Tank, D. W. (2017). Mapping of a non-spatial dimension by the hippocampal-entorhinal circuit. *Nature, 543,* 719–722.

Aronowitz, S. (2018). Memory is a modeling system. *Mind and Language, 34*(4), 483–502.

Baldassano, C., Hasson, U., & Norman, K. A. (2018). Representation of real-world event schemas during narrative perception. *Journal of Neuroscience, 38*(45), 9689–9699.

Behrens, T. E., Muller, T. H., Whittington, J. C., Mark, S., Baram, A. B., Stachenfeld, K. L., & Kurth-Nelson, Z. (2018). What is a cognitive map? Organizing knowledge for flexible behavior. *Neuron, 100*(2), 490–509.

Bellmund, J. L. S., Gardenfors, P., Moser, E. I., & Doeller, C. F. (2018). Navigating cognition: Spatial codes for human thinking. *Science, 362*(6415), eaat6766.

Bermudez, J. L. (1998). *The paradox of self-consciousness.* Cambridge, MA: MIT Press.

Bermudez, J. L. (2014). *Cognitive science: An introduction to the science of the mind* (2nd ed.). Cambridge: Cambridge University Press.

Camp, E. (2007). Thinking with maps. *Philosophical Perspectives, 21*(1), 145–182.

Camp, E. (2018). Why maps are not propositional. In A. Grzankowski & M. Montague (Eds.), *Non-propositional intentionality* (pp. 19–45). Oxford: Oxford University Press.

Carruthers, P. (2008). Meta-cognition in animals: A skeptical look. *Mind and Language, 23*(1), 58–89.

Clark, R. E., & Squire, L. R. (2013). Similarity in form and function of the hippocampus in rodents, monkeys, and humans. *Proceedings of the National Academy of Sciences of the United States of America, 110*(Supplement 2), 10365–10370.

Collins, A. M., & Loftus, E. F. (1975). A spreading-activation theory of semantic processing. *Psychological Review, 82*(6), 407–428.

Constantinescu, A. O., O'Reilly, J. X., & Behrens, T. E. (2016). Organizing conceptual knowledge in humans with a gridlike code. *Science, 352*(6292), 1464–1468.

Craver, C. F. (2007). *Explaining the brain: Mechanisms and the mosaic unity of neuroscience.* Oxford: Oxford University Press.

Dabaghian, Y., Brandt, V. L., & Frank, L. M. (2014). Reconceiving the hippocampal map as a topological template. *eLife, 3*, e03476.

Dechter, R., & Mateescu, R. (2007). AND/OR search spaces for graphical models. *Artificial Intelligence, 171*, 73–106.

Doeller, C. F., Barry, C., & Burgess, N. (2010). Evidence for grid cells in a human memory network. *Nature, 463*(7281), 657.

Dresler, M., Shirer, W. R., Konrad, B. N., Müller, N. C., Wagner, I. C., Fernández, G., . . . & Greicius, M. D. (2017). Mnemonic training reshapes brain networks to support superior memory. *Neuron, 93*(5), 1227–1235.

Dubrow, S., Rouhani, N., Niv, Y., & Norman, K. A. (2017). Does mental context drift or shift? *Current Opinion in Behavioral Sciences, 17*, 141–146.

Eichenbaum, H. (2014). Time cells in the hippocampus: A new dimension for mapping memories. *Nature Reviews Neuroscience, 15*, 732–744.

Ericsson, K. A. (2003). Exceptional memorizers: Made, not born. *Trends in Cognitive Sciences, 7*, 233–235.

Fodor, J. A. (1975). *The language of thought*. New York: Thomas Y. Crowell.

Fodor, J. A. (1987). *Psychosemantics*. Cambridge, MA: MIT Press.

Glymour, C. N. (2001). *The mind's arrows: Bayes nets and graphical causal models in psychology*. Cambridge, MA: MIT Press.

Goldberg, A. E. (2019). *Explain me this: Creativity, competition, and the partial productivity of constructions*. Princeton, NJ: Princeton University Press.

Gopnik, A., Glymour, C., Sobel, D. M., Schulz, L. E., Kushnir, T., & Danks, D. (2004). A theory of causal learning in children: Causal maps and Bayes nets. *Psychological Review, 111*(1), 3.

Gulliver, F. P. (1908). Orientation of maps. *Bulletin of the American Geographical Society, 40*(9), 538.

Hafting, T., Fyhn, M., Molden, S., Moser, M., & Moser, E. I. (2005). Microstructure of a spatial map in the entorhinal cortex. *Nature, 436*(7052), 801–806.

Hasselmo, M. E. (2008). Grid cell mechanisms and function: Contributions of entorhinal persistent spiking and phase resetting. *Hippocampus, 18*(12), 1213–1229.

Hasselmo, M. E. (2011). *How we remember: Brain mechanisms of episodic memory*. Cambridge, MA: MIT Press.

Høydal, Ø A., Skytøen, E. R., Moser, M., & Moser, E. I. (2019). Object-vector coding in the medial entorhinal cortex. *Nature, 568*(7752), 400–404.

Johnson-Laird, P. N. (1983). *Mental models: Towards a cognitive science of language, inference, and consciousness*. Cambridge, MA: Harvard University Press.

Kerr, N. H., & Neisser, U. (1983). Mental images of concealed objects: New evidence. *Journal of Experimental Psychology: Learning, Memory, and Cognition, 9*(2), 212.

Kuipers, B. (1982). The "map in the head" metaphor. *Environment and Behavior, 14*(2), 202–220.

Kuipers, B. (2007). An intellectual history of the spatial semantic hierarchy. In M. E. Jefferies & W.-K. Yeap (Eds.), *Robotics and cognitive approaches to spatial mapping* (pp. 243–264). Berlin: Springer.

Lombrozo, T. (2007). Simplicity and probability in causal explanation. *Cognitive Psychology, 55*(3), 232–257.

MacDonald, C. J., Lepage, K. Q., Eden, U. T., & Eichenbaum, H. (2011). Hippocampal "time cells" bridge the gap in memory for discontiguous events. *Neuron, 71*, 737–749.

Maguire, E. A., Frackowiak, R. S., & Frith, C. D. (1997). Recalling routes around London: Activation of the right hippocampus in taxi drivers. *Journal of Neuroscience, 17*(18), 7103–7110.

Maguire, E. A., Valentine, E. R., Wilding, J. M., & Kapur, N. (2003). Routes to remembering: The brains behind superior memory. *Nature Neuroscience, 6*, 90–95.

Mathis, A., Stemmler, M. B., & Herz, A. V. (2015). Probable nature of higher-dimensional symmetries underlying mammalian grid-cell activity patterns. *Elife, 4*, e05979.

Minchin, E. (2001). *Homer and the resources of memory: Some applications of cognitive theory to the Iliad and the Odyssey.* Oxford: Oxford University Press.

O'Keefe, J. (1991). An allocentric spatial model for the hippocampal cognitive map. *Hippocampus, 1*, 230–235.

O'Keefe, J., & Dostrovsky, J. (1971). The hippocampus as a spatial map. Preliminary evidence from unit activity in the freely-moving rat. *Brain Research, 34*(1), 171–175.

O'Keefe, J., & Nadel, L. (1978). *The hippocampus as a cognitive map.* Oxford: Clarendon Press.

Pylyshyn, Z. W. (1984). *Computation and cognition: Toward a foundation for cognitive science.* Cambridge, MA: MIT Press.

Raz, A., Packard, M. G., Alexander, G. M., Buhle, J. T., Zhu, H., Yu, S., & Peterson, B. S. (2009). A slice of π: An exploratory neuroimaging study of digit encoding and retrieval in a superior memorist. *Neurocase, 15*, 361–372.

Redish, A. D. (1999). *Beyond the cognitive map: From place cells to episodic memory.* Cambridge, MA: MIT Press.

Rescorla, M. (2009). Cognitive maps and the language of thought. *British Journal for the Philosophy of Science, 60*, 377–407.

Rescorla, M. (2017). Maps in the head? In K. Andrews and J. Beck (Eds.), *The Routledge handbook of philosophy of animal minds* (pp. 34–45). London: Routledge.

Roediger, H. L. (1980). Memory metaphors in cognitive psychology. *Memory and Cognition, 8*(3), 231–246.

Solstad, T., Boccara, C. N., Kropff, E., Moser, M., & Moser, E. I. (2008). Representation of geometric borders in the entorhinal cortex. *Science, 322*(5909), 1865–1868.

Stensola, T., Stensola, H., Moser, M. B., & Moser, E. I. (2015). Shearing-induced asymmetry in entorhinal grid cells. *Nature, 518*(7538), 20.

Taube, J. S., Muller, R. U., & Rank, J. B. (1990). Head-direction cells recorded from the postsubiculum in freely moving rats. *Journal of Neuroscience, 10*, 420–435.

Tavares, R. M., Mendelsohn, A., Grossman, Y., Williams, C. H., Shapiro, M., Trope, Y., & Schiller, D. (2015). A map for social navigation in the human brain. *Neuron, 87*(1), 231–243.

Tolman, E. C. (1948). Cognitive maps in rats and men. *Psychological Review, 55*(4), 189–208.

Wang, A. Y., & Thomas, M. H. (2000). Looking for long-term mnemonic effects on serial recall: The legacy of Simonides. *American Journal of Psychology, 113*, 331–340.

Wilding, J. M., & Valentine, E. R. (1997). *Superior memory*. Hove, UK: Psychology Press.

Wood, E. R., Dudchenko, P. A., & Eichenbaum, H. (1999). The global record of memory in hippocampal neuronal activity. *Nature, 397*, 613–616.

Yoon, J. S., Ericsson, K. A., & Donatelli, D. (2018). Effects of 30 years of disuse on exceptional memory performance. *Cognitive Science, 42*, 884–903.

Zacks, J. M., & Swallow, K. M. (2007). Event segmentation. *Current Directions in Psychological Science, 16*, 80–84.

Zhao, M. (2018). Human spatial representation: What we cannot learn from studies of rodent navigation. *Journal of Neurophysiology, 120*, 2453–2465.

12 Implications from the Philosophy of Concepts for the Neuroscience of Memory Systems

Anna Leshinskaya and Enoch Lambert

12.1 Introduction

It is a common assumption in cognitive science that the mind is furnished with elements called concepts. These constituents allow us to track observations and form thoughts about the same thing: if one believes that apples are tasty and observes that that's an apple, one can relate those propositions appropriately (Carey, 2009; Margolis & Laurence, 1999; Murphy, 2002; Rey, 1983; Smith & Medin, 1981). Psychologists seek to explain behavior hypothesized to rely on concepts, such as categorization, inference, and language comprehension. Cognitive neuroscientists seek to identify functionally distinct components of the mind and brain via what those components represent and compute (their characteristic features). Here, we ask how cognitive neuroscientists could probe conceptual representation as such a putative functional component.

Concepts are treated as a distinct topic of study within cognitive neuroscience (Hoffman, McClelland, & Lambon-Ralph, 2017; Mahon, 2015; A. Martin, 2007; Patterson, Nestor, & Rogers, 2007), but this literature is mostly silent on the question of what distinct cognitive or functional characteristic features concepts are thought to have. To be a well-defined empirical entity, concepts or conceptual memory must have operational criteria sufficiently precise to allow one to identify and measure them distinctly from other kinds of mental representations—notably, from other things stored in long-term memory.[1] Here, we seek out such criteria. Ideally, these criteria remain aligned with the findings and theoretical frameworks in psychology of concepts. Failing to find such criteria might lead one to reconsider what it means to have a cognitive neuroscience of concepts and whether this term is useful. This is the issue at stake here.

In the first half of this chapter, we review existing approaches in cognitive neuroscience for differentiating among kinds of long-term memories by virtue of their characteristic and measurable features, some of which could serve as candidates for operational criteria for concepts. For example, we discuss the difference between memories that aggregate across experience and those that pick out unique experiences,[2] and the difference between memories that are specific to a certain sensory channel and those that are channel invariant. However, we encounter a repeated problem: these distinctions do not neatly map onto the notion of concepts as it is often used in the psychology and neuroscience literature, and these criteria permit cases that would only reluctantly be called concepts by most researchers and laypersons (e.g., participants in psychology experiments). In our analysis of these cases, we propose that to account for both lay and researcher intuitions about concepts, we must appeal to a notion of sharedness: whether a memory refers to information that is also likely to be represented by others.

In the second half, we grapple with whether sharedness should be considered important in the scientific treatment of concepts rather than only describing a folk psychological intuition. In so doing, we raise the possibility that folk-psychological intuitions about memory may in fact play a role in a scientific theory of the brain, to the extent that they themselves may influence how memories are stored. Lastly, we propose an empirical path forward to settling this issue.

12.2   Preliminaries: Concepts as Representational Types

We take as a premise that concepts are (at least) one type of mental representation and are distinct from certain other distinguishable types. This view is a pillar for work in developmental psychology, which uses it to distinguish among hypotheses regarding infant cognition (Carey, 2009; Mandler, 2004) and in philosophical treatments, which discuss features distinguishing concepts from other kinds of representations (Margolis & Laurence, 1999; Millikan, 2017; Rey, 1983). It has also been a fundamental guiding framework in cognitive neuropsychology, which proposes distinct memory systems to account for patterns of impairment following neural damage (Caramazza et al., 1990; Eichenbaum & Cohen, 2001; Schacter & Tulving, 1994; Tulving, 1972, 1984; Warrington, 1975; Warrington & Taylor, 1978). For the distinctions among memory types that we delineate below, we draw on both theoretical and empirical justifications.

However, it must be noted that it is not a universally accepted premise in cognitive neuroscience that concepts are a type of mental representation. A different view is that concepts are a type of stimulus or task. This tradition studies representations of concepts, that is, mental representations about things with certain properties, rather than conceptual representations, that is, mental representations with particular properties (for reviews: Binder et al., 2009; Hoffman et al., 2017; A. Martin, 2007; McRae & Jones, 2013). In this representation of concepts view, there is no meaningful distinction among different representations evoked by the same "conceptual" stimulus. Thus, studying conceptual processing involves measuring the brain as it engages in processing the meaning of a word or picture relative to not accessing its meaning or relative to meaningless stimuli. For example, neural activity is measured as participants read and understand words relative to counting how many times the letter E appears in those words or reading pseudowords (Binder et al., 2009; Renoult, Irish, et al., 2019). Relatedly, different categories of words or images are contrasted (e.g., animals vs. tools), and the result is taken as a reflection of how animal and tools concepts are represented (A. Martin et al., 1996; Watson et al., 2013). All of the mental machinery engaged is relevant to the question of how we represent concepts, since the interest is in the set of processing mechanisms associated with certain classes of stimuli, not representations with certain qualities or of certain kinds.

In our view, a single stimulus—such as the word "apple"—can (at least conceivably) engage a multitude of different kinds of representations: we might retrieve a mental image of a bright red apple, knowledge of how apples grow, the lexical knowledge that its plural is "apples," a vivid childhood memory of picking apples with grandma, and even a salivation response in anticipation of eating one. Only some of these are conceptual. Simultaneously, a similar representation or process can be evoked by different stimuli: both the word apple and a classically conditioned auditory tone could evoke the anticipation of a reward or a salivation response. In our view, stimulus type cannot be assumed to provide a good classification scheme for the functional components of the mind and brain. Rather, it is an empirical matter to determine how mental representations are different or alike, which can be resolved by measuring properties of those mental representations. As we hope to make clear throughout the chapter, there are theoretical and empirical justifications for our view. Thus, we take as a starting point that there exists an important question regarding the different kinds of things stored in long-term memory, even if evoked by the same meaningful stimulus, and the questions we raise are how we should distinguish among them using the methods of cognitive neuroscience, and which ones best map on to the psychological notion of concept.

12.3   Approaches to Distinguishing among Types of Memory

Many cognitive neuroscientists have indeed taken up the task of distinguishing among types of long-term memory by virtue of their characteristic properties as measurable with the tools of neural recording or imaging. It is based on this work that we seek candidates for conceptual aspects of memory. We take the term "semantic memory" to be roughly equivalent to conceptual memory, and take concepts to be elements within semantic or conceptual memory.

We begin by reviewing three major well-operationalized criteria that have been used to distinguish among types of long-term memories: (1) experiential scope, or the extent to which a representation aggregates over specific experiences; (2) channel specificity, that is, whether a representation is specific to a sensory channel of input; and (3) its similarity profile, that is, whether a representation tracks closely with physical stimulus similarity or departs from it. We consider the idea that these distinctions can serve to operationalize the idea of conceptual or semantic representations, supplying the kind of operational definition cognitive neuroscience needs to study concepts as one or more functional neural components. Specifically, we consider the idea that conceptual representations could be ones that aggregate over more experience, are not specific to a sensory channel, and can depart from physical similarity space.

However, we also find that cognitive neuroscientists do not always use the terms "concept" or "semantic memory" to describe the components of memory that these approaches pick out, and that these operational definitions do not align with the way these terms are used in cognitive neuroscience or in cognitive science more broadly. Indeed, these criteria, alone and in combination, allow cases that are not typically considered concepts. While this might seem like only a terminological issue, it is in fact a deeper ontological one about what entities exist in the mind and how we should catalog mental phenomena. In the final section, we consider the latent intuition likely underlying these terminological choices, and whether it merits a revision to how memory systems are studied.

12.3.1 Experiential Scope

Suppose I believe that Szechuan peppers are spicy. How do I know that? Perhaps someone told me this once at a Chinese restaurant, pointing out the peppers on a plate, and I recall this specific episode vividly from a first-person perspective. Or perhaps I sampled them on my own and came to the conclusion that this must be generally true. Or perhaps it is a fact I know from repeated experiences, without recalling any individual experience. The difference between recalling unique personally experienced episodes and representations that are formed by aggregating over many such episodes has been an important distinction in cognitive neuroscience and neuropsychology.

Researchers have used a number of approaches to isolate memory of unique episodes (often termed "episodic memory") from our more general knowledge. Neuropsychological studies find that hippocampal damage disproportionately impairs patients' ability to recall individual events vividly, but that such patients typically retain general factual knowledge (Manns, Hopkins, & Squire, 2003; Mishkin, 1997; Nadel & Moscovitch, 1997; Schacter & Tulving, 1994; Sekeres, Winocur, & Moscovitch, 2018; Squire, Knowlton, & Musen, 1993; Tulving, 2002; Winocur et al., 2010). However, for recall of naturalistic lived experiences, these two kinds of memory can be difficult to isolate experimentally, as the opening vignette attests: when we don't actually know how an individual came to form their belief or what information allows them to attest it, it is difficult to know to what extent it encodes a single episode or aggregates across repeated experiences. To address this issue, one can distinguish individual from aggregated aspects of memory by probing learning in a controlled lab setting and recording neural activity in response to retrieving it. One can define episodic memory as those aspects of a memory that refer to unique aspects of a specific episode, and other kinds of memory as referring to regularities across many of them.

For example, suppose that one encountered Szechuan peppers on ten distinct occasions, and each of those times they were spicy but also each time served alongside different side dishes. Once, they were served with a yogurt dish—a memorable experience for the relief it gave to the palate afterwards. Recalling that particular individual pairing of Szechuan peppers with yogurt would be considered episodic because it is a unique encounter requiring individuating that experience from others, which did not feature yogurt. Recalling the regularity that Szechuan peppers are spicy without retrieving any individual episode but only by recalling aggregated experience is considered something else—something with a larger "experiential scope" (in the terminology used here—that is, the extent to which a representation aggregates over specific experiences).

There are both theoretical and empirical motivations for seeing the size of experiential scope as an important distinction. Some research finds that patients with episodic deficits appear able to learn some kinds of new representations that aggregate across repeated presentations of stimuli, though possibly in different ways than controls (Knowlton, Ramus, & Squire, 1992; Knowlton, Squire, & Gluck, 1994; Myers et al., 2003; Tulving, Hayman, & Macdonald, 1991; Verfaellie, Koseff, & Alexander, 2000; cf. Manns et al., 2003). Behavioral studies with nonhuman animals also support such distinctions. In an elegant approach, mice were given numerous trials of a water-maze task, in which they were placed in a pool too deep to stand and had to find a platform on which to rest (Richards et al., 2014). On each individual trial, the platform was in a different location, centered on a coordinate mean. On test trials, in which there is no platform, search behavior could be measured as reflecting the search for specific locations previously experienced or to the coordinate mean (which did not characterize any specific prior platform but rather the central tendency summarizing all of their prior experiences). On day 1 after learning, behavior was best characterized as reflecting specific locations experienced during learning, but after thirty days, mice went to their Cartesian average location, suggesting that memories might be transformed over time from unique to aggregated. Furthermore, neuroimaging work with humans finds that the hippocampus tends to represent episode-unique information, while other areas are more likely to aggregate (Dimsdale-Zucker et al., 2018; Tompary & Davachi, 2017; Wiggs, Weisberg, & Martin, 1999; but cf. Schapiro et al., 2016).

Finally, there are reasons to understand these as theoretically divergent functions of memory. One allows you to recall exactly what happened and when, while the other allows you to acquire a generalizable model of the world. There are compelling computational demonstrations indicating that the functions of storing episode-unique versus aggregated information inherently trade off in the kind of learning algorithms they require, and are thus best handled by at least partially distinct learning and memory systems (Eichenbaum, 2004; McClelland, McNaughton, & O'Reilly, 1995; O'Reilly & Rudy, 2001; Winocur et al., 2010). This is because one must use a slow learning rate to capture generalities that hold across experiences (not take any individual sample too seriously), which is inherently at odds with the fast learning rate that is required to recall the distinguishing features of a single episode. Having distinct systems with different learning rates allows us to learn both the rule and the exceptions governing experience.

The question we return to is whether aggregated memories are the right operationalization for concepts. Does the notion of concept or semantic memory from cognitive science map neatly onto the aggregated versus unique distinction used in neuroscience? On one hand, the idea of aggregation dovetails with the psychological literature on categorization, which shows that we tend to categorize entities on the basis of the typical features or characteristics of their category, being highly sensitivity to probabilities of a certain feature across encounters with that type (Rips, Shoben, & Smith, 1973; Rosch, 1975; for reviews, see Smith & Medin, 1981, and Murphy, 2002). For example, we readily judge that if something has wings, it probably flies but doesn't swim (even though penguins are an exception). Feature correlations across categories predict patterns of data on development and deterioration in semantic dementia. For example, categories that have more shared properties with their semantic neighbors are more robust to deterioration and yield different kinds of naming errors (Hodges & McCarthy, 1995; Rogers et al., 2004).

On the other hand, there are reasons to think that concepts are a more specific kind of representation than just any aggregate memory. Classically, Smith and Medin (1981) argued that psychological theories should distinguish between heuristics we use to identify members of a category ("identification procedures") and those we use to define inclusion in the category ("conceptual cores"). For example, we might use hairstyles and clothes to judge gender quickly, but we do not believe that these are defining or essential—even if they are extremely prevalent and diagnostic. The fact that participants can represent both of these facts simultaneously about a single category was elegantly demonstrated by Armstrong, Gleitman and Gleitman (1983), who showed that participants judge that some instances of the category "even number" are somehow better examples than others (e.g., 2 is better than 34), despite knowing the definition. Likewise, we might use appearance in Chinese restaurants to decide if an ambiguous vegetable is a Szechuan pepper, but resist the idea that being in a Chinese restaurant is important for it being one (Prasada & Dillingham, 2009). We may not even know what those defining features are, and yet we still believe they exist (Carey, 2009; Rey, 1983). In short, there is a disconnect between how we judge things to really be (definitional judgment) versus how we categorize them. While the latter might rely on aggregate representations, the former does something more—perhaps setting logical bounds of a concept—and is argued to be more important to concept meaning.

This distinction between categorization versus definition appears to inform not only everyday intuitions but also researchers' use of the term "concept." Representations simply summarizing experiences appear in a broad swath of literature not using the terms "concepts" or "semantic memory." For example, the literature on ensemble perception shows that we spontaneously average the features of sets of items, whether the sizes of circles or orientation of Gabor patches, and can report them after a delay with remarkable accuracy (Alvarez & Oliva, 2009; Ariely, 2001; Haberman & Whitney, 2009; Parkes et al., 2001). Aggregation over experience is prominent in reinforcement learning (Shohamy & Daw, 2015), speech segmentation and visual statistical learning (Chun & Turk-Browne, 2008; Saffran, Aslin, & Newport, 1996), natural scene statistics (Simoncelli & Olshausen, 2001), and a huge range of other cognitive domains. In few cases is the term "concept" invoked to describe these phenomena. Conversely, researchers in semantic memory do not take just any evidence about aggregated memory as relevant to their phenomena of study: their reviews do not cite learning experiments with Gabor patches, water maze platforms, or the distribution of phonemes in spoken language. A related fact was noted in a highly extensive review of the concepts literature in cognitive psychology, arguing that many of these omissions seem arbitrary (Murphy, 2002).

Thus, it would seem that the average location of water-maze platforms is not considered part of conceptual memory in the cognitive neuroscience of concepts in the same way as "apples are a fruit that tends to grow on trees", but there is no formal definition that clearly distinguishes these examples. It thus seems like there is a latent intuition driving these terminological uses. Before we probe this intuition further, and consider philosophical proposals to answering it, we evaluate two other criteria that may be important to characterizing concepts: channel specificity and stimulus similarity.

12.3.2 Channel Specificity

Neuropsychologists have long used specific criteria to probe impairments to semantic memory following degenerative disease (such as semantic dementia) or stroke (Caramazza, Berndt, & Brownell, 1982; Hodges, 1992; Shallice, 1988; Warrington, 1975). Such criteria have identified impairments that are consistent regardless of the sensory channel or stimulus modality (pictures, words, sounds) with which the patient is tested. Indeed, there are many cases of patients showing impairment across various tasks probing their knowledge, including matching pictures to their names, selecting sets of more versus less related words or pictures, or judging the typical attributes of categories. Such patients are relatively spared on measures of episodic memory, such as recalling autobiographical details of their lives (Hodges, 1992). Additionally, such patients can be contrasted with others who have deficits specific to representations proprietary to certain sensory channels. For example, damage to a structural description system is thought to be responsible for impairments to the ability to recognize unconventional views of objects, relative to recognizing them from conventional views, which attest to the sparing of the concept itself (Schacter & Tulving, 1994; Warrington & Taylor, 1978). Instead, the mapping between visual form and the conceptual system is what is affected. It is also thought to be implicated in modality-specific naming disorders, such as optic aphasia, in which patients cannot name objects when they are presented visually but can otherwise. Because they can name them when presented in tactile form or verbal description, it is not a deficit at the linguistic or semantic level. Simultaneously, it is not an impairment to basic visual processing, since they can perform appropriate actions with those objects (Beauvois, 1982; Caramazza et al., 1990; Riddoch & Humphreys, 1987). Therefore, the idea of a distinct neural system responsible for modality-invariant, non-autobiographical memory, and their double dissociates, is well grounded in neuropsychological data.

Neuroimaging approaches to semantic memory guided by the same principle have tracked neural responses that are common to different modalities of presentation of the same concept, such as the word "rabbit" and pictures of rabbits (Devereux et al., 2013; Fairhall & Caramazza, 2013; Simanova et al., 2012). However, these approaches can still reflect channel-specific representations because even representations proprietary to a sensory modality can be associatively retrieved from another cue. For example, a verbal cue can prime motor plans and vice versa (Fischer & Zwaan, 2008; Yee et al.,

2013), and likewise, whatever knowledge structure that allows one to map the visually specific, structural description of rabbit to the word "rabbit" can also be retrieved in reverse fashion when reading "rabbit." Another approach has been to study whether representations are preserved in cases where a sensory channel is missing from birth, such as congenital blindness (Bedny, Caramazza, Grossman, et al., 2008; Bedny, Caramazza, Pascual-Leone, et al., 2011; Peelen et al., 2013; Peelen, Romagno, & Caramazza, 2012; Striem-Amit et al., 2018; Wang et al., 2015). If a neural response to the word "rabbit" in the congenitally blind is similar to the responses in the sighted, it cannot be a representation that belongs only to the visual channel. While none of these approaches are impervious to criticism, the issue we focus on here is whether in ideal circumstances these characteristics are even sufficient. Are cross-modally accessible representations always conceptual? Are such criteria enough to distinguish conceptual representations from others?

One reason to think they are not is to consider the kinds of cases we raised in the earlier section on aggregation. We noted that representations such as the average location of water-maze platforms is not considered a concept by many researchers, and it does not seem that making this representation cross-modal helps. Suppose that the mice could find the platforms haptically when blindfolded. Even if their representation of location is not tied to vision specifically, the example does not seem intuitively more conceptual. It is also relevant to note that literature on cross-modal integration is not normally considered to be investigating concepts. For instance, there is a rich body of work reporting common responses to dynamic faces and vocal sounds in the posterior superior temporal sulcus (Deen et al., 2015). This area is not considered to represent concepts, and is instead described as an integrative audiovisual area. In sum, literatures on semantic memory and conceptual knowledge do not use just any cases of cross-modally accessible representations as part of the relevant phenomena. We suspend discussion of why, and why this should matter, until section 12.3.4.

12.3.3   Stimulus Similarity

A third approach that we consider measures different kinds of similarity relations among stimuli as they are represented in various neural areas. A cardinal feature of concepts, according to the psychological literature, is that similarity relations among concepts need not follow the surface or physical feature similarity of their referents (Carey, 2009; Gopnik & Meltzoff,

1997; Keil et al., 1998; Mandler, 2004). In a classic example, children will judge that an animal born a raccoon but transformed to look like a skunk is fundamentally still a raccoon despite having the superficial features of a skunk (Keil et al., 1998). This idea has been crucial in research on conceptual development and beyond (Murphy, 2002).

Departure from physical feature similarity has sometimes, but relatively rarely, been adopted in cognitive neuroscience. While it has been used in our work on semantic memory (Leshinskaya & Caramazza, 2015; Leshinskaya et al., 2017), its most extensive use has been in work on visual recognition and categorization where the term "concept" is only sometimes invoked (Bracci & Op de Beeck, 2016; Freedman et al., 2001; Haushofer, Livingstone, & Kanwisher, 2008; C. B. Martin & Barense, 2018; Miller et al., 2003; Proklova, Kaiser, & Peelen, 2016; Roy et al., 2010). Consider, for example, Freedman and colleagues (2001), who specifically probed representations that discretize a continuous feature space of visual stimuli, leading to a categorical structure. Macaques were taught category boundaries among artificial images, which were a continuously interpolated set of morphs between two reference images: a cat and a dog (Freedman et al., 2001). The category boundary in this parametric stimulus space was defined arbitrarily, designating everything on one side as cats and everything on the other as dogs. This made it possible to dissociate distance in the feature space from category membership by finding pairs of items that are equidistant in the feature space that are either within a boundary (are both cats) or straddle it (one is a cat; the other is a dog). Neural responses could then be analyzed as reflecting distance in the stimulus parameter space or distance according to the category boundary (crossing it or not). It was found that some neurons, specifically in the lateral prefrontal cortex (lPFC), exhibited categorical responses, while others (in the inferior temporal cortex) reflected the overall visual distance but were insensitive to the boundaries.

The above is an important finding in its own right, irrespective of whether one calls the categorical representation "conceptual." For the present discussion, however, we raise the question of whether it should be. Should this finding be considered as demonstrating that the lPFC is a semantic area, or something else? In spite of its use of an important characteristics of concepts from the psychological literature, work of this sort is not incorporated in many reviews of semantic memory and not used as criteria in an operationalization of concepts in cognitive neuroscience (Binder & Desai,

2011; Binder et al., 2009; A. Martin, 2007; Patterson et al., 2007; Renoult, Irish, et al., 2019). While this could be simply a methodological oversight, we consider below that a separate intuition guides researchers' use of this term, and ask whether this intuition should be followed or discarded.

12.4   Probing Intuitions about Concepts

In this section, we offer preliminary evidence that the operational characteristics examined thus far do not always guide cognitive neuroscientists' judgments about the notion of a concept. We suggest that judgments of the extent to which information is shared capture usage not covered by the operational characteristics.

We have considered three well-operationalized characteristics for distinguishing among kinds of representations in long-term memory that can and have been used in cognitive neuroscience: (1) whether a representation aggregates across experience or individuates specific episodes, (2) whether it is proprietary to a certain sensory input channel or is channel invariant, and (3) whether it can depart from physical stimulus similarity. These distinctions are well defined, cognitively meaningful, and work to distinguish distinct neural systems reliably. Many of them cohere with criteria set out in cognitive psychology for conceptual memory, even though each alone seemed not to capture everything the term "concept" is expected to capture. Below, we consider whether a combination of all of these criteria might serve as a good candidate in our search for an operational definition of "concept" for cognitive neuroscience. However, we argue that even as a combination, this set of criteria does not always guide cognitive neuroscientists' usage of the term "concept," and we further probe what intuition or theoretical commitment does guide it.

12.4.1   Combining the Operationalized Characteristics

Suppose a mouse is tasked with learning the location of water-maze platforms across repeated experiences, aggregating them into an average. The mouse can access this knowledge visually or haptically. Moreover, the mouse acquires two categories of water-maze platforms—ones that fall on the east side of the pool and ones that fall on the west side—and we find representations of categorical cross-modal representations of water-maze positions in their brains. Have we identified a neural substrate supporting conceptual memory, at least for a specific concept?

The participant in this hypothetical experiment is a mouse, but there are analogous examples from everyday human experience. Let's consider your long-term memory of the average route you take to work, the typical location of gas stations on the way there, the usual color of bicycles you own, or the typical contents of your grandmother's attic. All of these have (or can easily have) the set of characteristic features we just outlined but are not canonically conceptual. If our operational criteria for concepts include knowledge such as categories of water-maze platform locations and the typical content of one's grandmother's attic, then it would behoove concepts researchers to incorporate findings regarding such representations into their theoretical frameworks and reviews, and expand the scope of their experiments to include them. However, this does not appear to be the case. Despite meeting the representational criteria, the term "conceptual" is not typically used to denote these kinds of representations across the literature in cognitive neuroscience.

To substantiate our observations further regarding the use of the terms "concept" and "semantic memory" in the literature, we additionally queried cognitive neuroscientists of semantic memory directly. We sent a small survey to thirty-three researchers in the cognitive neuroscience of semantic memory. Of these, eleven completed the survey. Ten of these were either postdoctoral or professorial level scholars, and one was a graduate student. Although this is a small sample, we believe it is unbiased because the participants did not know the question under investigation or the position of the authors on it. Indeed, we had not discussed this topic with any of the survey participants.

Participants provided electronic consent in accordance with procedures approved by the Institutional Review Board of the University of California, Davis. Participation was entirely voluntary, and responses were anonymized; no compensation was provided. We first asked participants to indicate their background, and then presented the following instructions:

> In the following, each question will describe a mental representation or belief that a person might have. For each one, indicate the degree to which this representation would be typically studied under the domain of "semantic memory" or "conceptual knowledge." We are interested in your opinion about how these terms are used in your field.

Participants were then shown one phrase at a time describing a belief, and a slider response ranging from 1 ("definitely no") to 5 ("definitely yes") indicating the extent to which they agreed that this belief would be considered "semantic memory," "conceptual knowledge," or "something else." Phrases

were of three kinds: examples of highly canonical conceptual representations (such as "Dressers are a kind of furniture used to store clothes" and
"Grandmothers are women whose offspring also have offspring"), examples
of canonical episodic memories that are specific to a time and place ("One
time, I rode my bike along the river in town and found an abandoned
mill twenty miles away" and "The gas price today is $3.45/gallon"), and
examples that aggregate over experiences but are not canonically semantic
("The dresser in my bedroom is about two feet from the bed in the north-
west corner of the room, and I can usually navigate around it in the dark,"
"Most of the bicycles I've owned have been yellow," and "My grandmother
tends to store old photo albums in her attic but not heirlooms").

We present the results in table 12.1. Responses to "semantic memory"
and "conceptual knowledge" questions were highly correlated ($r=0.99$).
So, these were collapsed into a combined "semantic/conceptual" rating.
We found that the mean response on the semantic rating scale for canonical semantic examples ($M=4.67$) was significantly higher than for both
the canonical episodic examples ($M=2.66$, $t[10]=7.02$, $p<0.001$) and the
aggregate personal examples ($M=3.32$, $t[10]=5.63$, $p<0.001$). Correspondingly, responses on the "something else" rating scale were lower for the
canonical semantic items ($M=2.18$) than both the canonical episodic items
($M=4.27$, $t[10]=7.39$, $p<0.001$) and the aggregate non-canonical items
($M=3.25$, $t[10]=2.63$, $p=0.025$). The pattern of responses also reveals that
ratings for the aggregate non-canonical items were intermediate between
the canonical semantic and episodic items. Thus, the semantic rating was
indeed higher for the aggregate non-canonical items than for the canonical
episodic items ($t[10]=2.18$, $p=0.046$) and the rating for "something else"

Table 12.1
Results from a survey of eleven researchers in the cognitive neuroscience of semantic
memory

Item Type	Rating Type	
	Mean (SD) Rating on Semantic/Conceptual Scale (1–5)	Mean (SD) Rating on Something Else Scale (1–5)
Canonical semantic	4.67 (0.42)	2.18 (0.67)
Canonical episodic	2.66 (0.88)	4.27 (0.79)
Aggregate non-canonical	3.32 (0.91)	3.25 (1.28)

was lower ($t[10]=2.82$, $p=0.018$). Overall, this suggests that cognitive neuroscientists of semantic memory—at least, the ones queried here—classify
these aggregated experiences differently from both episodic and semantic
memory. They do not believe they are considered conceptual in the same
way as the canonical examples. Furthermore, they are less confident that
they are studied as "something else" relative to episodic examples. Overall,
this suggests that there is genuine uncertainty in cognitive neuroscience
of semantic memory about how such examples should be classified with
respect to existing ideas about memory systems. Although the small sample
may fail to capture intuitions of all cognitive neuroscientists, it reinforces
our observations from the literature regarding how these terms are used.
Altogether, we believe that something is missing from the definition of
semantic memory as it is currently actually used in the field.

There are two paths forward. One path is to look for additional criteria
that will help us classify the non-canonical examples and incorporate these
additional criteria into the operational definition of semantic memory.
The other path is to suppose that these in-practice use patterns of the term
"semantic memory" are not meaningful, or are erroneous, and should be
replaced with formal, operational definitions that make empirical predictions about brain and behavior. If intuitions about concepts do not coincide with relevant scientific notions in cognitive neuroscience, then these
intuitions have no place guiding terminology in the field. To adjudicate
between these paths, we consider what additional criteria might guide the
latent intuition behind the usage of the term "concept" and whether these
criteria may be scientifically useful.

12.4.2 Probing the Intuition

It is possible that there is no coherent way to distinguish between bona fide
examples of concepts, such as the knowledge that apples are a type of fruit
that grow on trees, and these seemingly odd, non-canonical examples, such
as the knowledge that my grandmother liked to store photo albums in her
attic. However, we propose that a coherent but latent intuition does guide
this distinction, and that the basis for this intuition turns on notions of
sharedness or intersubjectivity.

Historically, Tulving (1972) first proposed to distinguish semantic memory,
which stores concepts, from episodic memory, which stores personal experiences specific to a time and place. Episodic memory was thus characterized

by at least two features: specificity to time and place, and a personal or auto-biographical nature (Tulving, 2002). Both features have been used (some-times separately) in the operational definition of episodic memory since then (Nadel & Moscovitch, 1997; Renoult, Davidson, et al., 2012; Westmacott et al., 2004; Winocur et al., 2010). On the other hand, when semantic mem-ory has been operationalized, only the characteristic of generality of time and place have been used explicitly. The inverse of autobiographicality has not been used to operationalize it (though cf. Renoult, Davidson, et al., 2012).

We propose that this inverse could be called "sharedness" or "inter-subjectivity" and that it serves as one endpoint of a continuum opposite to autobiographical memory. This continuum would reflect the extent to which information describes one's own idiosyncratic and personal expe-rience or whether it is seemingly something that many others have also observed and know about. This idea is distinct from (though related to) generality or aggregation because autobiographical knowledge can span beyond a singular episode. For example, the knowledge of what I *typically* have for breakfast, or what my grandmother *tends* to store in her attic, are exactly the kinds of cases that this accommodates.

Figure 12.1 illustrates several examples of less and more autobiographi-cal information. At one extreme is the particular breakfast I had this morn-ing, alone at home. At the other is the idea of food eaten in the morning, which leaves open what exactly is eaten but increases its universality in human experience. No matter the country or culture, one could understand "breakfast" in this way. A simple litmus test for locating information along this dimension is by imagining what content is reliably conveyed by the

My breakfast this morning   My typical breakfast   Typical breakfast foods in my culture   Meal eaten in the morning

Autobiographical   *Intersubjective*

Figure 12.1
Hypothesized placement of examples of different kinds of memories in terms of their level of autobiographicality versus intersubjectivity/sharedness. (Images are open-source with free license). See color plate 3.

use of a word: when I say "breakfast" in conversation, I don't expect the word itself to convey anything about *my* breakfast that morning, or that I usually have breakfast in a sunny nook in my kitchen. I might expect others to understand that it is a meal eaten at the start of the day; and maybe that it contains typical breakfast foods in North America. Of course, this inference requires an understanding of what is common knowledge among speakers (like much of conversation).

This distinction between personal, autobiographical versus shared, intersubjective information seems to capture the intuitions and usage surrounding the word "concept" in cognitive neuroscience. For example, Hodges (1992) described semantic memory as knowledge that is "culturally shared," and Tulving (1972) described it as referring to the world rather than the self. The usage from these leading neuropsychologists has been influential; it is taken as the inherited view in modern work (e.g., Yee & Thompson-Schill, 2016).[3] It may be for this reason that cat–dog categories, locations of water-maze platforms, the contents of one's grandmother's attic, and other aggregated, categorical, cross-modal representations are not universally treated as conceptual. We argue that this is because they are idiosyncratic, personal, and lacking a sense of sharedness.

From the perspective of the observer or knower, a concern with sharedness is ultimately a concern with representing facts about the world that are independent of one's fallible view. A given observation may be a chance occurrence, and glitch of the senses, or an internally generated hallucination; it may be highly local and context dependent; or it may be simply incidental. Perhaps, the spiciness of a specific Szechuan pepper is because of my overly sensitive taste buds, a poor choice of specimen, contamination from hot sauce, or the peculiarities of my local Chinese restaurant. In these cases, storing the notion that Szechuan peppers are spicy as a fact in semantic memory would be a bad idea.

A similar principle might guide the intuitions documented in psychology experiments that clothing style is not part of the concept "boy" or that looking like a raccoon is not part of the concept "raccoon," as we described earlier. Judgments about more defining or essential properties are similarly concerned with representing what things really are. Even if we never obtain the ground truth, true properties are more likely to be shared across viewers, viewpoints, and incidentals of the environment. So, a concern with sharedness dovetails with these concerns also.

In the water-maze example, the tendency of platforms to be on the west side of the pool might be an accidental feature of the maze, itself an artificial setup specific to the lab. Whether other mazes have those features and whether other mice ever experience them is unknown to the mice. Yet, if the mice had evidence that they are not idiosyncratic, individual experiences, perhaps these representations would start to resemble more classic examples of concepts. One relevant cue would be if mice had ways of reaching the understanding that these mazes are a ubiquitous feature of mouse life. Suppose mouse A leaves the experiment chamber and returns to the cage, signaling to mouse B that it's a category 1 type day—the platforms will be mostly on the west side! They could come to believe these were real objects in the world, no longer purely autobiographical. We conjecture that the use of the term "concept" in cognitive neuroscience is implicitly guided by these considerations, similarly to how it guides laypersons in their judgments in psychology experiments.

The critical question we now raise is whether such considerations about sharedness versus autobiographicality should form part of the formal operational definition of the scientific term "concept" in cognitive neuroscience—that is, be used explicitly as a way to distinguish memory systems in neuroscientific experiments (which they have not been).

12.5   Proposals for Sharedness as Operational Criteria for Concepts

To address the question of whether sharedness should be part of the operational definition of conceptual memory, we turn to discussions in the philosophy and psychology of concepts regarding the importance of sharedness in a formal theory of concepts.

Two recent works on concepts, Ruth Millikan's *Beyond Concepts* (2017) and Susan Carey's *The Origin of Concepts* (2009), take opposing views on the extent to which cognitive science should treat sharedness as a critical property of conceptual representation. Their concern is in the objective fact of sharedness: whether concepts are, or cannot be, mental representations that are truly common among people, truly distinct from idiosyncratic bits of belief. This would seem relevant for the question of whether the scientific notion of "concept" should incorporate sharedness in its definition. Accepting Milikan's view would entail discarding this distinction and thus suggest a major ontology revision to cognitive neuroscience. Accepting Carey's view would require justifying an objective basis for distinguishing shared versus autobiographical content. We resolve this tension by offering a third view.

12.5.1 What Is Shared among Minds?

An influential philosophical position on concepts is that meanings, whether of words or mental representations such as concepts, are largely not in the head (Kripke, 1972; Putnam, 1973). A concept is a mental representation that points to a referent, but it need not have any special content apart from this referential relation. This is illustrated by a thought experiment that asks us to suppose that there exists a twin Earth in which you have a perfectly identical twin who has had all of the same experience as you (Putnam, 1973). The only difference in the twin world is that the molecular structure of water is XYZ rather than H_2O. Yet, to all available observations, the properties of twin water are identical to those of Earth water. The key intuition is that your thoughts about water and your twin's thoughts about water refer to different things, even though your experiences and knowledge about water are the same. This divergence in meaning is due to being embedded in environments in which these concepts point to different elements. By this argument, any property of a concept can be revised without changing its meaning. For example, it may turn out that all cats are really demons from Mars rather than mammals, and yet our concept "cat" would still point to the same things in the world. Thus, it is not by virtue of anything we believe about water or cats that these concepts point to their referents. Rather, reference or meaning obtains by virtue of a special kind of relation between a thing in the world and your mental representation of it.

If one accepts this argument, it is not clear what exactly in a mental representation is or need be shared across minds for concepts to mean the same thing—apart from their pointing to the same thing (where things may be physical, social, abstract, or otherwise). It does not, for purposes of meaning, matter what things we attribute to the concept "Szechuan pepper"—where one finds it, how it tastes, what our favorite dish is to put it in. The critical implication is that there is no objective fact of the matter as to which properties of Szechuan peppers are more or less important for the meaning of the concept "Szechuan pepper." This leaves us without an objective dividing line among the things in the head with respect to what something really is or means and, consequently, between personal belief and shared knowledge.

One response, then, is to throw in the towel—largely to accept that everything that is in the head is equally irrelevant to concept meaning, and to accept that there is no objective dividing line between meaning-relevant information and the rest of our seemingly local, subjective, and idiosyncratic memory: the average location of water-maze platforms and the typical contents of grandma's attic. We take this to be the implication of Millikan's (2017) arguments, as we elaborate below. The outcome of this view is that our intuitions about concepts as "shared" bits of knowledge are a relic of our folk psychology, and that they maybe should not be taken as defining the division between semantic memory and other things in long-term memory.

The alternative response is to defend the line between meaning-relevant, definitionally important, shared knowledge versus irrelevant, idiosyncratic, personal beliefs. This requires a concrete proposal for how to do so. We follow Carey's (2009) arguments for how and why we should attempt to draw this line with objective and formal criteria. If there really are such shared contents, then the scientific attempt to figure out what those are seems well motivated. We discuss whether these criteria can be practically applied.

A third, intermediate, view is to accept that there is no fact of the matter as to what mental content is central to conceptual meaning, but to propose that our meta-cognitive, intuitive concern with this makes the distinction psychologically real. This view, however, would suggest a significant departure for how cognitive neuroscience delineates memory systems, and we elaborate these implications in section 12.5.4.

12.5.2   Throw in the Towel: Unicepts and Non-Shared Content

Millikan (2017) suggests a revision to how we understand the mental representations typically called concepts; she offers instead the notions of unicepts and unitrackers, which, unlike concepts, have no mandate to be shared. While Millikan grants that mental representations should refer to the world, their function for the user is to track objects in the distal world, under the various sensory conditions in which they present themselves: to see the same thing as the same and different things as different. A unicept is a mental representation that binds together the various bits of information we might have about the same thing, and unitrackers flexibly connect unicepts to observations. For example, a unitracker for Szechuan pepper might be sensitive to the kind of restaurant or dish in which an observer has tended to find those peppers, to the extent that this information is useful. For unicepts and unitrackers to perform this role, they absolutely should and do reflect the idiosyncrasies, localities, and subject specificities of personal experience.

Despite having a highly experience-specific origin, these mental representations pick out enough commonality across individuals because there is systematicity to the information in the world. Millikan (2017) asks us to imagine the world as a many-dimensional space, where each dimension is a feature or property and each object is a point in this space. The result is clumpy, with dense areas separated by sparse gaps, because while some feature combinations are common, others are rare. Being a pepper and being served in a chicken dish are related. The same distal object thus has a "univocal" reason for creating its diverse sensory impressions, creating "many quite reliable ways to identify a human or a cat, an oak tree or an automobile or a laptop or a piano, as such" (p. 12)—a fact that helps us accumulate knowledge in constantly varying circumstances. When we talk to each other, we make pointers to the same clump, even if each of us has a slightly different view of it. What matters is that our mental representations share a referent to the same clump, and communication can work.

Millikan (2017) thus rejects the intuition that concepts must be the same across minds. On the basis of "meaning rationalism," she argues that communication, meaning, and reference need not be perfect but rather as evolved phenomena, and need only be as good as needed to be selected—and thus can fail even the vast majority of the time (e.g., mating acts in many species). Without speculating about failure rates in the use of concepts, this fact should already make us suspicious about intuitions concerning what must be present in the head for, say, humans to communicate accumulated knowledge across generations. In short, Millikan has developed and refined an elaborate and comprehensive theory about linguistic and mental intentionality and the lack of need for classical philosophical concepts.[4]

The upshot is a theory of mental content that does not posit a distinction between conceptual content and most of the rest of the information in memory.[5] "Information used to same-track is all of a kind, none of it more important, more defining, or more conceptual than the rest" (Milikan, 2017, p. 49). The stuff formerly taken to be concepts can be personal, local, and even temporary. In her example, "my glass at this party" is as much a concept as any other; a point we see as extending to today's water-maze platforms, and to imply that "typically found in Chinese restaurants" is part of the concept "Szechuan pepper," just as much as any other fact. If this does not conform to our intuitions, our intuitions are irrelevant. The upshot of this view for cognitive science is that trying to establish a conceptual versus non-conceptual boundary in memory may not be meaningful.

Millikan's (2017) take on mental representation is aligned with a number of approaches and observations in modern cognitive neuroscience. The idea of a clumpy correlated feature space stemming from the statistics of the world is an influential one, and can explain important phenomena in semantic system impairment (Capitani et al., 2009; McClelland & Rogers, 2003; Rogers et al., 2004, but cf. Caramazza et al., 1990, for arguments that it might not be the most parsimonious explanation). Recent neuroimaging work has suggested that the very organization of the cortex might be structured to facilitate the readout of correlated but diverse information. Konkle (2019) argues that having "content-channels" that represent a full diversity of features about the same kind of thing, from their typical basic forms to structural descriptions to more semantic knowledge, can facilitate their readout and extraction from observation by making their alignment explicit in cortical space. For example, animate things tend to have broadly curvy forms and tend to draw on foveal representations, while inanimate things tend to be boxy and extend more in to the periphery. All of these distinctions seem to be captured in aligned ways in overlapping parts of the ventral visual stream, so that curvy things, foveated things, and animate things all draw on overlapping cortex (Long, Yu, & Konkle, 2018). This can facilitate the use of this mutual information for recognition or, as Millikan would put it, "same-tracking." Perhaps, in neural space, representational types are too dense a mixture to cleave into concepts and the rest of knowledge cleanly.

Finally, some cognitive neuroscientists argue that because concepts are accumulated from one's potentially idiosyncratic personal experience, there is no reasonable way to separate a shared aspect from the rest, and the goal should not be to search for it (Casasanto & Lupyan, 2015; Yee & Thompson-Schill, 2016). They marshal substantial evidence that the information participants retrieve about the same referent varies to a large degree, influenced by current task, recent experience, accumulated experience, and individual cognitive differences, suggesting that the bulk of information we retrieve about the same thing is more variable than shared (Yee & Thompson-Schill, 2016). We have taken for granted that many different properties of the same referent are available, and the psychological evidence is compelling that their retrieval varies by situation: the spiciness of Szechuan peppers if anticipating eating it, the cuisine that uses it if seeking it out. However, the question remains whether we can systematically sort that information into aspects that are more idiosyncratic and more shared. Below, we review one additional proposal for how to do so.

12.5.3 Drawing the Line: Objective Criteria for Conceptual Content

Although not a direct response to Millikan, Carey (2009) argues that allowing idiosyncratic knowledge to be a part of concepts and not distinguishing between conceptual and idiosyncratic properties is a devastating move. Her rationale for this view, and approach to distinguishing these two forms of content, is as follows.

First, Carey (2009) notes that we must psychologically distinguish belief revision from conceptual change—operations that function differently in development. Conceptual change is slow and dramatic, and leads to previously unavailable representational resources that support new forms of inference. For example, children undergo a conceptual shift when they transition from being able to count to a finite number (initially one, two, or three) to a generative, productive understanding of the successor function (that one can always add one and obtain a new number ad infinitum). These enable radically distinct kinds of thoughts. In contrast, learning that Szechuan peppers typically appear in Chinese restaurants might be only a revision of belief—quick to learn with minimal impact on thought processes. There is thus an inferential origin story that concepts have that other beliefs do not.

Second, she motivates the distinction by arguments from Fodor and Lepore (1992) that if we do not draw that line, it would follow that *every* belief is dependent on every other. If we do not have a principle by which we determine which beliefs are important to understanding the concept "apple," we are committed to saying that all of them are, including our beliefs about camels and nuclear reactors. The consequences of this are philosophically objectionable. It would lead to absurd conclusions, for instance, that two people cannot have the belief that table salt is a kind of seasoning if they disagree about other things, such as salt is good for one's health. To avoid this consequent absurdity, dubbed "holism," some table salt content must be privileged in determining the concept of table salt. However, Fodor and Lepore (1992) argue that privileged content amounts to an analytic/synthetic distinction, which they take to be definitively debunked by Quine (1951). There are two general routes to countering Fodor and Lepore (1992). One can either deflate the supposedly absurd consequences of holism or defend a way of demarcating privileged, meaning-determining content. Carey (2009), following Block (1986), opts for the latter strategy.

Carey's (2009) proposal for how to draw this line is according to the causal processes by which we create new concepts: those concepts that were causally implicated in forming an initial new concept are relevant; others are not. Thus, there is not only a causal (referential) relation between a referent and a concept, but also a causal relation among concepts that are principled. Thus, we can draw the line between mental contents that are, and those which are not, relevant to a concept's meaning.

Could neuroscience take these concerns into consideration when mapping out memory systems? On theoretical grounds, Fodor and Lepore's (1992) argument does not seem to motivate drawing a distinction between *kinds* of knowledge: it simply argues for a need to distinguish which beliefs are relevant to which others, but not grounds for postulating distinct memory types, such as semantic and episodic. It is thus an argument about how to delimit the relations among beliefs but not that some beliefs X are always conceptual and other beliefs Y are always idiosyncratic. Another concern is that this distinction might not be empirically tractable. To trace the causal path between prior concepts and newly formed ones is currently beyond the capacities of empirical practice. Nonetheless, this challenge may be possible, and worthwhile, to meet in the future.

In the meantime, it could be possible to use participants' judgments about the status of different conceptual properties to make such distinctions. Prasada and Dillingham (2006, 2009) show that adults explicitly distinguish between properties that are equally prevalent for an object, but differ in their "principled connection" to that object's kind membership. For example, participants judge that the claim "dogs have four legs" is similar to the claim that "dogs, by virtue of being the kinds of things they are, are four-legged," but that being red, for a barn, amounts to only "barns, in general, are red"—but not that there is any meaningful relation between being a barn and being red. A range of related findings supports the idea that these distinctions are psychologically real and judged consistently. It is unclear whether this distinction maps onto the distinction between autobiographical and shared, but this could be measured also.

Finally, another important theoretical position on concepts (Machery, 2005, 2010) argues that psychological phenomena can reliably distinguish at least three kinds of representations typically called "concepts" by psychologists (Margolis & Laurence, 1999; Murphy, 2002). Sometimes, participants seem to use statistical summaries; other times, they retrieve individual exemplars and, yet other times, theory-like structures (such as causal models). If signatures of these phenomena are sufficiently stable and distinguishable, then a tripartite typology based on them is warranted. Furthermore, all three can be distinguished from "background," idiosyncratic knowledge by virtue of how automatically or prominently they are retrieved. While theorists such as Carey and Prasada would argue that more than ease of retrieval should characterize the distinction between concepts and idiosyncratic knowledge, the general idea of using dissociations among psychological tasks to determine a typology of long-term memory empirically is an appealing one.

In sum, it remains to be seen whether one can distinguish shared versus idiosyncratic aspects of knowledge objectively in ways that could be incorporated into neuroscientific operational criteria, and whether they would allow empirical traction on memory systems in the brain, but these paths also appear promising.

12.5.4 Biting the Meta-Cognitive Bullet

We propose a third solution, which simultaneously accepts that conceptual content is distinct from idiosyncratic personal experience but does not propose that this division is objective. Rather, this distinction is meaningful because concept users believe it is true and have ways of making it for themselves. The implication of this view is a very different way to approach the delineation of memory systems in cognitive neuroscience.

According to this perspective, the way memories are stored in the brain is influenced by the experiencer's own "meta-cognitive" assessment of whether the information they observed is personal and idiosyncratic versus intersubjective and shared. As inherently subjective observers, we can only make a best guess at to the extent to which some content is available to others. Nonetheless, many cues are available to help us do so, and these subjective judgments themselves could matter for how a memory is encoded.

Taking the example of Szechuan peppers, different circumstances of how I learned about their spiciness can support my own inferences about the intersubjectivity of this property: if a friend tells me that they are spicy, or if I read about this in an encyclopedia, I can rest assured that at least some other people experience them that way and conclude that it is probably a true fact about them. On the other extreme, if I sample Szechuan peppers just once, I could doubt whether my experience is either the same as anyone else's or factual at all. I might store the latter as an episodic memory and not integrate it into my knowledge of the world. But if I make the same observation repeatedly across different times and places, the chance that it is generally true increases. As we suggested earlier, we can use both consistency of observation and evidence of intersubjectivity as ways to infer whether an observation exists independently of our fallible sensory experience. This inference does not have to be made explicitly to exert an influence over how information is stored.

Not coincidentally, then, such cues coincide with many of the operational characteristics of memory reviewed through this chapter. Observations that are aggregated (consistent across multiple experiences) are less likely to be due to chance, and those that are cross-modally accessible are not likely relics of an idiosyncratic perceptual channel, and more likely to be common across observers with diverse perspectives (or even common with ourselves at another time, perhaps in a dark room). The concern with definitional, not just typical, features and with observations that reflect an underlying cause, rather than surface properties of stimuli, also reflect the concern with understanding what something *really is*. These criteria may thus operate at a psychological and neural level to distinguish between accidents of sensory noise and possibly real, shared referents in the world that exist independently of us. In this framework, there is a coherent way in which these different properties of memory distinguish our own observed experiences. The upshot is that we are able to use the characteristics of our own experiences to infer which observations are shared by others, and that this inference can itself impact the memory systems in which that information is stored. The empirical prediction that follows is that a person's own judgment about their experiences should affect how those experiences are neurally stored. If an observer has reason to believe that a certain observation is of a stable aspect of reality available to others, that observation would be used to update semantic memory. If that observation is instead only about their personal experience and not reflective of generally available facts, it would be stored in episodic memory. These judgments can be affected by any of a set of cues such as the ones described throughout this chapter. The way that these judgments are made by an observer would result in predictable and distinct neural and psychological outcomes for how that observation is stored in memory. The broader implication of such a finding would be that individuals' judgments about their own observations are a major determinant of how memories form. These remain open to empirical investigation.

12.6   General Conclusion

We have sought to identify how conceptual knowledge can be operationalized in cognitive neuroscience. We reviewed three characteristic features that have been successfully used to distinguish among types of long-term memory, including the extent to which a representation aggregates across experience, whether it is proprietary to a certain sensory input channel, and whether it can depart from physical stimulus similarity space. We found that these distinctions, while theoretically and empirically justified, did not map cleanly onto the notion of conceptual knowledge as it is used within cognitive neuroscience or in psychology: not all aggregated or cross-modal memories are considered concepts.

We argued that rather than just an error of terminology, this usage reflects an underlying intuition guiding the use of the term "concept" based on notions of sharedness. For example, many examples of memories that meet the operational criteria are ones that are personal and idiosyncratic—such as the typical contents of my grandmother's attic or my favorite color of bicycle. The guiding intuition that excludes such examples from the domain of semantic memory reflects historically influential ideas that characterize semantic memory as "shared" knowledge. Yet, this criterion has not itself become part of the formal operational definition for measuring semantic memory in cognitive neuroscience, and thus its empirical validity in predicting neural organization remains to be tested directly.

We sought here to determine whether sharedness should be considered part of the operational definition for semantic or conceptual memory by appealing to opposite theoretical positions on the topic. Millikan (2017) argues that sharedness is not a meaningful distinction for concepts, while Carey (2009) argues instead that it is essential. Taking Millikan's position entails largely discarding the semantic-episodic distinction in long-term memory and revising this ontology in cognitive neuroscience. Taking Carey's view requires justifying an objective basis for distinguishing between shared versus autobiographical/idiosyncratic content, but proposals for how neuroscientists could test such divisions remain to be developed.

We offered a third view, which argues that individual observers are capable of making the distinction between shared and autobiographical aspects of their own observations themselves subjectively. We predict that these subjective judgments themselves influence how observations are allocated among memory systems. For observers, these judgments are motivated by their ultimate concern with establishing whether their observations reflect intersubjective reality, and they can rationally make use of a variety of cues to guide their judgments. By systematically manipulating such cues experimentally, we expect to be able to affect whether observers classify an observation as personal or shared and, consequently, how that observation is allocated to their own memory systems. Overall, we believe that the theoretical alternatives we have outlined offer robust empirical predictions, and testing them has the potential to redefine our theories of memory systems.

Notes

1. "Concepts" in cognitive neuroscience are thought to be part of long-term (semantic) memory (e.g., Schacter & Tulving, 1994). It is possible that there are other cognitive domains in which concepts are a distinguishable part, but we focus on memory here.

2. We use "experience" broadly here: experiences need not be sensory, but also based in thought, language, and imagination.

3. These authors challenge that view; we turn to their arguments in section 12.4.2.

4. Mark Wilson's (2008) work against classical conceptions of concepts shows that in actual science, they do not even behave in the way many philosophers have thought they needed to in order for our paradigm of knowledge to be successful. Nonetheless, Millikan's theory was met with a number of objections and counterintuitive consequences. Rather than address them all here, we instead highlight the power of her motivating basics and the subsequent opportunity to develop alternatives within a "Millikan tradition." Approaching cognition as being confronted with some basic problems to solve and considering signs and communication systems as evolved solutions to such problems is an underexplored and potentially powerful way of addressing traditional philosophical issues about content—or, perhaps in some cases, learning to recognize false puzzles and dilemmas they have been taken to raise.

5. Millikan does, however, draw a distinction between those observations for which we do not form unicepts for referents which we do not care to same-track as such. Her examples are perceptual constancy (seeing the color of the wall as constant despite

differences in illumination), phonemes (categories of speech sounds), and oriented edges such as Gabor patches. These do not involve unicepts because we do not care to track them as such—to gather information *about* edges, *about* phonemes, or *about* colors. This may or may not be psychologically true. This distinction between things we track and things we do not could map onto the functional divisions we earlier described between semantic memory and the structural description system, but it would not draw the same line between semantic and episodic memory.

References

Alvarez, G. A., & Oliva, A. (2009). Spatial ensemble statistics are efficient codes that can be represented with reduced attention. *Proceedings of the National Academy of Sciences of the United States of America, 106*(18), 7345–7350.

Ariely, D. (2001). Seeing sets: Representation by statistical properties. *Psychological Science, 12*(2), 157–162.

Armstrong, S. L., Gleitman, L. R., & Gleitman, H. (1983). What some concepts might not be. *Cognition, 13(3),* 263–308.

Beauvois, M.-F. (1982). Optic aphasia: A process of interaction between vision and language. *Philosophical Transactions of the Royal Society B: Biological Sciences, 298*(1089), 35–47.

Bedny, M., Caramazza, A., Grossman, E., Pascual-Leone, A., & Saxe, R. R. (2008). Concepts are more than percepts: The case of action verbs. *Journal of Neuroscience, 28*(44), 11347–11353.

Bedny, M., Caramazza, A., Pascual-Leone, A., & Saxe, R. R. (2011). Typical neural representations of action verbs develop without vision. *Cerebral Cortex, 22*(2), 286–293.

Binder, J. R., & Desai, R. H. (2011). The neurobiology of semantic memory. *Trends in Cognitive Sciences, 15*(11), 527–536.

Binder, J. R., Desai, R. H., Graves, W. W., & Conant, L. L. (2009). Where is the semantic system? A critical review and meta-analysis of 120 functional neuroimaging studies. *Cerebral Cortex, 19*(12), 2767–2796.

Block, N. (1986). Advertisement for a semantics for psychology. *Midwest Studies in Philosophy, 10*(1986), 615–678.

Bracci, S., & Op de Beeck, H. (2016). Dissociations and associations between shape and category representations in the two visual pathways. *Journal of Neuroscience, 36*(2), 432–444.

Capitani, E., Laiacona, M., Pagani, R., Capasso, R., Zampetti, P., & Miceli, G. (2009). Posterior cerebral artery infarcts and semantic category dissociations: A study of 28 patients. *Brain, 132,* 965–981.

Caramazza, A., Berndt, R. S., & Brownell, H. H. (1982). The semantic deficit hypothesis: Perceptual parsing and object classification by aphasic patients. *Brain and Language, 15*(1), 161–189.

Caramazza, A., Hillis, A. E., Rapp, B. C., & Romani, C. (1990). The multiple semantics hypothesis: Multiple confusions? *Cognitive Neuropsychology, 7*(3), 161–189.

Carey, S. (2009). *Origin of concepts.* Oxford: Oxford University Press.

Casasanto, D., & Lupyan, G. (2015). All concepts are ad hoc concepts. In E. Margolis & S. Laurence (Eds.), *The conceptual mind: New directions in the study of concepts* (pp. 543–566). Cambridge, MA: MIT Press.

Chun, M. M., & Turk-Browne, N. B. (2008). Associative learning mechanisms in vision. In S. J. Luck & A. Hollingworth (Eds.), *Visual memory* (pp. 209–245). Oxford: Oxford University Press.

Deen, B., Koldewyn, K., Kanwisher, N., & Saxe, R. (2015). Functional organization of social perception and cognition in the superior temporal sulcus. *Cerebral Cortex, 25*(11), 4596–4609.

Devereux, B. J., Clarke, A., Marouchos, A., & Tyler, L. K. (2013). Representational similarity analysis reveals commonalities and differences in the semantic processing of words and objects. *Journal of Neuroscience, 33*(48), 18906–18916.

Dimsdale-Zucker, H. R., Ritchey, M., Ekstrom, A. D., Yonelinas, A. P., & Ranganath, C. (2018). CA1 and CA3 differentially support spontaneous retrieval of episodic contexts within human hippocampal subfields. *Nature Communications, 9*(1), 294.

Eichenbaum, H. B. (2004). Hippocampus: Cognitive processes and neural representations that underlie declarative memory. *Neuron, 44*(1), 109–120.

Eichenbaum, H. B., & Cohen, N. J. (2001). *From conditioning to conscious recollection: Memory systems of the brain.* Oxford: Oxford University Press.

Fairhall, S. L., & Caramazza, A. (2013). Brain regions that represent abstract conceptual knowledge. *Journal of Neuroscience, 33*(25), 10552–10558.

Fischer, M. H., & Zwaan, R. (2008). Embodied language: A review of the role of the motor system in language comprehension. *Quarterly Journal of Experimental Psychology, 61*(6), 825–850.

Fodor, J., & Lepore, E. (1992). *Holism: A shopper's guide.* Cambridge, MA: Blackwell.

Freedman, D. J., Riesenhuber, M., Poggio, T., & Miller, E. K. (2001). Categorical representation of visual stimuli in the primate prefrontal cortex. *Science, 291*, 312–316.

Gopnik, A., & Meltzoff, A. N. (1997). *Words, thoughts, and theories.* Cambridge, MA: MIT Press.

Haberman, J. M., & Whitney, D. (2009). Averaging facial expression over time. *Journal of Vision, 9*(2009), 1–13.

Haushofer, J., Livingstone, M. S., & Kanwisher, N. G. (2008). Multivariate patterns in object-selective cortex dissociate perceptual and physical shape similarity. *PLoS Biology, 6*(7), 1459–1647.

Hodges, J. R. (1992). Semantic dementia: Progressive fluent aphasia with temporal lobe atrophy. *Brain, 115,* 1783–1806.

Hodges, J. R., & McCarthy, R. A. (1995). Loss of remote memory: A cognitive neuropsychological perspective. *Current Opinion in Neurobiology, 5*(2), 178–183.

Hoffman, P., McClelland, J. L., & Lambon-Ralph, M. A. (2017). Concepts, control and context. *Psychological Review, 125*(3), 293–328.

Keil, F. C., Smith, W. C., Simons, D. J., & Levin, D. T. (1998). Two dogmas of conceptual empiricism: Implications for hybrid models of the structure of knowledge. *Cognition, 65*(2–3), 103–135.

Knowlton, B. J., Ramus, S. J., & Squire, L. R. (1992). Intact artificial grammar learning in amnesia: Dissociation of category-level knowledge and explicit memory for specific instances. *Psychological Science, 3*(3), 172–179.

Knowlton, B. J., Squire, L. R., & Gluck, M. A. (1994). Probabilistic classification learning in amnesia. *Learning and Memory, 1*(2), 106–120.

Konkle, T. (2019). Emergence of multiple retinotopic maps without a feature hierarchy. *Journal of Vision, 19* (10), 90a–90a.

Kripke, S. (1972). *Naming and necessity.* Malden, MA: Blackwell.

Leshinskaya, A., & Caramazza, A. (2015). Abstract categories of functions in anterior parietal lobe. *Neuropsychologia, 76,* 27–40.

Leshinskaya, A., Contreras, J. M., Caramazza, A., & Mitchell, J. P. (2017). Neural representations of belief concepts: A representational similarity approach to social semantics. *Cerebral Cortex, 27,* 344–357.

Long, B., Yu, C.-P., & Konkle, T. (2018). Mid-level visual features underlie the high-level categorical organization of the ventral stream. *Proceedings of the National Academy of Sciences of the United States of America, 115*(38), E9015–E9024.

Machery, E. (2005). Concepts are not a natural kind. *Philosophy of Science, 72*(3), 444–467.

Machery, E. (2010). Précis of doing without concepts. *Behavioral and Brain Sciences, 33,* 195–244.

Mahon, B. Z. (2015). Missed connections: A connectivity constrained account of the representation and organization of object concepts. In E. Margolis & S. Laurence

(Eds.), *The conceptual mind: New directions in the study of concepts* (pp. 79–116). Cambridge, MA: MIT Press.

Mandler, J. M. (2004). *The foundations of mind: The origins of conceptual thought.* New York: Oxford University Press.

Manns, J. R., Hopkins, R. O., & Squire, L. R. (2003). Semantic memory and the human hippocampus. *Neuron, 38*(1), 127–133.

Margolis, E., & Laurence, S. (1999). Introduction. In S. Laurence & E. Margolis (Eds.), *Concepts: Core readings* (pp. 3–81). Cambridge, MA: MIT Press.

Martin, A. (2007). The representation of object concepts in the brain. *Annual Review of Psychology, 58,* 25–45.

Martin, A., Wiggs, C. L., Ungerleider, L. G., & Haxby, J. V. (1996). Neural correlates of category specific knowledge. *Nature, 379,* 649–652.

Martin, C. B., & Barense, M. D. (2018). Integrative and distinctive coding of visual and conceptual object features in the ventral visual stream. *ELife, 7,* e31873.

McClelland, J. L., McNaughton, B. L., & O'Reilly, R. C. (1995). Why there are complementary learning systems in the hippocampus and neo-cortex: Insights from the successes and failures of connectionists models of learning and memory. *Psychological Review, 102*(3), 419–457.

McClelland, J. L., & Rogers, T. T. (2003). The parallel distributed processing approach to semantic cognition. *Nature Reviews Neuroscience, 4*(4), 310–322.

McRae, K., & Jones, M. (2013). Semantic memory. In D. Reisberg (Ed.), *Oxford handbook of cognitive psychology* (pp. 206–219). New York: Oxford University Press.

Miller, E. K., Nieder, A., Freedman, D. J., & Wallis, J. D. (2003). Neural correlates of categories and concepts. *Current Opinion in Neurobiology, 13,* 198–203.

Millikan, R. G. (2017). *Beyond concepts: Unicepts, language, and natural information.* Oxford: Oxford University Press.

Mishkin, M. (1997). Hierarchical organization of cognitive memory. *Philosophical Transactions of the Royal Society B: Biological Sciences, 352*(1360), 1461–1467.

Murphy, G. L. (2002). *Big book of concepts.* Cambridge, MA: MIT Press.

Myers, C. E., Shohamy, D., Gluck, M. A., Grossman, S., Kluger, A., Ferris, S., . . . Schwartz, R. (2003). Dissociating hippocampal versus basal ganglia contributions to learning and transfer. *Journal of Cognitive Neuroscience, 115*(2), 185–193.

Nadel, L., & Moscovitch, M. (1997). Memory consolidation and the hippocampal complex. *Cognitive Neuroscience, 7,* 217–227.

O'Reilly, R. C., & Rudy, J. W. (2001). Conjunctive representations in learning and memory: Principles of cortical and hippocampal function. *Psychological Review, 108*(1), 83–95.

Parkes, L., Lund, J., Angelucci, A., Solomon, J. A., & Morgan, M. (2001). Compulsory averaging of crowded orientation signals in human vision. *Nature Neuroscience, 4*(7), 739–744.

Patterson, K., Nestor, P. J., & Rogers, T. T. (2007). Where do you know what you know? The representation of semantic knowledge in the human brain. *Nature Reviews Neuroscience, 8*, 976–989.

Peelen, M. V., Bracci, S., Lu, X., He, C., Caramazza, A., & Bi, Y. (2013). Tool selectivity in left occipitotemporal cortex develops without vision. *Journal of Cognitive Neuroscience, 25*(8), 1225–1234.

Peelen, M. V., Romagno, D., & Caramazza, A. (2012). Independent representations of verbs and actions in left lateral temporal cortex. *Journal of Cognitive Neuroscience, 24*(10), 2096–2107.

Prasada, S., & Dillingham, E. M. (2006). Principled and statistical connections in common sense conception. *Cognition, 99*(1), 73–112.

Prasada, S., & Dillingham, E. M. (2009). Representation of principled connections: A window onto the formal aspect of common sense conception. *Cognitive Science, 33*(3), 401–448.

Proklova, D., Kaiser, D., & Peelen, M. V. (2016). Disentangling representations of object shape and object category in human visual cortex: The animate–inanimate distinction. *Journal of Cognitive Neuroscience, 28*(5), 680–692.

Putnam, H. (1973). Meaning and reference. *The Journal of Philosophy, 70*(9), 699–711.

Quine, W. V. O. (1951). Two dogmas of empiricism. *Philosophical Review, 60*(1), 20–43.

Renoult, L., Davidson, P. S. R., Palombo, D. J., Moscovitch, M., & Levine, B. (2012). Personal semantics: At the crossroads of semantic and episodic memory. *Trends in Cognitive Sciences, 16*(11), 550–558.

Renoult, L., Irish, M., Moscovitch, M., & Rugg, M. D. (2019). From knowing to remembering: The semantic–episodic distinction. *Trends in Cognitive Sciences, 23*(12), 1041–1057.

Rey, G. (1983). Concepts and stereotypes. *Cognition, 15*, 237–262.

Richards, B. A., Xia, F., Santoro, A., Husse, J., Woodin, M. A., Josselyn, S. A., & Frankland, P. W. (2014). Patterns across multiple memories are identified over time. *Nature Neuroscience, 17*(7), 981–986.

Riddoch, M. J., & Humphreys, G. W. (1987). Visual object processing in optic aphasia: A case of semantic access agnosia. *Cognitive Neuropsychology, 4*(2), 131–185.

Rips, L. J., Shoben, E. J., & Smith, E. E. (1973). Semantic distance and the verification of semantic relations. *Journal of Verbal Learning and Verbal Behavior, 12*(1), 1–20.

Rogers, T. T., Lambon-Ralph, M. A., Garrard, P., Bozeat, S., McClelland, J. L., Hodges, J. R., & Patterson, K. (2004). Structure and deterioration of semantic memory: A neuropsychological and computational investigation. *Psychological Review, 111*(1), 205–235.

Rosch, E. (1975). Cognitive representations of semantic categories. *Journal of Experimental Psychology: General, 3*, 192–233.

Roy, J. E., Riesenhuber, M., Poggio, T., & Miller, E. K. (2010). Prefrontal cortex activity during flexible categorization. *Journal of Neuroscience, 30*(25), 8519–8528.

Saffran, J. R., Aslin, R. N., & Newport, E. L. (1996). Statistical learning by eight-month-old infants. *Science, 274*(5294), 1926–1928.

Schacter, D. L., & Tulving, E. (1994). *Memory systems*. Cambridge, MA: MIT Press.

Schapiro, A. C., Turk-Browne, N. B., Norman, K. A., & Botvinick, M. M. (2016). Statistical learning of temporal community structure in the hippocampus. *Hippocampus, 26*(1), 3–8.

Sekeres, M. J., Winocur, G., & Moscovitch, M. (2018). The hippocampus and related neocortical structures in memory transformation. *Neuroscience Letters, 680*, 39–53.

Shallice, T. (1988). *From neuropsychology to mental structure*. Cambridge: Cambridge University Press.

Shohamy, D., & Daw, N. D. (2015). Integrating memories to guide decisions. *Current Opinion in Behavioral Sciences, 5*, 85–90.

Simanova, I., Hagoort, P., Oostenveld, R., & van Gerven, M. A. J. (2012). Modality-independent decoding of semantic information from the human brain. *Cerebral Cortex, 24*(2), 426–434.

Simoncelli, E. P., & Olshausen, B. A. (2001). Natural image statistics and neural representation. *Annual Review of Neuroscience, 24*, 1193–1216.

Smith, E. E., & Medin, D. L. (1981). *Categories and concepts*. Cambridge, MA: Harvard University Press.

Squire, L. R., Knowlton, B. J., & Musen, G. (1993). The structure and organization of memory. *Annual Review of Psychology, 44*(1), 453–495.

Striem-Amit, E., Wang, X., Bi, Y., & Caramazza, A. (2018). Neural representation of visual concepts in people born blind. *Nature Communications, 9*(1), 5250.

Tompary, A., & Davachi, L. (2017). Consolidation promotes the emergence of representational overlap in the hippocampus and medial prefrontal cortex. *Neuron, 96*(1), 228–241.e5.

Tulving, E. (1972). Episodic and semantic memory. In E. Tulving & W. Donaldson (Eds.), *Organization of memory* (pp. 381–402). New York: Academic Press.

Tulving, E. (1984). Précis of elements of episodic memory. *Behavioral and Brain Sciences, 7*(2), 223–238.

Tulving, E. (2002). Episodic memory: From mind to brain. *Annual Review of Psychology, 53,* 1–25.

Tulving, E., Hayman, C. A. G., & Macdonald, C. A. (1991). Long-lasting perceptual priming and semantic learning in amnesia: A case experiment. *Journal of Experimental Psychology: Learning, Memory, and Cognition, 17*(4), 595–617.

Verfaellie, M., Koseff, P., & Alexander, M. P. (2000). Acquisition of novel semantic information in amnesia: Effects of lesion location. *Neuropsychologia, 38*(4), 484–492.

Wang, X., Peelen, M. V., Han, Z., He, C., Caramazza, A., & Bi, Y. (2015). How visual is the visual cortex? Comparing connectional and functional fingerprints between congenitally blind and sighted individuals. *Journal of Neuroscience, 35*(36), 12545–12559.

Warrington, E. K. (1975). The selective impairment of semantic memory. *The Quarterly Journal of Experimental Psychology, 27*(4), 635–657.

Warrington, E. K., & Taylor, A. M. (1978). Two categorical stages of object recognition. *Perception, 7,* 395–401.

Watson, C. E., Cardillo, E., Ianni, G., & Chatterjee, A. (2013). Action concepts in the brain: An activation-likelihood estimation meta-analysis. *Journal of Cognitive Neuroscience, 25*(8), 1191–1205.

Westmacott, R., Black, S. E., Freedman, M., & Moscovitch, M. (2004). The contribution of autobiographical significance to semantic memory: Evidence from Alzheimer's disease, semantic dementia, and amnesia. *Neuropsychologia, 42*(1), 25–48.

Wiggs, C. L., Weisberg, J., & Martin, A. (1999). Neural correlates of semantic and episodic memory retrieval. *Neuropsychologia, 37*(1), 103–118.

Wilson, M. (2008). *Wandering significance: An essay on conceptual behavior.* New York: Oxford University Press.

Winocur, G., Moscovitch, M., Rosenbaum, R. S., & Sekeres, M. J. (2010). An investigation of the effects of hippocampal lesions in rats on pre- and postoperatively acquired spatial memory in a complex environment. *Hippocampus, 20*(12), 1350–1365.

Yee, E., Chrysikou, E. G., Hoffman, E., & Thompson-Schill, S. L. (2013). Manual experience shapes object representations. *Psychological Science, 24*(6), 909–919.

Yee, E., & Thompson-Schill, S. L. (2016). Putting concepts into context. *Psychonomic Bulletin and Review, 23,* 1015–1027.

13   The Scientific Study of Passive Thinking: Methods of Mind-Wandering Research

Samuel Murray, Zachary C. Irving, and Kristina Krasich

13.1   Introduction

Chances are, at some point when you're reading this, your eyes will move across the page while your mind is elsewhere. This is likely true even if you really want to stay focused on what we have to say. People's ability to remain vigilant toward any one thing is remarkably flawed—the mind is rich with internal thoughts, concerns, simulation, and feelings that can pull our attention away. Of course, sometimes you can actively shift attention. If you get bored, you might look at your phone to check email or social media. This kind of attention shift is an active mental phenomenon as opposed to the more passive mental phenomenon of having one's attention drift away.

Mental life involves a dynamic coordination of these active and passive elements. You can *decide* whether to go to the store before or after work, *deliberate* on what to purchase, and *intentionally* buy the fruit instead of the chips (active phenomena), but you cannot easily manage your *desire* to eat chips or your *belief* that eating them will make you happy (passive phenomena). With regards to your attention span (assuming you are still with us), you may have decided to read this chapter, but that doesn't mean your desires to keep up with work or friends won't distract you.

Instead of reading carefully throughout, your mind will sometimes wander. It might wander to an upcoming test, a dinner out with friends last weekend, or a song you recently heard. Wherever it wanders, your mind will be wandering away from whatever we're saying. This isn't something you'll decide to do, but you will catch yourself doing it from time to time. For this reason, mind wandering falls on the *passive* side of the active/passive divide. This isn't a definition. Instead, it's an example that illustrates the phenomenon's passivity.

Special methods are often required to measure passive phenomena such as mind wandering or dreaming (Windt, 2015; Irving, 2018). In particular, the sciences of passive thinking often require self-reports or retrospective assessments of the content or character of one's mental states. However, philosophers and cognitive scientists have raised nontrivial epistemic concerns about self-report. It is therefore unclear whether we have an adequate methodology to study mind wandering empirically.

Our solution to this problem proceeds through a metaphysical account of mind wandering. We explain how mind wandering fits into the wider fabric of human agency, which makes sense of the causes and conditions of mind wandering. These can be leveraged into a (limited) defense of the self-report methods used to study mind wandering.

Our chapter has six parts. We first describe the central role of self-report in the rapid expansion of mind-wandering research over the last twenty years (section 13.2). Next, we argue that the passivity of mind wandering explains why self-report is necessary for its study (section 13.3), which may raise skeptical worries about the veracity of mind-wandering research (section 13.4). We then consider whether objective methods (section 13.5) or studies of intentional mind wandering (section 13.6) can obviate the need for self-report (spoiler: they can't). Finally, we propose a metaphysical solution to the epistemic problems of self-report (section 13.7).

13.2 Methodological Innovations

Two methodological innovations explain why mind wandering came to prominence in cognitive psychology. First was the discovery of the so-called default mode network—a set of brain regions associated with task-independent activity. Of equal importance was the revival of self-report methods to measure the wandering mind. In this section, we discuss why these two innovations are central to the history of mind-wandering research.

Twenty years ago, almost nobody in cognitive psychology or neuroscience talked about mind wandering (notable exceptions include Antrobus, 1968; Giambra, 1995; Wegner, 1997). Presently, each of the last five years has seen more than 100 articles published on mind wandering. One might assume that the reasons for this shift concern the importance of mind wandering itself. Mind wandering occupies a significant portion of our waking thoughts (Kane, Brown, et al., 2007; Seli, Beaty, et al., 2019). It is associated with a range of costs, including higher rates of car crashes (Yanko & Spalek, 2014; Gil-Jardiné et al., 2017), occupational accidents (Warm, Parasuraman, & Matthews, 2008), and general negative affect (Killingsworth & Gilbert, 2010).[1] It also has benefits for self-control (Gorgolewski et al., 2014), planning (Baird, Smallwood, & Schooler, 2011), and creativity (Preiss et al., 2016; Gable, Hopper, & Schooler, 2019; for a review, see Smallwood & Schooler, 2015). But the importance of mind wandering cannot explain its *increase* in prominence within the scientific community. After all, twenty years ago, mind wandering was just as pervasive in everyday life and had the same costs and benefits. Why, then, have the last two decades ushered in so much new research on mind wandering? Two methodological innovations deserve the lion's share of credit.

One innovation was the discovery of the so-called default mode network. In the early 2000s, researchers discovered a functionally connected set of brain regions that become considerably more active during moments of rest or inactivity than when subjects perform tasks (Raichle, 2015).[2] This helped to spur interest in studying what the brain is doing when it's not engaged in a task. In other words, what is the brain doing when it's "resting"? This marked an exciting departure from the norm in cognitive psychology and neuroscience, which was to study exclusively the cognitive processes that support task performance (Irving, 2018; Callard, Smallwood, & Margulies, 2011).

Over time, researchers realized that the default mode network subserves spontaneous internally directed cognition (Buckner, Andrews-Hanna, & Schacter, 2008) and that default mode activity increases when people's minds are wandering (Christoff, Gordon, et al., 2009). Initially, these results suggested that mind wandering might reflect a default state of human cognition: the unperturbed stream of thought. That suggestion turned out to be overly simplistic due to evidence that the default network can support goal-directed cognition (Spreng et al., 2010) and that other networks (notably the executive control network) are active during mind wandering (Fox et al., 2015; see Klein, 2012, for a philosophical discussion).[3] Still, the discovery of the default mode network helped mind wandering emerge as a research topic. It set the stage for observational experiments that were primarily concerned with studying what the brain does when one is not explicitly engaged in goal-directed thinking.

Mind-wandering science also benefited from a second methodological innovation: the development and refinement of self-report measures. Most studies of mind wandering in the lab and everyday life use a self-report method called retrospective thought sampling (Smallwood & Schooler, 2015). In these studies, participants are periodically interrupted as they perform tasks in the lab or go about their daily lives. They are then given a thought probe that asks questions about their immediately preceding experiences (figure 13.1). For example, one influential study asked subjects whether they agreed that "At the time of the beep [the thought probe], my mind had wandered to something other than what I was doing" (Kane, Brown, et al., 2007). Subjects who answered "yes" were classified as mind wandering. The study of mind wandering, then, leaned heavily on participants making retrospective judgments about their mental state just prior to being probed.[4]

Scientists have relied on these self-report methods to discover many characteristics of mind wandering, including (but not limited to) its frequency, costs, benefits, role in education, and its relationship to working memory, affect, episodic thinking, mindfulness, and the stream of thought (for reviews, see Smallwood & Schooler, 2015; Christoff, Irving, et al., 2016). Indeed, the vast majority of our knowledge of mind wandering is owed (at least in part) to self-report. In the next section, we explain why.

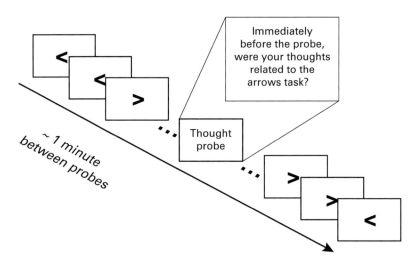

Figure 13.1
Retrospective thought sampling in a laboratory study (example). Subjects are interrupted by a thought probe on average once a minute. They are then asked whether their immediately preceding thoughts are related to the laboratory tasks. If not, their thoughts are classified as mind wandering.
Source: Reproduced with permission from Irving and Glasser (2020).

13.3   Self-Report and Passive Phenomena

Self-report is central to mind-wandering research. In this section, we explain why this is due to the passivity of mind wandering. Specifically, we build on an argument from Irving (2018). Irving notes that cognitive psychologists typically study a cognitive process by giving subjects a voluntary task that activates that process. But because mind wandering is passive, no voluntary task initiates mind wandering. So, psychologists need a task-free method to study mind wandering, and presently the best task-free method is self-report.

Here is our expanded version of Irving's argument:

1. If a cognitive process or state ϕ cannot[5] be voluntarily initiated and does not reliably subserve the performance of a task, then the psychological study of ϕ requires self-report.
2. Mind wandering cannot be voluntarily initiated.
3. Mind wandering does not reliably subserve the performance of a task.
4. Therefore, the psychological study of mind wandering requires self-report.

The remainder of this section offers support for these premises.

13.3.1   Processes and Reports (Premise 1)

Cognitive psychology often relies on experimental tasks in part to avoid the need for self-report. Let's say that a cognitive psychologist wants to study a cognitive process or state ϕ. She will typically design an experimental task the performance of which requires some behavioral response τ, where τ is known to activate ϕ. Our experimenter then has no need to ask subjects to self-report on whether they use process ϕ in the experiment, since we already know that anyone who performs τ activates process ϕ. We call this approach "task-based psychology" (for an example, see our discussion of the go/no-go paradigm in the next paragraph).

Tasks can activate a psychological process or state in two ways, depending on how directly one can control the process or state. First, some tasks exploit the fact that subjects can voluntarily initiate a psychological process or state. We'll say that an agent can voluntarily initiate ϕ if and only if she can bring about the occurrence of ϕ immediately by deciding, choosing, or willing to make ϕ obtain (see Adams, 1985, p. 8; cf. van Inwagen, 1989, p. 410). One example of the use of voluntary initiation is how psychologists use a go/no-go task to study inhibitory processes (Lappin & Eriksen,

1966; Logan, 2015). Subjects respond quickly to targets (e.g., "go" signs) and withhold their response when they receive a stop signal (e.g., "stop" signs). Because the task is timed, subjects typically begin to respond as soon as they see a stimulus, even if it is a stop signal. In stop trials, they then have to inhibit their response. Psychologists use go/no-go tasks to study the psychological process of inhibiting an ongoing action. Subjects can voluntarily initiate this sort of inhibition. So, anyone who completes the go/no-go task will voluntarily activate the process in question.

Psychologists cannot use this direct method to study passive states and processes such as beliefs, desires, or dreaming. Agents cannot voluntarily initiate passive processes or directly be in some passive state because they can neither come to be in these states nor bring about the occurrence of these processes immediately. At most, agents can indirectly be in some passive state or bring about the occurrence of a passive process φ by performing *another* action that they know is likely to bring it about that φ. Suppose you don't now desire to eat a tomato, for example. You cannot simply decide or choose to have that desire in a way that immediately and directly causes you to want a tomato. You can do things that make the desire more likely, such as looking at pictures of caprese salads. But this amounts to only indirectly controlling your desires.

Task-based psychologists can sometimes use indirect methods to study passive states and processes. Sometimes, a passive state or process φ will subserve the voluntary performance of a task τ. If so, psychologists can ask subjects to voluntarily perform τ in order to activate φ indirectly. Consider, for example, how psychologists use spatial cueing paradigms (Posner, 1980) to study the effects of participant's beliefs on visual attention. Beliefs are passive, insofar as one cannot directly choose or decide to believe something (van Fraassen, 1984). But spatial cueing tasks manipulate subjects' beliefs about the location of a target stimulus. Participants are assigned to one of three conditions. In one condition, a peripheral cue is presented indicating the location of the target stimulus (the congruent condition). In another condition, the peripheral cue is presented in a location that differs from the target stimulus (the incongruent condition). In the control condition, no peripheral cue is presented. Fixation occurs much more quickly in congruent conditions relative to incongruent conditions (Israel, Jolicoeur, & Cohen, 2018). This suggests that participants with true beliefs about the target location outperform subjects with false beliefs.

Belief, though passive, can be studied within the standard task-based experimental paradigm because we understand something about how belief produces action. Part of what it is to be a belief is to play a certain *causal* role in the production of action and a *normative* role in the explanation of action (Stalnaker, 1984, pp. 4–5, 82). Paradigmatically, the desire G, coupled with the belief that A-ing is a means to realize G, will cause one to begin A-ing (Audi, 1979). When we know about someone's actions and desires, we can therefore use this schema to infer someone's beliefs. For example, visual fixation is an action (or a component of an action) that subjects use to complete an experimental task. So, we can reliably infer that subjects form beliefs about the best location for visual fixation. Crucially, we can make this inference even though belief is passive because we know how beliefs tend to subserve action.

We argue that there are two conditions under which psychologists can use task-based methods to study a cognitive process or state ϕ: (1) when subjects can voluntarily initiate ϕ, and (2) when ϕ reliably subserves the performance of a task that one can voluntarily initiate. If neither of those conditions obtain, psychologists must use the methods of task-free psychology: self-report.

13.3.2   Tasks and Mind Wandering (Premises 2 and 3)

Premise 2 in our argument says that one cannot voluntarily initiate mind wandering. This is intuitive. People cannot make their minds wander at the drop of a hat, and they often catch their minds wandering (just as you can catch yourself dozing off). People can put themselves in a position that makes mind wandering likely, such as reading a boring book. But at the end of the day, you must let mind wandering come to you.

Premise 3 says that mind wandering does not reliably subserve the voluntary performance of a task. Unlike belief, researchers cannot ask subjects to perform tasks that require mind wandering because there are no such tasks. Our argument for this premise differs, depending on how you define mind wandering (cf. Irving, 2018, for similar arguments).

Most psychologists define mind wandering as task-unrelated thought. In a laboratory, this means that our wandering thoughts are (by definition) unrelated to the task assigned by the experimenter. But if thoughts subserve an experimental task, then they are related to that task. By definition, any thoughts that subserve the experimental task are not mind wandering—at least according to the standard operationalization (Smallwood & Schooler, 2015).

Others define mind wandering in terms of its dynamics rather than its relationship to an ongoing task (Irving, 2016; Christoff, Irving, et al., 2016; Sripada, 2018). Dynamic views focus on how mind wandering unfolds over time, meandering from one topic to another. Such views contrast mind wandering with goal-directed thinking, which remains focused on the agent's task. Irving (2016) appeals to guidance to explain the dynamic difference between mind wandering and directed thinking. During goal-directed thinking, for example, the agent guides her attention to remain on task-relevant stimuli. During mind wandering, in contrast, the agent's attention is unguided and thus free to meander from topic to topic.

We can now explain why no task can reliably recruit mind wandering. Part of what it means to perform a task is to guide your attention to thoughts that are relevant to that task. So, any thoughts that subserve an agent's task will be thoughts toward which she guides her attention. Therefore, those thoughts will not constitute mind wandering.

We now have an argument that explains why mind-wandering research relies on self-report. Agents cannot voluntarily initiate mind wandering, and nor can mind wandering reliably subserve the performance of a task. The methods of task-based psychology are therefore ill suited to study the wandering mind. Mind-wandering researchers must instead rely on task-free methods—namely, self-report.

13.4   Skepticism about Self-Report

Self-report is central to the science of mind wandering. At bottom, mind-wandering research fundamentally requires that we ask people to reflect on what's going on in their minds. You might think this is a serious cause for concern. Empirical evidence suggests that people are notoriously unreliable when they make judgments about their own minds. The situation is particularly bad when we use survey responses as data points. Survey responses are subject to framing effects (Sinnott-Armstrong, 2008), social desirability biases (Robins & John, 1997), and simple misunderstanding of the questions being asked (Cullen, 2010). Moreover, people are bad at remembering their responses to surveys, raising the question of whether responding to surveys activates reliable, reasons-responsive processes. Hall, Johansson, and Strandberg (2012) showed this when they reversed people's responses to a moral and political opinion survey and got people to defend their reversed positions!

Surveys about experience are notoriously unreliable. Fortunately, such surveys differ from thought sampling methodology in important ways. Thought probes do not interrogate *why* you are mind wandering, just *whether* you are. So, while people exhibit unreliability in making process assessments ("Why are you in a particular mental state?"), mind-wandering research requires state assessment ("Are you in this particular mental state?"). Furthermore, thought sampling questions do not ask about obscure technical concepts such as knowledge or freedom. Rather, they ask about a pervasive and familiar experience: whether your mind was wandering. Finally, thought-sampling methods ask about your immediately preceding experiences, rather than asking you to report on what your experiences are like in general. Such methods may therefore avoid the distorting effects of memory (Windt, 2016).

We have empirical reasons, though, to trust self-reports of mind wandering. Mind-wandering research shows surprising consistency across subjects in self-reports of mind wandering (see Kane, Smeekens, et al., in preparation). For instance, there is emerging evidence of convergent validity between self-reports and third-party assessments of mind wandering (Mills, Rafaelli, et al., 2018). Self-reports of mind wandering consistently correlate with various indirect measures of mind wandering. Some of these are behavioral, such as performance errors (McVay & Kane, 2009), comprehension failures (Schooler, Reichle, & Halpern, 2004), eye movements (Uzzaman & Joordens, 2011), and changes in response variability (Seli, Smallwood, et al., 2015). Other measures are physiological, including evoked response potentials (Smallwood, Beach, et al., 2008), pupil dilation (Franklin, Mrazek, et al., 2013), eye blink rate (Smilek, Carriere, & Cheyne, 2010), and changes in brain activity (Christoff, Gordon, et al., 2009).

Of course, none of this is decisive. Global skeptical arguments about even the simplest experiential reports (Schwitzgebel, 2008) are notoriously hard to defuse (see Sosa, 1994). Furthermore, there are critics of self-report within the science of mind wandering (see Weinstein, 2018). One issue concerns the distorting effects of thought probes. Studies that present probes too often may disrupt the natural flow of thought in at least four ways. First, probes may reorient attention to task demands and reduce mind wandering (Seli, Carriere, Levene, et al., 2013; although see Robison, Miller, & Unsworth, 2019). Second, hyper-probing can lead to increased meta-awareness of one's thoughts, which itself alters the occurrence of mind wandering (Zedelius, Broadway, & Schooler, 2015). Third, this increased meta-awareness is a problem in itself: it inflates assessments of the degree to which mind wandering is accompanied by meta-awareness. Finally, laboratory tasks are often structurally dissimilar from real-world tasks, meaning that rates of mind wandering in the lab are difficult to translate into predictions of rates of mind wandering in the real world (see Murray et al., 2020).

Another methodological problem concerns how we define mind wandering. The vast majority of self-report studies operationally define mind wandering as task-unrelated thought (Mills, Rafaelli, et al., 2018). Indeed, many task-unrelated thoughts overlap with mind wandering. If your mind wanders from topic to topic as your eyes scan this page, for example, your thoughts are unrelated to the task of reading. Yet, the standard view has come under fire from multiple fronts. Task-unrelated thought is a broad and heterogeneous category (Irving, 2016; Christoff, Irving, et al., 2016). While your eyes mindlessly scan the page, for example, you might diligently do mental math for your statistics class. Or you might obsessively ruminate on a fight you just had with your friend. Such focused thinking seems to contrast with the meandering characteristic of mind wandering. Furthermore, it's not clear how the standard view handles cases of mind wandering in the absence of a task. Your mind can wander while you are resting on the beach, but then you don't have a task to wander away from (Seli, Kane, et al., 2018; Irving, 2016).

Cognitive scientists have recently proposed alternatives to the standard definition of mind wandering, but it is still an open question how amenable those new theories are to self-report. The dynamic view says that mind wandering is relatively unconstrained thought, which freely meanders from topic to topic (Irving, 2016; Christoff, Irving, et al., 2016; Sripada, 2018; see section 13.2). Yet, the empirical study of the dynamic stream of thought is still in its infancy, and ongoing work tells a nuanced story. One study found that reports of freely moving and task-unrelated thought are orthogonal in daily life and that the former independently predicts affect (Mills, Rafaelli, et al., 2018). Another found that unique electrophysiological signatures are differentially associated with freely moving, directed, and task-unrelated thought (Kam et al., 2021). However, yet another study failed to find predicted relationships between freely moving thought and various clinical pathologies (OCD, depression, and anxiety), as well as finding a negative correlation between freely moving thought and creative idea generation (Smith et al., unpublished manuscript). Additional studies also suggest that measures of thought dynamics are redundant with measures of task unrelatedness (O'Neill et al., 2020). This might suggest that the probes used to assess freely moving thought do not yet measure constructs that are readily interpretable.

Others argue that mind wandering is a heterogeneous construct with multiple overlapping attributes that no single instance of mind wandering fully exemplifies (Seli, Kane, et al., 2018). This argument depends on the (contentious) claim that no single definition of mind wandering picks out an extensionally adequate concept. On this basis, some have begun to propose alternative subjective measures of mind wandering that reflect the inherent heterogeneity of the construct (Murray et al., 2020).

Our review suggests that psychologists and philosophers have both worried about our current self-report measures of mind wandering. But the solutions to these problems almost exclusively involve the refinement of self-reports. You may wonder: Can we do better?

13.5   Intentional Mind Wandering

We find ourselves in a pickle. On the one hand, we argue that the scientific study of mind wandering requires self-report. On the other hand, there are several reasons to be skeptical of self-report methods. You may therefore ask whether we can break out of the cycle of self-report.

Our master argument in section 13.2 suggests that mind-wandering research must rely on self-report. Recall that our argument is as follows:

1. If a cognitive process or state ϕ cannot be voluntarily initiated and does not reliably subserve the performance of a task, then the psychological study of ϕ requires self-report.

2. Mind wandering cannot be voluntarily initiated.

3. Mind wandering does not reliably subserve the performance of a task.

4. Therefore, the psychological study of mind wandering requires self-report.

If you think mind-wandering research can break the cycle—that is, can do without self-report—then there must be something wrong with this argument. In the next two sections, we consider two objections to our argument. The first targets Premises 2 and 3, arguing that so-called intentional mind wandering can either be voluntarily initiated or subserve a task. The second targets Premise 1, on the grounds that objective measures allow us to study mind wandering without tasks or self-report. We turn first to the objection from intentional mind wandering.

Premises 2 and 3 state that no voluntary task can activate mind wandering, either directly or indirectly. You might resist this claim based on empirical evidence for intentional mind wandering. People frequently report intentional mind wandering—that is, letting their minds wander on purpose—both in the laboratory (Seli, Risko, Smilek, & Schacter, 2016) and in everyday life (Kane, Brown, et al., 2007). Empirical research suggests that everyday thinking recognizes the possibility of intentional mind wandering (Irving et al., 2020). Furthermore, intentional and unintentional mind wandering are differentially associated with independent state- and trait-level variables (Seli, Risko, & Smilek, 2016).

The possibility of intentional mind wandering might seem intuitive. While you're reading this chapter, for example, you might defiantly turn your head away from the page and let your thoughts drift from topic to topic. You might then argue that intentional mind wandering is a voluntary task that activates mind wandering. If so, then task-based methods may be able to measure mind wandering after all (contra Premises 2 and 3).

Our response to this objection depends on how you define mind wandering. Consider the task-unrelated thought theory of mind wandering. On this view, intentional mind wandering would consist in you intending to have task-unrelated thoughts. However, in intending to have such thoughts, you acquire a task: letting your mind wander. So, your thoughts are task related—not mind wandering. Intentions generate tasks. So, it's impossible to intend to think task-unrelated thoughts (see Murray & Krasich, 2020).

Of course, alternative characterizations of mind wandering might have the conceptual tools to explain intentional mind wandering. Some have suggested that intentional mind wandering might reflect maintaining variable streams of thought through meta-control (Irving, under revision). This is similar to contemplative techniques used by surrealist painters (Green, 2010) and meditators (Lutz et al., 2008).

Unfortunately, this interesting proposal is limited because the ability to maintain a wandering stream of thoughts requires a specialized process—meta-control—and employing it likely requires specialized training. Meta-controlled mind wandering may therefore differ in important ways from mind wandering in the absence of meta-control. However, there is no way to assess this without the use of self-report. Again, we find that self-report is indispensable for the scientific study of mind wandering.

Our discussion thus far shows the limits of methods that require direct voluntary control over mind wandering. If mind wandering is task-unrelated thought, such methods are conceptually incoherent. If mind wandering is unguided thought, such methods cannot tell us about mind wandering in general unless we use self-report. But perhaps we will have more luck with methods that exploit indirect voluntary control over mind wandering.

It is possible to control mind wandering indirectly: you can perform some *other* voluntary task τ because you know that τ reliably leads to mind wandering. Suppose that after a long day of studying, you want to let your mind wander to clear your head. You might do this by taking a shower or washing the dishes, which are the kinds of boring tasks that reliably lead to mind wandering (Mason et al., 2007; Smallwood, Fitzgerald, et al., 2009). You might assume that indirect voluntary control provides a route to the task-based study of mind wandering. Rather than ask subjects whether their mind was wandering, for example, you might simply give them the kind of boring task that reliably leads to mind wandering.

Baird, Smallwood, Mrazek, and colleagues' (2012) influential study of mind wandering and creativity employs this kind of indirect method. They tested whether mind wandering can lead to creative insights. This idea is intuitive: when you're stuck on a problem, for example, you might make progress by letting your mind wander in the shower. Baird, Smallwood, Mrazek, and colleagues (2012) tested this idea by having subjects solve a creative problem, take a break, and then return to the same problem. During the break, subjects in the experimental condition performed an easy task known to induce mind wandering. Subjects in one control condition performed a difficult task known to reduce mind wandering. Baird, Smallwood, Mrazek, and colleagues found that subjects who performed the easy task were more creative after the break than those who performed the difficult task. So, they concluded that mind wandering facilitates creativity.

But Baird, Smallwood, Mrazek, and colleagues' indirect method makes it difficult to interpret their results (Irving, 2018). It is possible that easy tasks facilitate creativity more than difficult tasks *because* they lead to higher rates of mind wandering. But it is also possible that the effect of task difficulty on creativity bypasses mind wandering entirely. Easy tasks might lead to relaxation, whereas difficult tasks might lead to frustration. And this difference in affective state might be what drives creativity. This is not a problem that indirect task-based methods can solve. Baird, Smallwood, Mrazek, and colleagues' subjects do not voluntarily initiate mind wandering or perform a task that mind wandering subserves. Rather, they perform a task τ that has two likely effects: mind wandering and creativity. Indirect methods alone cannot determine whether the effect on creativity is due to τ or mediated by mind wandering. To test for this mediation, we would need to ask subjects whether their minds are wandering during the break, and see whether those self-reported rates of mind wandering predict creativity—that is, we would need to rely on self-report.

Our discussion in this section shows that mind-wandering researchers cannot use intentional mind wandering to make do without self-report. A careful study of intentional mind wandering might lead to the development of direct and indirect methods to study the wandering mind. But those methods at best complement self-report; they are not a replacement.

13.6   Establishing Objective Measures

We cannot exploit intentional mind wandering to ground a task-based science of the wandering mind. Premise 1 of our master argument states, roughly, that self-report is the best (perhaps the only) alternative to task-based methods. If so, it follows that mind-wandering research must rely on self-report. You might object that there is a third alternative to task-based methods and self-report. Tasks and reports are useful because they reliably indicate that a process such as mind wandering is active. If we can devise alternative indicators for mind wandering, then we can do without tasks or reports. This hope animates the search for objective measures of mind wandering.

Seemingly toward this search, researchers have found various objective markers that correlate with reports of mind wandering, such as motor response times, task accuracy, and eye movements. One hope is that those objective markers can obviate the need for self-report. Rather than ask subjects whether their mind was wandering, the dream is simply to determine this answer by looking at behavior.

After a careful review of objective methods, we argue that we cannot replace self-report for two reasons. First, current findings are mixed and likely impacted by idiosyncrasies of ongoing task demands and strategies. As such, an indisputable set of mind wandering–specific behaviors has yet to be established. Second, self-report is still used to establish objective methods. Thus, objective methods actually extend, rather than replace, self-reports in the study of mind wandering.

13.6.1 Response Times and Performance Accuracy

Many laboratory studies of mind wandering have integrated thought probes into the Sustained Attention to Response Task (SART; Robertson et al., 1997), which is a task historically used to study vigilance. The SART requires participants to respond to nearly all stimuli and withhold responses to infrequent targets. Errors of commission and faster responses times (RT) are generally taken as evidence of vigilance lapses—that is, it is supposed that participants perform the task more quickly because they rely on their prepotent responses to stimuli rather than making on-the-fly adjustments, and errors of commission are thought to occur due to the associated attenuated response inhibition.

In many studies, self-reported mind wandering during the SART is also correlated with worse task performance and faster RT (e.g., McVay & Kane, 2011; Kane & McVay, 2012; Thomson, Besner, & Smilek, 2015). These findings seem consistent with frameworks of mind wandering that characterize it in terms of executive control failures (McVay & Kane, 2010) or lapses in vigilance (Spruyt et al., 2019). The link with faster RT is particularly intriguing as a potential real-time, thought-probe, independent index of mind wandering: with faster RT, the more likely it is that a participant's mind is currently wandering.

Unfortunately, this framework is too simple and, at times, inaccurate. Indeed, a few studies have found that self-reported mind wandering was associated with significantly *slower* RT (Baird, Smallwood, Lutz, et al., 2014; Bastian & Sackur, 2013). Factors contributing to these discrepancies are unclear but may include a variety of interrelated factors pertaining to differences in thought content, progression, and meta-awareness of mind wandering (Bastian & Sackur, 2013). Another possibility is that RT variability, rather than speed, is actually a better indicator of mind wandering (Bastian & Sackur, 2013; Seli, Cheyne, & Smilek, 2013; Thomson et al., 2014), but others have failed to find this effect (e.g., McVay & Kane, 2011). Therefore, the collective evidence does not eliminate the link between mind wandering and RT in the SART, but it does call into question RT as a reliable indicator of mind wandering across experimental conditions. In other words, the link between RT and mind wandering seems highly influenced by specific task parameters, which challenges the utility of using it as a single reliable behavioral index of mind wandering.

A tempting hypothesis, then, is that changes in RT relative to task performance could improve predictions of mind wandering. Supportive evidence of this has shown that as an iterated task progressed, the frequency of reported mind wandering increased, the observed differences in associated RT become more robust, and accuracy decreased (e.g., Krimsky et al., 2017).

Even this framework, however, is not without its challenges, especially because people can perform simple tasks accurately enough even while mind wandering, such as mind wandering while driving home from work. For example, Brosowsky and colleagues (forthcoming) measured rates of mind wandering during an implicit learning task. They found that as participants learned the task, RTs became faster and depth of mind wandering increased. Performance, however, improved throughout the task (consistent with the task becoming learned). Thus, some tasks are such that as they become automatized, mind wandering does not interfere with their performance. RTs and performance are uninformative indicators of mind wandering in these situations.

Furthermore, people exhibit signs of strategic mind wandering, especially in predictable or familiar task environments (Seli, Carriere, Wammes, et al., 2018). Seli and colleagues had participants watch a clock hand that moved in discrete steps, completing one revolution every twenty seconds. The instructions were simple: press the space bar on the keyboard whenever the clock hand reached the "12" (upright) position. Using thought probes to measure the occurrence of mind wandering, they found that mind wandering was significantly more likely to occur when the clock hand was in the second or third quadrant ("3"–"9") than the first or fourth quadrants. People began mind wandering when they knew they didn't need to pay attention, and they returned to the task when they knew they should. Accordingly, there was no relationship between the rates of reported mind wandering and task accuracy.

Collectively, findings from Seli, Smilek, et al. (2018) suggest that people can continue monitoring the task and remain aware of the task environment even when they report mind wandering. Again, we see that the link between mind wandering and potential behavioral indices is highly influenced by specific task parameters and might not be a reliable indicator of mind wandering across contexts.

13.6.2 Gaze Behaviors

A growing body of research has investigated eye movements as a potential behavioral index of mind wandering. Motivating this approach, eye movements are closely linked to the visual processing priorities of the visual system (e.g., Just & Carpenter, 1976; Kowler et al., 1995). This is in part due to the structural and functional limitations of the visual system in virtue of the anatomy of the eye, the organization of neurons in the primary visual cortex, and strict capacity limits on attention and working memory. People tend to look wherever they are attending (rare exceptions include covert shifts of attention) in such a way that best serves ongoing task goals.

During mind wandering, however, the visual system becomes, to some degree, perceptually decoupled from sensory inputs (Schooler et al., 2011; Smallwood, 2013). Neuroscientific measures support this idea. For instance, electroencephalography studies have shown that self-reported mind wandering is associated with attenuated P1 event-related potential (ERP) component amplitude (Baird, Smallwood, Lutz, et al., 2014; Kam et al., 2011; Smallwood, Beach, et al., 2008)—the ERP component associated with low-level visual processing (Hillyard et al., 1973). Mind wandering is also associated with attenuated cognitive processing of external stimuli, as indicated by an attenuated P3 ERP component amplitude (Barron et al., 2011). Considered together, these findings indicate a reduction in the cortical processing of external visual information associated with self-reported mind wandering.

Changes in visual processing during mind wandering, then, should result in corresponding changes in gaze behaviors. Accordingly, research has identified a number of changes in gaze behaviors associated with mind wandering, although results have been relatively mixed. To illustrate, as one of the first studies to investigate mind wandering–related gaze behaviors, Reichle, Reineberg, and Schooler (2010) showed that the eyes tended to fixate on words for a longer duration of time before reports of mind wandering compared to reports of attentive reading. These longer fixations were not related to word length or frequency as they typically are during attentive reading (Juhasz & Rayner, 2006; Rayner & Duffy, 1986; Reichle, Rayner, & Pollatsek, 2003). This finding suggests that the observed longer fixations durations before reports of mind wandering reflected perceptual decoupling, corroborating the self-reports.

Reichle and colleagues (2010) suggests that longer fixation durations might be a promising objective measure of mind wandering, especially because this relationship has been replicated in other work using reading tasks (e.g., Faber, Bixler, & D'Mello, 2018; Foulsham, Farley, & Kingstone, 2013; Frank et al., 2015; Steindorf & Rummel, 2020; although see Smilek, Carriere, & Cheyne, 2010; Uzzaman & Joordens, 2011) as well as scene-viewing tasks (Krasich et al., 2018; Zhang, Anderson, & Miller, 2020). The challenge, though, is that many contemporary frameworks of eye movements consider longer fixation durations a marker of *increased* visual processing (e.g., Choi et al., 2017; Coco, Nuthmann, & Dimigen, 2020; Henderson, Choi, Luke, & Schmidt, 2018; Luke et al., 2018), which is supported by neuroscientific evidence in reading (Henderson, Choi, Luke, & Desai, 2015) and scene viewing (Henderson & Choi, 2015). Therefore, it would seem as though the relationship between mind wandering and fixation duration— and perhaps gaze control in general—might also prove to be idiosyncratic.

To test this idea, Faber, Krasich, and colleagues (2020) asked participants to complete a battery of computer-based cognitive tasks while their eye movements were measured. These tasks included the SART, listening to an audiobook (while looking at a central fixation), reading a narrative story, comprehending an illustrated text, viewing visual scenes, watching a recorded academic lecturer, and watching a narrative film. As such, the tasks varied across the spatial extent of the visual stimuli as well as the visual and semantic processing demands. Accordingly, these tasks should demand very different gaze patterns during attentive viewing, and the purpose of the study was to examine whether the relationship between mind wandering and eye movements would also vary by task.

The findings showed just this contextual variance. Specifically, in tasks that required extensive sampling of the visual field (i.e., scene viewing, comprehending an illustrated text, and narrative reading), fewer fixations were made prior to self-reported mind wandering compared to reported attentive viewing. Depending on the task, these fixations were also longer and more spatially dispersed. Conversely, in tasks that required more centrally focused gaze (i.e., the SART, listening to an audiobook while looking at a central fixation, and watching a recorded academic lecture), mind wandering was associated with shorter and more dispersed fixations as well as larger saccades. These findings support the idea that the relationship between mind wandering and gaze behavior varies according to the idiosyncrasies of the task.

These findings pose a challenge for initiatives attempting to use gaze patterns as a behavioral index of mind wandering. It is probably no surprise, then, that predictive modeling methods (such as those described in Yarkoni & Westfall, 2017) have yet to identify a common set of gaze parameters that can be used across *all* tasks. Still, modeling eye movements to predict mind wandering *within* a task has some potential. Essentially, this approach considers a variety of changes in eye movements associated with reports of mind wandering in concert.

For instance, Faber, Bixler, and colleagues (2018) developed a gaze behavior–based, machine-learned model of mind wandering utilizing reading data from 132 undergraduate students across two universities (data set from Kopp, D'Mello, & Mills, 2015). Specifically, this "mind-wandering detector" trained a supervised classification model of gaze behaviors associated with self-caught mind wandering (participants pressed a key on the keyboard when they caught themselves mind wandering).[6] The model included sixty-two global (content-independent) gaze features, such as the number, duration, and dispersion of fixations; the number of saccades (ballistic eye movements between fixations); and the number of blinks.

The model was validated using a leave-one-reader-out cross-validation method that trained the model on data from $n - 1$ participants and tested the model on data from the remaining participant until all 132 served as a "test" participant. The model (sequential minimization optimization) showed a weighted precision of 72.2 percent and a weighted recall of 67.4 percent. Stated simply, this mind-wandering detector could accurately (though not perfectly) predict self-reported mind wandering from gaze behaviors. Others have used similar approaches to predict the likelihood of ongoing mind wandering offline (e.g., Bixler & D'Mello, 2016; Brishtel et al., 2020) and in real time (Mills, Gregg, et al., 2020) during reading.

Outside of the context of reading, Hutt and colleagues (2019) adopted an offline-to-online classification-verification approach to predict the likelihood of ongoing mind wandering within an artificial intelligence tutoring system (ITS). These authors first gathered data from 135 high school students who completed the ITS and responded to pseudo-randomly distributed thought probes (probe-caught mind wandering) as to whether they were on- or off-task at a given moment. The authors then used Bayesian networks to classify gaze behaviors associated with mind wandering (from a set of fifty-seven global gaze parameters, eighty content-specific parameters, and eight features related to the human–computer interactions) and a leave-several-students-out cross-validation scheme (67 percent of students were used in the training set and 33 percent were assigned to the test set) for fifteen iterations. Findings showed accuracies (mind wandering $F_1 = 0.59$) were better than chance ($F_1 = 0.24$), and this model could generalize to data collected from a controlled laboratory study.

Live mind-wandering detection was then tested on a new sample of thirty-nine high school students. Students completed the same ITS while mind-wandering probabilities were generated. Two types of thought probes were distributed throughout the learning session: (1) a probe that was triggered by the mind-wandering detector used to assess hits and false alarms, and (2) pseudo-random probes used to identify missed mind-wandering episodes. Findings showed that mind-wandering detection ($F_1 = 0.40$) was above chance ($F_1 = 0.24$). Collective evidence from Hutt and colleagues (2019) again indicates that accurate (though still imperfect) real-time mind-wandering detection can be achieved by training classification algorithms on a collection of gaze behaviors linked to self-reported mind wandering.

13.6.3 Limitations of Objective Methods

Classifying eye movements to detect ongoing mind wandering seems to promise an objective measure of mind wandering: online machine learning classifiers may allow us to detect ongoing mind wandering without disrupting the subject's task performance or spontaneous stream of thoughts. This would ease worries about how self-reports disrupt the thoughts they are designed to measure.

Yet, even these mind-wandering detectors do not remove the need for self-report. First, we argued that objective measures of mind wandering are not always available and that they require unique classification for each task context. Moreover, this so-called objective method for identifying mind wandering still has a subjective core. Machine learning classifiers are initially trained on data about self-reported mind wandering. Researchers then validate a classifier on the basis of how well it detects said self-reported mind wandering. Self-report is therefore still the epistemic foundation of objective measures of mind wandering (Irving, 2018).

Classifiers inherit many of the other challenges of self-report methods as well. For instance, even supposing that the classifiers perfectly detect self-reported mind wandering (which they don't), they will be trained to detect the experimenter-imposed operational definition of mind wandering, which is most frequently task-unrelated thought. But we've already seen that various researchers have criticized this definition (section 13.3). Therefore, we can develop a more nuanced classifier—that detects the dynamics of thought, for example—only *after* we develop more nuanced self-report measures of mind wandering.

Our more fundamental point is that wholehearted skepticism about self-report is not an option for mind wandering research. In section 13.3, we noted that philosophers and cognitive scientists have both questioned whether self-report can *ever* be a reliable method. We think the success of mind-wandering science generates a powerful response to this brand of skepticism. However, if skepticism still tempts you, objective methods should provide you no relief, since those methods are only as good as the self-reports that they are designed to track.

13.7 Functional Justification of Self-Report

We've shown that mind-wandering science cannot break out of the cycle of self-report, but that does not mean we must trust self-report blindly. Instead, we have empirical and philosophical reasons to be confident in self-reports about mind wandering. We have already reviewed some of the empirical reasons for confidence (section 13.4). Self-reports of mind wandering are remarkably consistent in their contents and their ability to pick out behavior, neural activation, and third-party reports. If self-reports were spurious, we should predict that this consistency would not arise.

We will now provide a philosophical argument for the reliability of self-reports. Recall earlier our discussion of how the functional role of belief plays an important role in supporting inferences about it in experimental contexts. We believe that understanding the functional role of mind wandering will be similarly helpful in vindicating some of the self-report methods used to measure mind wandering. Our argument has the following structure. We have independent reason to believe two leading (and compatible) theories about the function of mind wandering (related to plural goal pursuit and exploration). We note that each of those theories makes predictions about the contexts in which mind wandering should be most prevalent. Self-reports confirm those predictions. The fact that our best theories and self-report methods converge is reason to trust both theory and method.

Let us start by reviewing two theories about the function of mind wandering.

13.7.1   Plural Goal Pursuit

The mind's tendency to wander is both salient and puzzling. Studies indicate that we spend a lot of time mind wandering (Seli, Beaty, et al., 2019). Perhaps this is obvious from your ordinary experience. But take a step back, and this statistic is shocking. When the world is a dynamic, unpredictable, and dangerous place, why would any creature spend half their lives being inattentive? What function could mind wandering have that explains its prominence in our lives?

The first theory is that mind wandering enables scattered agents like us to plan our futures. Human beings have hierarchically structured sets of goals, commitments, and projects that they aim to complete over timescales of days, weeks, months, and even years. This requires balancing many different demands that these place on our time. For example, one must balance the demands of being a teacher, being a parent, being a friend, and so on. Navigating social space requires living up to the different goals and expectations associated with the various roles one occupies.

In addition to having many goals, people have relatively limited capacities. Some of these limitations are physical. The particularities of embodiment preclude us from moving in two opposite directions simultaneously. However, some of these limitations are psychological. Foveal vision does not extend much beyond about 5° of arc. Working memory has severe capacity limits (Logie, 2011), items stored in working memory decay rapidly (Baddeley & Hitch, 1974; Oberauer & Kliegel, 2006), and self-control exhibits depletion effects over short timescales (Dang, 2018). These limits present obstacles to plural goal pursuit.

In order to do a lot with a little, people scatter their agency over time (see Murray, 2020). Non-scattered agents act sequentially, and they can only start a new activity upon completing or abandoning their current activity. Scattered agents, on the other hand, are capable of being engaged simultaneously in various projects without explicitly acting on any particular project then and there. In this way, scattered agents act in such a way that their actions have distinct (non-overlapping) temporal parts (see Sorensen, 1985). While scattering action is an efficient solution to the problem of plural goal pursuit under conditions of computational limitation, scattered agency raises unique engineering problems. People must now balance focus on their present activities with keeping an eye on acting in the future.

Balancing consists in dynamically altering the allocation of psychological resources to pursue multiple goals fluidly at the same time. At any point in time, there will be multiple, jointly incompatible ways of allocating one's psychological resources to pursuing distinct tasks. For example, do you focus on answering emails, editing a manuscript, or diving into some grading? You can't do all three at once, and you've got to do them all eventually. The marginal utility of focusing on any task diminishes over time. This creates pressure to switch to other tasks, the phenomenological correlate of which might be the experience of effort (Shenhav et al., 2017). However, the possibility of switching to a disengaged plural pursuit state can be beneficial, where people think about various goals without thinking too hard about any particular goal. This is especially true when current task performance is unlikely to improve with increased or sustained focus. Mind wandering's first function is arguably to enable this disengaged plural pursuit state.[7] On this view, mind wandering drifts between various contents, though it is predominantly drawn to one's goals and concerns (cf. Klinger, 2013; Irving, 2016). Mind wandering, then, reflects a way of managing limited mental resources for scattered agency.

13.7.2 The Function of Exploration

Mind wandering's second function may be to help us navigate the explore–exploit trade-off (Sripada, 2018; Irving, 2019). Imagine that you are beginning to write an essay. Before you put pen to paper, you are faced with a choice. Should you pursue one of the ideas that you have already come up with, or should you explore a new idea? Explorers take a risk: you might waste precious time. But you might strike gold. This decision is an example of a fundamental trade-off between two aims of a cognitive agent: to explore for new ideas, or to exploit the ideas we already have in order to get things done.

Humans arguably have different modes of thought that are tailored to the demands of exploitation and exploration (Sripada, 2018). Exploitative modes of thinking remain focused on our goals or personal concerns for extended periods of time and shield out distractions. You engage in exploitative thought, for example, when you successfully guide your attention to this chapter for long enough to get through a few pages. Exploitative thinking is necessary for limited agents like us. At any time, you can only focus on a tiny subset of all the relevant information. To make this subset count, you must be able to focus on what you think is relevant (e.g., this chapter) and ignore almost limitless things that might distract you.

Exploitation is an incomplete strategy, however, because you can be wrong about what is in fact relevant to you. Imagine that you dutifully guide your attention to this chapter, and nothing but, for a whole day. We the authors would be thrilled! But we recognize that you might miss out. Perhaps you would forget about other commitments that you have made. Or you might simply fail to expand the sphere of what you are reading and thinking about. By constantly guiding your attention to what you *think* is relevant, you risk being caught in a cognitive bubble, where you never notice the information that would lead you to expand your own point of view.

Exploratory modes of thought such as mind wandering may help us burst cognitive bubbles. When your mind drifts between topics, you may think of ideas that seem strange, useless, or irrelevant. This is not a bug of mind wandering but rather a feature, since seemingly useless information can turn out to be crucial. Seemingly irrelevant ideas may be just what you need to expand your point of view. Mind wandering, then, may be an inbuilt mode of cognitive exploration that helps us take risks and search for new ideas. Most of the time, this exploration may prove useless. But sometimes it will strike gold.

13.7.3 Triangulation of Function and Method

We have sketched two theories about the functions of mind wandering. First, mind wandering supports plural goal pursuit. Second, mind wandering supports cognitive exploration. Despite their differences, these theories generate common a priori predictions about the circumstances likely to elicit mind wandering.

Mind wandering should occur when we are bored or under-stimulated. Both of our functional theories contrast mind wandering with more focused forms of thinking, which occur when we are pursuing a single important goal (e.g., writing an exam or running from a tiger). We should therefore expect mind wandering to occur when people are bored, as boredom reflects a perceived lack of important information to extract from the task environment. Relatedly, we should expect strategic mind wandering in familiar and predictable task environments. Such environments are structured in ways where task performance does not benefit from sustained focus on the task (Kane, Gross, et al., 2017). When engaged in boring or predictable tasks, we can free our cognitive resources to focus on the type of plural goal pursuit and/or exploration characteristic of mind wandering (Geana et al., 2016).

Results from self-report studies are consistent with our a priori predictions about the circumstances likely to elicit mind wandering. For example, mind wandering is more frequent when people are bored or under-stimulated, or when they are doing something either too easy or too difficult (Eastwood et al., 2012). People report that mind wandering is more pleasant when they are doing something unengaging, which suggests that mind wandering relieves boredom (Mooneyham & Schooler, 2013). The conditions that we expect to elicit mind wandering turn out to be the conditions where we observe a lot of mind wandering.

The fact that our best functional theories and self-report methods converge is reason to trust both theory and method. Consider an analogy. Imagine that we have two very different methods to estimate a quantity, such as the density of gold. Can we trust that our methods are accurate? Well, suppose we find that each method is internally consistent: they yield the same measurements at different times and for different samples of gold. Suppose further that the methods are consistent with each other: both estimate the density at 19.32 cm^3. Unless we have reason to think otherwise, the best explanation of this internal and external consistency is that (1) the density of gold is 19.32 cm^3, and (2) both methods can accurately estimate this density. The fact that diverse methods triangulate onto a single answer gives us reason to trust both methods.

We think a similar form of triangulation gives us reason to trust self-report methods in mind-wandering science. Self-report methods are internally consistent, at least about the result that variables such as boredom elevate mind wandering. Self-report methods are also consistent with independently plausible theories about the function of mind wandering. Put colloquially, one reason to believe self-reports about mind wandering is that they make sense.

Put more precisely, philosophical argumentation and self-report are both reliable methods for forming beliefs about mind wandering. These methods triangulate onto common claims about mind wandering—for example that boring tasks elevate mind wandering. Unless we have reason to think otherwise, the best explanation for this triangulation is (1) that mind wandering occurs more frequently in boring and undemanding tasks and (2) that philosophical argumentation and self-report are both accurate.

We have focused on predictions about boredom. But triangulation has much broader application in the science of mind wandering. We can generate more consensus predictions from our best theories of the function of mind wandering. We can then see whether self-report methods yield results that make sense, given these predictions. If they do, this gives us defeasible reason to trust that self-reports about mind wandering are accurate. Mind-wandering science may be unable to do without self-report. But our trust in self-report can be based in theory rather than blindness.

13.8   Conclusion

Mind-wandering science cannot do without self-report. At best, objective measures complement but do not replace self-report methods. This is not a defect of mind-wandering research but rather a reflection of the kind of cognitive state that mind wandering is: a passive manifestation of agency. We suspect that there are many other passive manifestations of agency that are inaccessible to purely task-based methodologies. As such, research on mind wandering provides valuable lessons on how to refine self-report methodologies and use them in conjunction with more standard methods in cognitive psychology and neuroscience.

Skeptics about the reliability of self-report may take this as reason to doubt mind-wandering science. Our response to the skeptic appeals to the metaphysics of mind wandering to partially vindicate self-report methods. This response highlights how theoretical models play an important role in verifying the accuracy of experimental methods. When the outputs of these methods (data) converge with the predictions of the model, this is crucial evidence for the validity of empirical methods. Convergence, then, also provides evidence in favor of the model. The justificatory pathway between theory and data runs both ways. This runs contrary to a tendency in cognitive psychology to separate theory and data, with data being considered the only source of evidence. Theories, on this view, might make predictions that guide the acquisition of data, but theories themselves are not considered sources of evidence. Our argument depends on this being wrong. Instead, the theoretical content of and argument for a model play important roles both in verifying each of these methods and in the model itself. This is consistent with a model-based view of scientific methodology that began to flourish as a reaction against the logical positivism of the twentieth century (see Kuhn, 1962; Feyerabend, 1969; Tal, 2011).

While this is a longer story than we have space here to tell, we want to highlight the fact that, on our view, *philosophical theorizing* plays an important role in generating evidence insofar as such theorizing can produce models. Our argument is one example of this. The functional characterizations of mind wandering offered above are drawn from philosophical reflections on mind wandering, attention, and action. This suggests that philosophers have a more active role to play in empirical inquiry than simply double-checking the inferences that cognitive scientists make.

Notes

1. The Killingsworth and Gilbert (2010) study enjoys some prominence in the mind-wandering literature for being one of the first large-scale experience sampling studies of mind wandering published in a prestigious venue. The paper is cited mainly for two findings: (1) the results that 30–50 percent of waking thoughts are mind wandering, and (2) the association between mind wandering and negative affect. However, both results have been challenged recently. Some studies have shown that the frequency of mind wandering varies as a function of the response options provided to indicate mind wandering. Based on how self-reports are interpreted, mind wandering can constitute anywhere from 18 to 60 percent of waking thoughts (Seli, Beaty, et al., 2019). Moreover, the association with negative affect fails to distinguish between different kinds of mind wandering (intentional/unintentional) and does not consider relevant trait-level moderators (e.g., mindfulness; see Wang et al., 2017). Reanalysis of the Killingsworth and Gilbert (2010) data shows that accounting for interest in the content of off-task thoughts reveals a positive association between mind wandering and positive affect (Franklin, Broadway, et al., 2013). This shows that the results of the Killingsworth and Gilbert (2010) study should be viewed within the context of recent developments and criticisms.

2. Shulman and colleagues (1997) provided the initial characterization of the default mode network by contrasting activations in control states from activations in task states in positron emission tomography studies. Follow-up studies confirmed the identity of the network (Binder et al., 1999; Mazoyer et al., 2001).

3. Nonhuman animals, including marmosets and macaques, also have a network that is organized similarly to the default network (for a review, see Buckner & DiNicola, 2019). This may also point to a dissociation between mind wandering and the

default network if there are independent reasons to deny that nonhuman animals mind wander.

4. There are many different kinds of self-report used throughout the behavioral sciences. We are interested in the kind of self-report described here—namely, making a judgment about one's mental states on the basis of retrospective assessments. All subsequent reference to self-report picks out this subset of self-report measures.

5. The modal scope includes nomological possibility. In section 13.4, we explain why creatures like us in worlds like this cannot voluntarily initiate certain activities. Perhaps there are worlds and creatures that differ on these grounds. We think it is the job of theologians and poets (rather than this chapter) to consider these possibilities.

6. Faber, Bixler, and D'Mello (2018) actually tested a few different classification algorithms. The best model was a sequential minimization optimization algorithm, which is an implementation of a support vector machine classifier.

7. The term "disengaged" serves to distinguish mind wandering from multitasking, the latter of which is an engaged or committed form of plural goal pursuit. Roughly, in multitasking, someone explicitly commits to focusing on several tasks simultaneously. In mind wandering, several tasks can at any one time be at the forefront of one's mind, but there is no explicit commitment to keep any single task (or set of tasks) in focus.

References

Adams, R. M. (1985). Involuntary sins. *The Philosophical Review, 94*(1), 3–31.

Antrobus, J. S. (1968). Information theory and stimulus-independent thought. *British Journal of Psychology, 59*, 423–430.

Audi, R. (1979). Wants and intentions in the explanation of action. *Journal for the Theory of Social Behaviour, 9*(3), 227–249.

Baddeley, A. D., & Hitch, G. (1974). Working memory. In G. H. Bower (Ed.), *The psychology of learning and motivation* (pp. 47–89). Cambridge, MA: Academic Press.

Baird, B., Smallwood, J., Lutz, A., & Schooler, J. W. (2014). The decoupled mind: Mind-wandering disrupts cortical phase-locking to perceptual events. *Journal of Cognitive Neuroscience, 26*, 2596–2607.

Baird, B., Smallwood, J., Mrazek, M. D., Kam, J. W., Franklin, M. S., & Schooler, J. W. (2012). Inspired by distraction: Mind wandering facilitates creative incubation. *Psychological Science, 23*, 1117–1122.

Baird, B., Smallwood, J., & Schooler, J. W. (2011). Back to the future: Autobiographical planning and the functionality of mind wandering. *Consciousness and Cognition, 20*, 1604–1611.

Barron, E., Riby, L. M., Greer, J., & Smallwood, J. (2011). Absorbed in thought: The effect of mind wandering on the processing of relevant and irrelevant events. *Psychological Science, 22*, 596–601.

Bastian, M., & Sackur, J. (2013). Mind wandering at the fingertips: Automatic parsing of subjective states based on response time variability. *Frontiers in Psychology, 4*, 573.

Binder, J. R., Frost, J. A., Hammeke, T. A., Bellgown, P. S. F., Rao, S. M., & Cox, R. W. (1999). Conceptual processing during the conscious resting state: A functional MRI study. *Journal of Cognitive Neuroscience, 11*(1), 80–93.

Bixler, R., & D'Mello, S. (2016). Automatic gaze-based user-independent detection of mind wandering during computerized reading. *User Modeling and User-Adapted Interaction, 26*(1), 33–68.

Brishtel, I., Khan, A. A., Schmidt, T., Dingler, T., Ishimaru, S., & Dengel, A. (2020). Mind wandering in a multimodal reading setting: Behavior analysis and automatic detection using eye-tracking and an EDA sensor. *Sensors, 20*(9), 2546.

Brosowsky, N. P., Murray, S., Schooler, J. W., & Seli, P. (forthcoming). Attention need not always apply: Mind wandering impedes explicit but not implicit sequence learning. *Cognition*. Advance online publication. doi:10.1016/j.cognition.2020.104530.

Buckner, R. L., Andrews-Hanna, J. R., & Schacter, D. L. (2008). The brain's default network: Anatomy, function, and relevance to disease. *Annals of the New York Academy of Sciences, 1124*, 1–38.

Buckner, R.L. and DiNicola, L.M. (2019). The brain's default network: Updated anatomy, physiology and evolving insights. *Nature Reviews Neuroscience, 20*(10), 593–608. doi: 10.1038/s41583-019-0212-7.

Callard, F., Smallwood, J., & Margulies, D. S. (2011). Default positions: How neuroscience's historical legacy has hampered investigation of the resting mind. *Frontiers in Psychology, 3*, 321.

Choi, W., Lowder, M. W., Ferreira, F., Swaab, T. Y., & Henderson, J. M. (2017). Effects of word predictability and preview lexicality on eye movements during reading: A comparison between young and older adults. *Psychology and Aging, 32*(3), 232.

Christoff, K., Gordon, A. M., Smallwood, J., Smith, R., & Schooler, J. W. (2009). Experience sampling during fMRI reveals default network and executive system contributions to mind wandering. *Proceedings of the National Academy of Sciences of the United States of America, 106*, 8719–8724.

Christoff, K., Irving, Z. C., Fox, K. C. R., Spreng, R. N., & Andrews-Hanna, J. R. (2016). Mind wandering as spontaneous thought: A dynamic framework. *Nature Reviews Neuroscience, 17*(11), 718–731.

Coco, M. I., Nuthmann, A., & Dimigen, O. (2020). Fixation-related brain potentials during semantic integration of object–scene information. *Journal of Cognitive Neuroscience, 32*(4), 571–589.

Cullen, S. (2010). Survey-driven romanticism. *Review of Philosophy and Psychology, 1*, 275–296.

Dang, J. (2018). An updated meta-analysis of the ego depletion effect. *Psychological Research, 82*(4), 645–651.

Eastwood, J. D., Frischen, A., Fenske, M. J., & Smilek, D. (2012). The unengaged mind: Defining boredom in terms of attention. *Perspectives on Psychological Science, 7*(5), 482–495.

Faber, M., Bixler, R., & D'Mello, S. K. (2018). An automated behavioral measure of mind wandering during computerized reading. *Behavior Research Methods, 50*, 134–150.

Faber, M., Krasich, K., Bixler, R. E., Brockmole, J. R., & D'Mello, S. K. (2020). The eye–mind wandering link: Identifying gaze indices of mind wandering across tasks. *Journal of Experimental Psychology: Human Perception and Performance, 46*(10), 1201–1221.

Feyerabend, P. K. (1969). Science without experience. In P. K. Feyerabend (Ed.), *Realism, rationalism, and scientific method* (Vol. 1: Philosophical Papers; pp. 132–136). Cambridge: Cambridge University Press.

Foulsham, T., Farley, J., & Kingstone, A. (2013). Mind wandering in sentence reading: Decoupling the link between mind and eye. *Canadian Journal of Experimental Psychology, 67*, 51–59.

Fox, K. C. R., Spreng, R. N., Ellamil, M., Andrews-Hanna, J. R., & Christoff, K. (2015). The wandering brain: Meta-analysis of functional neuroimaging studies of mind wandering and related spontaneous thought processes. *NeuroImage, 111*, 611–621.

Frank, D. J., Nara, B., Zavagnin, M., Touron, D. R., & Kane, M. J. (2015). Validating older adults' reports of less mind wandering: An examination of eye-movements and dispositional influences. *Psychology and Aging, 30*(2), 266–278.

Franklin, M. S., Broadway, J. M., Mrazek, M. D., Smallwood, J., & Schooler, J. W. (2013). Window to the wandering mind: Pupillometry of spontaneous thought while reading. *Quarterly Journal of Experimental Psychology, 66*, 2289–2294.

Franklin, M. S., Mrazek, M. D., Anderson, C. L., Smallwood, J., Kingstone, A., & Schooler, J. W. (2013). The silver lining of a mind in the clouds: Interesting musings are associated with positive mood while mind wandering. *Frontiers in Psychology, 4*, 583.

Gable, S. L., Hopper, E. A., & Schooler, J. W. (2019). When the muses strike: Creative ideas of physicists and writers routinely occur during mind wandering. *Psychological Science, 30*(3), 396–404.

Geana, A., Wilson, R., Daw, N. D., & Cohen, J. (2016). Boredom, information-seeking and exploration. In A. Papafragou, D. Grodner, D. Mirman, & J. C. Trueswell (Eds.), *Proceedings of the Cognitive Science Society* (pp. 1751–1756). Seattle, WA: Cognitive Science Society.

Giambra, L. M. (1995). A laboratory method for investigating influences on switching attention to task-unrelated imagery and thought. *Consciousness and Cognition, 4*(1), 1–21.

Gil-Jardiné, C., Née, M., Lagarde, E., Schooler, J., Contrand, B., Orriols, L., & Galera, C. (2017). The distracted mind on the wheel: Overall propensity to mind wandering is associated with road crash responsibility. *PLoS One, 12*(8), e0181327.

Gorgolewski, K. J., Lurie, D., Urchs, S., Kipping, J. A., Craddock, R. C., . . . Smallwood, J. (2014). A correspondence between individual differences in the brain's intrinsic functional architecture and the content and form of self-generated thoughts. *PLoS One, 9*(5), e97176.

Green, S. (2010). Consciousness and Ian McEwan's *Saturday*: "What Henry Knows." *English Studies, 91*(1), 58–73.

Hall, L., Johansson, P., & Strandberg, T. (2012). Lifting the veil of morality: Choice blindness and attitude reversals on a self-transforming survey. *PLoS One, 7*(9), e45457.

Henderson, J. M., & Choi, W. (2015). Neural correlates of fixation duration during real-world scene viewing: Evidence from fixation-related (FIRE) fMRI. *Journal of Cognitive Neuroscience, 27*(6), 1137–1145.

Henderson, J. M., Choi, W., Luke, S. G., & Desai, R. H. (2015). Neural correlates of fixation duration in natural reading: Evidence from fixation-related fMRI. *NeuroImage, 119*, 390–397.

Henderson, J. M., Choi, W., Luke, S. G., & Schmidt, J. (2018). Neural correlates of individual differences in fixation duration during natural reading. *Quarterly Journal of Experimental Psychology, 71*(1), 314–323.

Hillyard, S. A., Hink, R. F., Schwent, V. L., & Picton, T. W. (1973). Electrical signs of selective attention in the human brain. *Science, 182*(4108), 177–180.

Hutt, S., Krasich, K., Mills, C., Bosch, N., White, S., Brockmole, J. R., & D'Mello, S. K. (2019). Automated gaze-based mind wandering detection during computerized learning in classrooms. *User Modeling and User-Adapted Interaction, 29*, 821–867.

Irving, Z. C. (2016). Mind wandering is unguided attention: Accounting for the "purposeful" wanderer. *Philosophical Studies, 173*, 547–571.

Irving, Z. C. (2018). Psychology off tasks. *Journal of Consciousness Studies, 25*(5–6), 63–84.

Irving, Z. C. (2019). Attention norms in Siegel's *The Rationality of Perception. Ratio, 32*(1), 84–91.

Irving, Z. C. (resubmitted). Drifting and directed minds: The significance of mind wandering for mental action. Retrieved from https://philpapers.org/rec/IRVDAD

Irving, Z. C. and Glasser, A. (2020). Mind-wandering: A philosophical guide. *Philosophy Compass,* 15:e12644. https://doi.org/10.1111/phc3.12644.

Irving, Z. C., Glasser, A., Gopnik, A., & Sripada, C. (2020). What does "mind wandering" mean to the folk? An empirical investigation. *Cognitive Science, 44*(10), e12908.

Israel, M. M., Jolicoeur, P., & Cohen, A. (2018). The effect of pre-cueing on spatial attention across perception and action. *Psychonomic Bulletin and Review, 25*(2), 1840–1846.

Juhasz, B. J., & Rayner, K. (2006). The role of age of acquisition and word frequency in reading: Evidence from eye fixation durations. *Visual Cognition, 13*(7–8), 846–863.

Just, M. A., & Carpenter, P. A. (1976). Eye fixations and cognitive processes. *Cognitive Psychology, 8*(4), 441–480.

Kam, J. W. Y., Dao, E., Farley, J., Fitzpatrick, K., Smallwood, J., Schooler, J. W., & Handy, T. C. (2011). Slow fluctuations in attentional control of sensory cortex. *Journal of Cognitive Neuroscience, 23*(2), 460–470.

Kam, J.W.Y., Irving, Z.C., Mills, C., Patel, S., Gopnik, A., and Knight, R.T. (2021) Distinct electrophysiological signatures of task-unrelated and dynamic thoughts. *Proceedings of the National Academy of Sciences, 118*(4), e2011796118. Doi: 10.1073/pnas.2011796118.

Kane, M. J., Brown, L. H., McVay, J. C., Silvia, P. J., Myin-Germeys, I., & Kwapil, T. R. (2007). For whom the mind wanders, and when: An experience-sampling study of working memory and executive control in daily life. *Psychological Science, 18*(7), 614–621.

Kane, M. J., Gross, G. M., Chun, C. A., Smeekens, B. A., Meier, M. E., Silvia, P. J., & Kwapli, T. R. (2017). For whom the mind wanders, and when, varies across laboratory and daily-life settings. *Psychological Science, 28*(9), 1271–1289.

Kane, M. J., & McVay, J. C. (2012). What mind wandering reveals about executive-control abilities and failures. *Current Directions in Psychological Science, 21*, 348–54.

Kane, M. J., Smeekens, B. A., Meier, M. E., Welhaf, M. S., & Phillips, N. E. (in preparation). Testing the construct validity of competing measurement approaches to probed mind-wandering reports. Retrieved from https://psyarxiv.com/te9yc/

Killingsworth, M. A., & Gilbert, D. T. (2010). A wandering mind is an unhappy mind. *Science, 330*(6006), 932.

Klein, C. (2012). Cognitive ontology and region- versus network-oriented analyses. *Philosophy of Science, 79*(5), 952–960.

Klinger, E. (2013). Goal commitments and the content of thoughts and dreams: Basic principles. *Frontiers in Psychology, 4*, 415.

Kopp, K., D'Mello, S., & Mills, C. (2015). Influencing the occurrence of mind wandering while reading. *Consciousness and Cognition, 34*, 52–62.

Kowler, E., Anderson, E., Dosher, B., & Blaser, E. (1995). The role of attention in the programming of saccades. *Vision Research, 35*(13), 1897–1916.

Krasich, K., McManus, R., Hutt, S., Faber, M., D'Mello, S. K., & Brockmole, J. R. (2018). Gaze-based signatures of mind wandering during real-world scene processing. *Journal of Experimental Psychology: General, 147*(8), 1111–1124.

Krimsky, M., Forster, D. E., Llabre, M. M., & Jha, A. P. (2017). The influence of time on task on mind wandering and visual working memory. *Cognition, 169*, 84–90.

Kuhn, T. S. (1962). *The structure of scientific revolutions.* Chicago, IL: University of Chicago Press.

Lappin, J. S., & Eriksen, C. W. (1966). Use of a delayed signal to stop a visual reaction-time response. *Journal of Experimental Psychology, 72*(6), 805.

Logan, G. D. (2015). The point of no return: A fundamental limit on the ability to control thought and action. *Quarterly Journal of Experimental Psychology, 68*, 833–857.

Logie, R. H. (2011). The functional organization and capacity limits of working memory. *Current Directions in Psychological Science, 20*(4), 240–245.

Luke, S.G., Darowski, E.S., and Gale, S.D. (2018). Predicting eye-movement characteristics across multiple tasks from working memory and executive control. *Memory & Cognition, 46*, 826-839.

Lutz, A., Slagter, H. A., Dunne, J. D., & Davidson, R. J. (2008). Attention regulation and monitoring in meditation. *Trends in Cognitive Sciences, 12*(4), 163–169.

Mason, M. F., Norton, M. I., Van Horn, J. D., Wegner, D. M., Grafton, S. T., & Macrae, C. N. (2007). Wandering minds: The default network and stimulus-independent thought. *Science, 315*, 393–395.

Mazoyer, P., Zago, L., Mallet, E., Bricogne, S., Etard, O., Houde, O., . . . Tzourio-Mazoyer, N. (2001). Cortical networks for working memory and executive function sustain the conscious resting state in man. *Brain Research Bulletin, 54*, 287–298.

McVay, J. C., & Kane, M. J. (2009). Conducting the train of thought: Working memory capacity, goal neglect, and mind wandering in an executive-control task. *Journal of Experimental Psychology: Learning, Memory, and Cognition, 35*, 196–204.

McVay, J. C., & Kane, M. J. (2010). Does mind wandering reflect executive function or executive failure? Comment on Smallwood and Schooler 2006 and Watkins 2008. *Psychological Bulletin, 136*, 188–197.

McVay, J. C., & Kane, M. J. (2011). Why does working memory capacity predict variation in reading comprehension? On the influence of mind wandering and executive attention. *Journal of Experimental Psychology: General, 141*, 302–320.

Mills, C., Gregg, J., Bixler, R., & D'Mello, S. K. (2020). Eye-Mind Reader: An intelligent reading interface that promotes long-term comprehension by detecting and responding to mind wandering. *Human–Computer Interaction*. Advance online publication. doi:10.1080/07370024.2020.1716762.

Mills, C., Rafaelli, Q., Irving, Z. C., Stan, D., & Christoff, K. (2018). Is an off-task mind a freely-moving mind? Examining the relationship between different dimensions of thought. *Consciousness and Cognition, 58*, 20–33.

Mooneyham, B. W., & Schooler, J. W. (2013). The costs and benefits of mind wandering: A review. *Canadian Journal of Experimental Psychology, 67*, 11–18.

Murray, S. (2020). The place of the trace: Negligence and responsibility. *Review of Philosophy and Psychology, 11*, 39–52.

Murray, S., & Krasich, K. (2020). Can the mind wander intentionally? *Mind and Language*. Advance online publication. doi:10.1111/mila.12332.

Murray, S., Krasich, K., Schooler, J. W., & Seli, P. (2020). What's in a task? Complications in the study of the task-unrelated-thought (TUT) conception of mind wandering. *Perspectives on Psychological Science, 15*(3), 572–588.

Oberauer, K., & Kliegl, R. (2006). A formal model of capacity limits in working memory. *Journal of Memory and Language, 55*(4), 601–626.

O'Neill, K., Smith, A. P., Smilek, D., & Seli, P. (2020). Dissociating the freely-moving thought dimension of mind-wandering from the intentionality and task-unrelated thought dimensions. *Psychological Research*. Advance online publication. doi:10.1007/s00426-020-01419-9.

Posner, M. I. (1980). Orienting of attention. *Quarterly Journal of Experimental Psychology, 31*(1), 3–25.

Preiss, D. D., Cosmelli, D., Grau, V., & Ortiz, D. (2016). Examining the influence of mind wandering and metacognition on creativity in university and vocational students. *Learning and Individual Differences, 51*, 417–426.

Raichle, M. E. (2015). The brain's default mode network. *Annual Review of Neuroscience, 38*, 433–447.

Rayner, K., & Duffy, S. A. (1986). Lexical complexity and fixation times in reading: Effects of word frequency, verb complexity, and lexical ambiguity. *Memory and Cognition, 14*, 191–201.

Reichle, E. D., Rayner, K., & Pollatsek, A. (2003). The E-Z reader model of eye-movement control in reading: Comparisons to other models. *Behavioral and Brain Sciences, 26*(4), 445–476.

Reichle, E. D., Reineberg, A. E., & Schooler, J. W. (2010). Eye movements during mindless reading. *Psychological Science, 21*(9), 1300–1310.

Robertson, I. H., Manly, T., Andrade, J., Baddeley, B. T., & Yiend, J. (1997). "Oops!": Performance correlates of everyday attentional failures in traumatic brain injured and normal subjects. *Neuropsychologia, 35*, 747–758.

Robins, R. W., & John, O. P. (1997). The quest for self-insight: Theory and research on accuracy and bias in self-perception. In R. Hogan, J. A. Johnson, & S. R. Briggs (Eds.), *Handbook of personality psychology* (pp. 649–679). Cambridge, MA: Academic Press.

Robison, M. K., Miller, A. L., & Unsworth, N. (2019). Examining the effects of probe frequency, response options, and framing within the thought-probe method. *Behavior Research Methods, 51*(1), 398–408.

Schooler, J. W., Reichle, E. D., & Halpern, D. V. (2004). Zoning out while reading: Evidence for dissociations between experience and metaconsciousness. In D. T. Levin (Ed.), *Thinking and seeing: Visual metacognition in adults and children* (pp. 203–226). Cambridge, MA: MIT Press.

Schooler, J. W., Smallwood, J., Christoff, K., Handy, T. C., Reichle, E. D., & Sayette, M. A. (2011). Meta-awareness, perceptual decoupling and the wandering mind. *Trends in Cognitive Sciences, 15*(7), 319–326.

Schwitzgebel, E. (2008). The unreliability of naïve introspection. *Philosophical Review, 117*(2), 245–273.

Seli, P., Beaty, R. E., Cheyne, J. A., Smilek, D., Oakman, J., & Schacter, D. L. (2019). How pervasive is mind wandering, really? *Consciousness and Cognition, 66*, 74–78.

Seli, P., Carriere, J. S. A., Levene, M., & Smilek, D. (2013). How few and far between? Examining the effects of probe rate on self-reported mind wandering. *Frontiers in Psychology, 4*, 430.

Seli, P., Carriere, J. S. A., Wammes, J. D., Risko, E. F., Schacter, D. L., & Smilek, D. (2018). On the clock: Evidence for the rapid and strategic modulation of mind wandering. *Psychological Science, 29*(8), 1247–1256.

Seli, P., Cheyne, J. A., & Smilek, D. (2013). Wandering minds and wavering rhythms: Linking mind wandering and behavioral variability. *Journal of Experimental Psychology: Human Perception and Performance, 39*, 1–5.

Seli, P., Kane, M. J., Smallwood, J., Schacter, D. L., Maillet, D., Schooler, J. W., & Smilek, D. (2018). Mind wandering as a natural kind: A family-resemblances view. *Trends in Cognitive Sciences, 22*(6), 479–490.

Seli, P., Risko, E. F., & Smilek, D. (2016). On the necessity of distinguishing between unintentional and intentional mind wandering. *Psychological Science, 27*(5), 685–691.

Seli, P., Risko, E. F., Smilek, D., & Schacter, D. L. (2016). Mind wandering with and without intention. *Trends in Cognitive Sciences, 20*(8), 605–617.

Seli, P., Smallwood, J., Cheyne, J. A., & Smilek, D. (2015). On the relation of mind wandering and ADHD symptomatology. *Psychonomic Bulletin and Review, 22*(3), 629–636.

Shenhav, A., Musslick, S., Lieder, F., Kool, W., Griffiths, T. L., Cohen, J. D., & Botvinick, M. M. (2017). Toward a rational and mechanistic account of mental effort. *Annual Review of Neurosciences, 40*, 9–124.

Shulman, G. L., Corbetta, M., Fiez, J. A., Buckner, R. L., Miezin, F. M., Raichle, M. F., & Petersen, S. F. (1997). Searching for activations that generalize over tasks. *Human Brain Mapping, 5*(4), 317–322.

Sinnott-Armstrong, W. (2008). Framing moral intuitions. In W. Sinnott-Armstrong (Ed.), *Moral psychology. Vol. 2. The cognitive science of morality: Intuition and diversity* (pp. 47–76). Cambridge, MA: MIT Press.

Smallwood, J. (2013). Distinguishing how from why the mind wanders: A process-occurrence framework for self-generated thought. *Psychological Bulletin, 139*, 519–535.

Smallwood, J., Beach, E., Schooler, J. W., & Handy, T. C. (2008). Going AWOL in the brain: Mind wandering reduces cortical analysis of external events. *Journal of Cognitive Neuroscience, 20*, 458–469.

Smallwood, J., Fitzgerald, A., Miles, L. K., & Phillips, L. H. (2009). Shifting moods, wandering minds: Negative moods lead the mind to wander. *Emotion, 9*, 271–276.

Smallwood, J., & Schooler, J. W. (2015). The science of mind wandering: Empirically navigating the stream of consciousness. *Annual Reviews in Psychology, 66*, 487–518.

Smilek, D., Carriere, J. S. A., & Cheyne, J. A. (2010). Out of mind, out of sight: Eye blinking as indicator and embodiment of mind wandering. *Psychological Science, 21*(6), 786–789.

Smith, A., Brosowsky, N. P., Murray, S., & Seli, P. (under review). Testing the predictions of the dynamic framework of mind wandering.

Sorensen, R. A. (1985). Self-deception and scattered events. *Mind, 94*(373), 64–69.

Sosa, E. (1994). Philosophical Scepticism. *Aristotelian Society Supplementary Volume, 68*(1), 263-307.

Spreng, N., Stevens, W. D., Chamberlain, J. P., Gilmore, A. W., & Schacter, D. L. (2010). Default network activity, coupled with the frontoparietal control network, supports goal-directed cognition. *NeuroImage, 53*(1), 303–317.

Spruyt, K., Herbillon, V., Putois, B., Franco, P., & Lachaux, J.-P. (2019). Mind-wandering, or the allocation of attentional resources, is sleep-driven across childhood. *Scientific Reports, 9*(1), 1269.

Sripada, C. (2018). An exploration/exploitation tradeoff between mind wandering and task-directed thinking. In K. C. R. Fox & K. Christoff (Eds.), *Oxford handbook of spontaneous thought and creativity* (pp. 23–34). New York: Oxford University Press.

Stalnaker, R. (1984). *Inquiry*. Cambridge, MA: MIT Press.

Steindorf, L., & Rummel, J. (2020). Do your eyes give you away? A validation study of eye-movement measures used as indicators for mindless reading. *Behavioral Research Methods, 52*(1), 162–176.

Tal, E. (2011). How accurate is the standard second? *Philosophy of Science, 78*(5), 1082–1096.

Thomson, D. R., Besner, D., & Smilek, D. (2015). A resource-control account of sustained attention: Evidence from mind wandering and vigilance paradigms. *Perspectives in Psychological Science, 10*(1), 82–96.

Thomson, D. R., Seli, P., Besner, D., & Smilek, D. (2014). On the link between mind wandering and task performance over time. *Consciousness and Cognition, 27*, 14–26.

Uzzaman, S., & Joordens, S. (2011). The eyes know what you are thinking: Eye movements as an objective measure of mind wandering. *Consciousness and Cognition, 20*(4), 1882–1886.

van Fraassen, B. (1984). Belief and the will. *The Journal of Philosophy, 81*(5), 235–256.

van Inwagen, P. (1989). When is the will free? *Philosophical Perspectives, 3*, 399–422.

Wang, Y., Xu, W., Zhuang, C., & Liu, X. (2017). Does mind wandering mediate the association between mindfulness and negative mood? A preliminary study. *Psychological Reports, 120*(1), 118–129.

Warm, J. S., Parasuraman, R., & Matthews, G. (2008). Vigilance requires hard mental work and is stressful. *Human Factors, 50*(3), 433–441.

Wegner, D. M. (1997). Why the mind wanders. In J. D. Cohen & J. W. Schooler (Eds.), *Scientific approaches to consciousness* (pp. 295–315). Mahwah, NJ: Lawrence Erlbaum Associates.

Weinstein, Y. (2018). Mind-wandering, how do I measure thee with probes? Let me count the ways. *Behavior Research Methods, 50*(2), 642–661.

Windt, J. (2015). *Dreaming: A conceptual framework for philosophy of mind and empirical research*. Cambridge, MA: MIT Press.

Windt, J. (2016). Dreaming, imagining, and first-person methods in philosophy: Commentary on Evan Thompson's "Waking, Dreaming, Being." *Philosophy East and West, 66*(3), 959–981.

Yanko, M. R., & Spalek, T. M. (2014). Driving with the wandering mind: The effect that mind wandering has on driving performance. *Human Factors, 56*(2), 260–269.

Yarkoni, T., & Westfall, J. (2017). Choosing prediction over explanation in psychology: Lesson from machine learning. *Perspectives on Psychological Science, 12*(6), 1100–1122.

Zedelius, C. M., Broadway, J. M., & Schooler, J. W. (2015). Motivating meta-awareness of mind wandering: A way to catch the mind in flight? *Consciousness and Cognition, 36,* 44–53.

Zhang, H., Anderson, N.C., & Miller, K.F. (2020). Refixation patterns of mind-wandering during real-world scene perception. *Journal of Experimental Psychology: Human Perception and Performance, 47*(1), 36-52.

14 Neuroscience and Cognitive Ontology: A Case for Pluralism

Joseph McCaffrey and Jessey Wright

14.1 Introduction

Developing taxonomies—systems for classifying phenomena or objects of study—is an important scientific activity. By categorizing and classifying phenomena, scientists can more effectively organize research programs, and can even discover new phenomena and relations between them. Cognitive psychology seeks to explain and predict human behavior. The mental domain is thus organized into categories such as attention, episodic memory, reward prediction, and cognitive control. Collectively, these categories comprise a "cognitive ontology"—a set of categories taken to reflect the mind's organization (Price & Friston, 2005; Poldrack, 2010; Klein, 2012; Anderson, 2015). Cognitive neuroscience aims to understand how different brain structures carry out these psychological capacities.

This chapter examines the role of cognitive neuroscience in shaping our cognitive ontology. The term "cognitive ontology" sometimes denotes the informatics concept of a database mapping the domain of mental entities and the relationships between them (Price & Friston, 2005; Poldrack, Kittur, et al., 2011). Elsewhere, cognitive ontology refers more loosely to the taxonomy of mental kinds adopted by contemporary psychology. Worries about our cognitive ontology manifest as concerns about the terminology used to refer to cognitive phenomena, the mental kinds those terms and concepts refer to, or as conflicts between views about the structure of cognition itself (Janssen, Klein, & Slors., 2017).[1] These concerns are interrelated. For instance, a clear and shared ontology is important for fruitfully engaging in debates about the structure of cognition or bringing empirical results to bear on such debates. After all, scientists cannot integrate findings from different research projects if they do not, at minimum, use terms and concepts in similar ways.

How do we know if we have the right concepts for mapping the mind? On one hand, psychology exhibits conceptual progress over time. For example, most psychologists would agree that response inhibition, a psychological construct related to impulse control and measured using tasks such as the go/no-go task (Morein-Zamir & Robbins, 2015), is a better psychological category than Sigmund Freud's id or pleasure principle (Freud, 1989).[2] On the other hand, there are numerous debates about whether particular elements and concepts deserve to be in our cognitive ontology. Researchers debate how many types of memory there are (Tulving, 2007), whether emotions such as fear and anger are real mental kinds (Barrett, 2006; Lindquist et al., 2012), whether task switching is a unique cognitive control element (Lenartowicz et al., 2010), and so forth. In many domains of psychology (social, affective, cognitive, etc.), there are debates about the foundational terms and concepts used to describe our mental abilities, and about how these terms and concepts align with the experimental tasks used to study them in the laboratory (Sabb et al., 2008, Scarantino & Griffiths, 2011; Sullivan, 2016, 2017; Quesque & Rossetti, 2020).

Recently, philosophers and cognitive scientists have expressed a more global concern that perhaps our entire cognitive ontology (or large territories of it) needs substantial revision (Price & Friston, 2005; Lenartowicz et al., 2010; Poldrack, 2010; Anderson, 2015). In this chapter, we present and critique the main arguments motivating large-scale cognitive ontology revision in the philosophical and neuroscientific literature. The first argument for large-scale ontological revision is methodological and conceptual. It asserts that psychological categories are folksy, unscientific, and otherwise carelessly deployed. Not only should we expect radical ontological revision as neuroscience makes progress, but the looseness of the ontology itself may be holding cognitive neuroscience back (Poldrack, 2010; Bunzl, Hanson, & Poldrack, 2010). The second argument for cognitive ontology revision is empirical. This argument alleges that neuroimaging data—for example, functional magnetic resonance imaging (fMRI), positron emission tomography, and magnetic electroencephalography studies—suggest that the processes specified in our cognitive ontology do not map onto structural or functional features of the brain (e.g., Lenartowicz et al., 2010).

Advocates for this position claim these findings challenge the status of cognitive terms as demarcating scientifically legitimate kinds.

We have three goals for this chapter. The first goal is to familiarize readers with the main issues and debates about cognitive ontology revision (section 14.2). The second goal is to show that while the master arguments for large-scale ontology revision highlight important theoretical and methodological issues in the mind–brain sciences (we agree that neuroscience will contribute to taxonomic revision in psychology and that psychologists should systematically clarify their conceptual categories), our existing ontology is not necessarily the problem (section 14.3). The third goal is to challenge the widespread assumption of taxonomic monism in the cognitive ontology literature. Most researchers working on this issue seem to hold that (1) each item in a scientifically valid cognitive ontology will map to a specific item in the neural ontology (the collection of categories for organizing and describing parts or functions of the brain), and (2) there exists one correct (or best) cognitive ontology. We think these assumptions are mistaken and instead echo and reinforce calls for taxonomic pluralism (Danks, 2015; Hochstein, 2016; Sullivan, 2016, 2017). Taxonomic pluralism, applied to these issues, is the view that cognitive neuroscience will need different ontologies for achieving different research goals and for providing different perspectives on the phenomena of interest. After advocating for taxonomic pluralism, we devote the end of this chapter to examining what that might look like for cognitive neuroscience (section 14.4).

14.2   The Case for Cognitive Ontology Revision

The question of how to classify mental kinds has deep historical roots in philosophy and psychology. The ancient Greek philosopher Plato argued that the soul has three parts: the rational part, the spirited part, and the appetitive part (Plato, *Republic IV*, 431a1 ff). Scottish philosopher Thomas Reid (1710–1796) recognized dozens more "mental faculties" such as memory, abstraction, judgment, and imagination (Brooks, 1976). Disputes about how to classify mental kinds are similarly longstanding. Nineteenth-century phrenologist Franz Gall (1758–1828), who advocated twenty-seven mental faculties (e.g., memory for words, sense of places, poetic talent, love of offspring, etc.), chastised philosophers for developing their mental taxonomies through internal reflection instead of scientific investigation (Gall, 1835).[3]

The recent debate about cognitive ontology brings these perennial questions about classifying mental kinds into the contemporary era of big data, open science initiatives, and brain imaging (Poldrack, 2010; Poldrack, Kittur, et al., 2011; Poldrack & Yarkoni, 2016). This emerging literature has three recurring themes. The first is a concern that our current psychological categories and the practices surrounding them may be holding back scientific research (Poldrack, Kittur, et al., 2011; Sullivan, 2016; Quesque & Rossetti, 2020). The second is that developing cognitive ontologies will improve science by making conceptual disagreements explicit and permitting researchers from disparate laboratories and research traditions (e.g., psychiatry and cognitive neuroscience) to integrate their findings (Price & Friston, 2005; Sabb et al., 2008; Poldrack, Kittur, et al., 2011; Sullivan, 2017). The third concerns how our cognitive ontology aligns with neuroimaging data (Anderson, 2010, 2014; Klein, 2012; Rathkopf, 2013; McCaffrey, 2015; Viola, 2017; Khalidi, 2017). Some propose using neuroimaging to determine what elements belong in our cognitive ontology (Lenartowicz et al., 2010; Poldrack 2010). This raises important questions about "the proper role of neuroscientific evidence in determining what mental entities there are" (Viola, 2017, p. 163).

There are varying views about how neuroscience might transform our cognitive ontology. Some researchers envision modest reform, while others envision a radical overhaul and a new beginning (Anderson, 2015). There are many research agendas with different theoretical orientations, experimental methods, and analytic tools. Some initiatives are primarily aimed at charting our existing cognitive ontology to achieve greater terminological and conceptual consistency across laboratories and fields (e.g., Poldrack, Kittur, et al., 2011; Quesque & Rossetti, 2020). Other work calls for testing our existing ontology using brain data (e.g., Lenartowicz et al., 2010; Lindquist et al., 2012), or using brain data to mine novel relationships between tasks and brain activation patterns (Poldrack, Halchenko, et al., 2009; Anderson, Kinnison, & Pessoa, 2013; Yeo et al., 2015). Despite these differences, we can extract an overarching narrative from advocates of cognitive ontology revision (Price & Friston, 2005; Poldrack, 2010; Anderson, 2014, 2015). According to this narrative, our current cognitive ontology is systematically ill suited for explaining human behavior or for predicting patterns of brain activation in experiments *because* it was developed in absence of consideration of the brain. We should rectify this by testing, refining, and potentially rebuilding our cognitive ontology using the tools of neuroscience (Lenartowicz et al., 2010; Poldrack, 2010; Anderson, 2015). As Michael Anderson (2015) puts it, "the brain can (and should) act as one arbiter of the psychologically real" (p. 70).

There are two main arguments for this position. The first argument for revision is that we should distrust our current ontology because it contains concepts that are provisional, vague, or borrowed from prescientific or "folk" psychology (Bunzl et al., 2010; Poldrack, 2010). As cognitive neuroscientist Russel Poldrack laments, "the fundamental problem is our stone age psychological ontology" (Bunzl et al., 2010, p. 54). Moreover, cognitive psychologists often deploy these concepts heterogeneously and imprecisely (Poldrack, Kittur, et al., 2011; Quesque & Rossetti, 2020) or fail to analyze the relationships rigorously between tasks (activities performed by human subjects in a laboratory) and the mental processes those tasks recruit (Figdor, 2011; Sullivan, 2017). The second argument for revision is the empirical observation that neuroimaging studies show virtually no one-to-one mappings between brain structures or activation patterns and mental processes (Price & Friston, 2005; Anderson, 2010; Poldrack, 2010). This implies that our current taxonomy of mental processes fails to map onto the neural structures underlying cognition. We discuss these arguments in turn.

14.2.1 Why Our Cognitive Ontology Is Mistaken: Conceptual Woes

The first major argument for large-scale cognitive ontology revision is that we have good antecedent conceptual reasons for doubting our existing cognitive ontology. According to this proposal, our cognitive ontology contains numerous items that are vague, ill defined, or borrowed from folk psychology (Bunzl et al., 2010; Poldrack, 2010). Furthermore, different laboratory groups frequently use cognitive terms and concepts interchangeably (Poldrack, Kittur, et al., 2011; Quesque & Rossetti, 2020) and/or fail to provide adequate analyses of the tasks used to study cognition and the cognitive processes those tasks allegedly recruit (Figdor, 2011; Swick, Ashley, & Turken, 2011; Sullivan, 2016, 2017). In other words, psychology may have many flawed concepts and a set of embedded social practices that are unconducive to weeding them out. We call these issues "conceptual woes."

Psychologists constantly debate the validity of their concepts. Good psychological concepts are supposed to explain current behavior, predict future behavior or behavior in different contexts, and be distinct from one another (Feest, 2020). For example, the social psychology concept "grit" (i.e., perseverance toward long-term goals when faced with obstacles) is not intended to *describe* the fact that some people succeed in spite of difficulties but to *explain* in part why they do and to *predict* who is likely to (Duckworth et al., 2007). For proponents, grit should be in our cognitive ontology because (1) its measures (e.g., the Grit Scale) are independent of, and have independent predictive value compared to, measures for other personality traits (e.g., self-control), and (2) the construct predicts patterns of performance in multiple laboratory tasks and in real-world behavior. Common reasons for jettisoning psychological concepts include a lack of ecological validity (laboratory measures do not predict real-world behavior) and or redundancy with existing concepts. For example, Vazsonyi and colleagues (2019) found that scales for measuring grit and self-control have basically the same power for predicting success—thus, beyond terminology, "grit might be indistinguishable from self-control" (p. 224).

One would hope that the routine, iterative process of construct validation (Sullivan, 2016; Feest, 2020) would converge on a plausible cognitive ontology. But many have recently argued that current practices in cognitive psychology are unconducive to this goal of conceptual progress (Poldrack, Kittur, et al., 2011; Sullivan, 2017; Quesque & Rossetti, 2020). One worry is a lack of consistency in the terms and concepts used to describe cognitive processes. A second concern is a lack of consistency in the tasks used to study those putative processes and a lack of engagement with whether those tasks actually tap into the process in question. A final concern is that the terms and concepts psychologists employ are borrowed from prescientific folk psychology or other outdated psychological theories (Poldrack, 2010).

First, there is significant heterogeneity in the terms and concepts used to describe cognitive processes (Sullivan, 2017). Poldrack and Yarkoni (2016) claim that unlike other sciences, "cognitive neuroscience is awash in a sea of conflicting terms and concepts" (p. 588). Psychologists frequently use different terms to refer to the same cognitive process. For example, social cognition researchers use a motley of terms such as "mindreading," "mentalizing," "theory of mind," "perspective taking," "cognitive empathy," and so on to denote the same hypothetical construct (Quesque & Rosetti, 2020). The same term can also have wildly diverging meanings—for example, the term "empathy" has more than forty different definitions and refers to at least nine putative cognitive constructs (Quesque & Rossetti, 2020). Poldrack, Kittur, and colleagues (2011) highlight the diverging meanings of "working memory," which is sometimes construed as manipulating information in an online fashion, sometimes merely holding that information, and other times as memory for "temporally varying" aspects of a task (p. 1).

Second, there is significant heterogeneity regarding the experimental tasks used to study cognitive concepts. Researchers frequently use different experimental tasks to measure the same cognitive process—for example, the go/no-go task (a person is given a standing rule to respond on certain trials and hold back on others), and the stop-signal task (a person is instructed to respond on trials until a stop signal) are both used to measure response inhibition (Swick et al., 2011; Morein-Zamir & Robbins, 2015). Likewise, researchers sometimes use the same task to measure different putative cognitive processes. In different laboratories, the Stroop task is used to measure response inhibition, response selection, and conflict detection.

Conceptual progress in cognitive neuroscience requires linking task performance in experimental settings (e.g., performance on a go/no-go task) to cognitive processes (e.g., response inhibition) and functional neuroanatomy (the brain areas involved in the task). These links must be made within individual experiments but also across various laboratories with different experimental protocols (Sullivan, 2016). Psychologists' tendency to use terms, concepts, and experimental tasks interchangeably creates barriers to integrating findings across different laboratories and subdisciplines. If one group uses the Stroop task to study "conflict detection" and another group uses it to study "response inhibition," then there could be surprising neuroanatomical and behavioral overlap between these (allegedly) distinct cognitive kinds. Avoiding this problem requires task analysis—that is, articulating and testing different theories of the cognitive processes involved in the task (Sullivan, 2016; Poldrack, 2010).

In some cases, researchers have likely not undertaken these task analyses because there is no widely used repository for keeping track of these proposed connections. Where these relationships are explicitly tested, the data are troubling. For example, Swick and colleagues (2011) report that two popular response inhibition tasks—the go/no-go task and the stop-signal task—elicit distinct patterns of brain activity. This challenges whether these tasks are truly equivalent measures of the same neurocognitive process. Poldrack, Kittur, and colleagues (2011) summarize this dire situation as follows: "this lack of consistency in the way that tasks and concepts are treated in the literature makes it difficult to draw meaningful inferences from existing literature and limits the cumulative value of the knowledge represented in this literature" (p. 2).

Finally, some worry that our cognitive ontology is suspect because it derives from folk psychology and other prescientific cognitive theories (Bunzl et al., 2010; Poldrack, 2010). Poldrack (2010, p. 754) poses a vivid thought experiment: suppose phrenologist Franz Gall had fMRI at his disposal. Poldrack (2010, p. 754) claims Gall could have performed fMRI investigations of his phrenological faculties (e.g., sense of property, poetic talent) under contemporary labels for actual neuroimaging experiments (e.g., hoarding behavior, generation of creative vs. uncreative narrative). The narrow warning here is we should not reify a cognitive kind in our ontology merely because something in the brain "activates" during tasks meant to recruit it. The broader concern is that we should distrust our cognitive ontology because it comes from folk psychology rather than neuroscience. We revisit this claim in section 14.3.

14.2.2   Why Our Cognitive Ontology Is Mistaken:
No One-to-One Mappings

The second main argument for large-scale taxonomic revision is the empirical observation that our current cognitive kinds do not perform well at predicting or explaining why particular brain structures are, or are not, involved in various cognitive tasks.[4] Some claim that if our cognitive ontology is correct, we should expect one-to-one mappings between those categories and brain structures—for example, regions, networks, or activation patterns (Price & Friston, 2005; Poldrack, 2010). As Poldrack (2010) puts it, "correctness of the ontology would be reflected in selective associations between structures and functions" (p. 754). The assumption here is that since the brain carries out cognition, we should expect to map items in a cognitive ontology onto brain structures. Likewise, we should be suspicious if our ontology contains an item that cannot be mapped onto some element of our neural ontology. But here we have a major problem: "a review of the neuroimaging literature suggests that selective association between mental processes and brain structures is currently impossible to find" (Poldrack, 2010, p. 274). We call this the problem of no one-to-one mappings.

A traditional view holds that each brain structure is associated with a single mental process (e.g., the amygdala is a "fear area"). Indeed, the assumption that "each mental entity should correspond to a single neural entity (and vice versa) is deeply embedded in cognitive neuroscience" (Viola & Zanin, 2017, p. 947). However, these simple mappings are not what neuroimaging studies show. Instead, there are one-to-many mappings in which each brain structure is associated with multiple cognitive processes (Poldrack, 2006; Anderson, 2010; Klein, 2012; McCaffrey, 2015; Burnston, 2016) and many-to-one mappings in which each cognitive process is associated with several different brain areas (Pessoa, 2014). The term "many-to-one" encompasses both the idea that cognitive processes map onto brain networks rather than regions (Pessoa, 2014) and the concept of neural degeneracy—that is, different brain structures can perform the same cognitive process (Price & Friston, 2002). Brain structures are typically associated with multiple cognitive processes, and cognitive processes are typically associated with multiple brain structures.

One way forward is to revise our cognitive ontology. Several researchers have proposed that if we altered our cognitive ontology to capture better what brain structures actually contribute to cognition—for example, in computational or mechanistic terms—we would achieve selective or one-to-one mappings (Price & Friston, 2005; Poldrack, 2010; Shine, Eisenberg, & Poldrack, 2016). The case of the so-called visual word form area exemplifies this proposal. Price and Friston (2005) claim that while the posterior lateral fusiform gyrus (plFG) is reputed to process the visual form of written characters, it performs (or contributes to) many other functions such as reading by touch and processing non-word visual stimuli. They argue that the plFG is not a word area or a visual area, but rather one that performs a kind of sensorimotor integration (p. 267).

Broca's area, once thought to be specialized for speech production (i.e., control of muscles involved in articulating speech), is implicated in myriad functions, including tool use, comprehending arm gestures, syntax comprehension, and even musical syntax (Tettamanti & Weniger, 2006; Gentilucci & Volta, 2008). Tettamanti and Weniger (2006) claim that Broca's involvement in these seemingly disparate cognitive processes is explained by the fact that all of them involve representing hierarchical structures—that is, Broca's area computes the kinds of hierarchical structures that are common to musical syntax, comprehending grammatical rules, and orchestrating speech.

There are proposals to revise the functions of many regions beyond the traditional psychological labels commonly ascribed to them (McCaffrey, 2015). Examples include claims that the pIFG is for sensorimotor integration (Price & Friston, 2005), Broca's area is a hierarchical processor (Tettamanti & Weniger, 2006), the anterior insula is a salience detector (Menon & Uddin, 2010), the parahippocampal cortex functions in contextual processing (Aminoff, Kveraga, & Bar, 2013), and so on. For these authors, while our current ontology fails to find one-to-one mappings between brain structures and cognitive processes, we can revise our ontology to achieve them. Proponents of large-scale cognitive ontology revision (e.g., Poldrack, Kittur, et al., 2011; Anderson, 2014) think we should revise the bulk of our cognitive ontology in this way. Klein (2012) objects that this proposal results in vague functional labels that are poor substitutes for cognitive terms, worrying that sensorimotor integration could describe what any part of the brain does, and hence fails to make specific predictions about the pIFG. While we agree in practice, in principle, we think terms such as "sensorimotor integration" or "salience detection" are placeholders for some as yet discovered either mechanistic or computational contribution to cognition.[5]

Some resist this conclusion, suggesting that these misaligned mappings will realign if we focus on the correct brain structures. Some authors have responded to the apparent multi-functionality of brain areas by arguing that we need to zoom in by dividing the region into functionally distinct areas (Scholz et al., 2009) or zoom out to map functions onto large-scale brain networks (Pessoa, 2014) to achieve one-to-one mappings. Increasing precision about our neural ontology is important, since there is no guarantee that traditional anatomical divisions (e.g., Brodmann areas) are the most cognitively interesting ones. But these simple fixes are unlikely to achieve neat one-to-one mappings, since multi-functionality is observed at many scales of brain organization.[6] Large-scale brain networks also appear to be multifunctional just as individual regions are (Pessoa, 2014; Viola, 2017).

Some authors attempt to resolve the problem of no one-to-one mappings by revising our assumptions about neural functioning rather than or in addition to our cognitive ontology (Klein, 2012; Anderson, 2014; McCaffrey, 2015; Burnston, 2016; Viola, 2017; Hutto, Peeters, & Segundo-Ortin, 2017). In particular, we should abandon the idea that each brain structure performs a single function. For example, Burnston (2016) argues that brain regions perform different functions and computations in a context-sensitive fashion. The middle temporal visual area (MT) is usually thought to process motion. However, as Burnston notes, MT is also involved in processing color, fine depth, texture, and other visual properties. While a proponent of computational specificity (e.g., Shine et al., 2016) might argue that we need to revise our cognitive ontology to include some common computation MT is performing, Burnston argues that models of MT suggest different computations underlying these distinct visual properties.

We examine these calls for neurofunctional revision more closely in section 14.3. The main point, for revisionists, is that failures to map elements of our cognitive ontology neatly onto the brain supply evidence that our cognitive ontology is flawed. Thus, we need to revise our cognitive ontology if we hope to predict brain activation patterns on the basis of cognitive functioning or to describe correctly what brain regions contribute to cognition (Price & Friston, 2005; Poldrack, 2010; Rathkopf, 2013).

14.2.3   Proposed Remedies

Here, we briefly review some work from this burgeoning literature that makes concrete recommendations for resolving the problems with our cognitive ontology that were outlined above (Price & Friston, 2005; Sabb et al., 2008; Lenartowicz et al., 2010; Poldrack, Kittur, et al., 2011; Anderson et al., 2013; Yeo et al., 2015; Poldrack & Yarkoni, 2016).

Efforts to articulate our existing cognitive ontology range from local, informal pleas to use terms and concepts consistently—for example, Quesque and Rossetti (2020) in social cognition research—to formal, global efforts to chart our whole ontology in a searchable database. For example, the *Cognitive Atlas* (Poldrack, Kittur, et al., 2011) is a wiki-inspired infrastructure that aims to depict the current ontology of cognitive neuroscience. One goal of the *Cognitive Atlas*, which currently charts 868 cognitive concepts, 775 tasks, and proposed relationships between them (is a, is part of, is measured by, etc.), is to facilitate data aggregation and collaboration between different laboratory groups by providing "a systematic characterization of the broad range of cognitive processes" (Poldrack, Kittur, et al., 2011, p. 2).

Some researchers have tested cognitive ontologies directly against fMRI data in a top-down fashion. By "top-down," we mean that they first articulated a cognitive ontology and then tested its elements using neuroimaging methods. In a widely discussed study, Lenartowicz and colleagues (2010) tested an ontology of cognitive control against fMRI data. They performed a meta-analysis of hundreds of fMRI studies of cognitive control constructs such as working memory, response selection, response inhibition, and task switching. They trained a machine learning classifier on this data set to see if it could reliably discriminate patterns of brain activation associated with each construct. If the classifier can discriminate between the brain activation patterns for two cognitive kinds (e.g., working memory and response inhibition), this suggests they correspond to distinct neural processes. If the classifier cannot discriminate the patterns, this suggests the cognitive kinds may be different labels for the same neural process. Interestingly, Lenartowicz and colleagues (2010) report that the classifier can readily distinguish these cognitive control concepts, with the exception of task switching. The classifier was unable to reliably discriminate task switching from response inhibition and response selection. This, the authors argue, suggests that task switching is not a distinct neural process and should be removed from our cognitive ontology. Task switching may exist "only in the minds of cognitive scientists" (p. 690).

Others have used data-driven approaches to mine for novel conceptual categories that link patterns of brain activation to cognitive tasks (Poldrack, Halchenko, et al., 2009; Anderson et al., 2013; Yeo et al., 2015). Data-driven approaches often begin with data drawn from databases of task-elicited brain activity such as *Neurosynth* (Yarkoni et al., 2011) or *BrainMap* (Laird et al., 2011). Information about the cognitive processes is set aside, and the brain activation data are grouped and analyzed using dimensionality reduction techniques such as factor analysis, principal components analysis, or multidimensional scaling. Various machine learning methods, including classification (Yeo et al., 2015) and neural networks (Poldrack, Halchenko, et al., 2009), have also been used to explore alternative ways of categorizing neural data. This ideally results in a novel way of grouping together the brain data that either echoes existing categories or inspires new cognitive classifications and terms.

For example, Yeo and colleagues (2015) performed an analysis of roughly 10,000 fMRI experiments in the *BrainMap* database (www.brainmap.org). They built a mathematical model (an author-topic hierarchical Bayesian model) linking the fMRI activation patterns to eighty-three task categories (e.g., N-back task, visual pursuit task, etc.) via latent variables (which correspond to cognitive components). The model estimated the probability that a task would recruit a cognitive component and the probability that a cognitive component would activate a particular region (set of volumetric pixels or "voxels"). The model was built to formalize mathematically the plausible notions that: (1) tasks can recruit a number of different cognitive components, (2) each cognitive component can participate in multiple tasks, and (3) cognitive components can map onto distributed brain regions. The model uncovered twelve hypothetical cognitive components that predicted their data set. These components were "discovered" by the model and do not correspond to traditional cognitive kinds. As Yeo and colleagues (2015) write, they "have refrained from explicitly labeling the cognitive components in order to not bias the readers' interpretation" (p. 3661).

Data-driven approaches such as that those briefly discussed above are attractive vehicles for a conceptual revolution because they allow for the adoption of an agnostic stance toward the cognitive ontology. That is, the focus of these approaches is on identifying categories that have the potential to solve the problems motivating ontology revision—both conceptual woes and no one-to-one mappings—without running headlong into them. A machine learning classifier or an exploratory factor analysis procedure can, in theory, categorize data without referencing or considering the cognitive capacities and tasks used to produce it in the first place. In the next section, we will critically evaluate the prospects of large-scale ontology revision.

14.3 A Knot with Three Ends: Assessing the Case for Large-Scale Taxonomic Revision

So far, we have presented arguments for revising our cognitive ontology using the tools of neuroscience (Price & Friston, 2005; Poldrack, 2010; Anderson, 2015). Next, we evaluate the case for large-scale cognitive ontology revision. Rather than pick apart individual studies, we will address the central philosophical arguments motivating revision. We claim while the arguments presented in section 14.2 provide an important theoretical and methodological critique of cognitive neuroscience and a vital source of inspiration for moving forward, they are not universally damaging to our existing ontology. We propose that the major problems raised by revisionists—(1) our cognitive ontology is in doubt, (2) our neurofunctional ontology is in doubt, and (3) psychology needs more terminological, conceptual, and experimental rigor—are so intertwined that blaming a lack of progress specifically on our cognitive ontology is premature.

First, we address whether our cognitive ontology *must* be flawed because it descends from folk psychology rather than neuroscience (e.g., Bunzl et al., 2010). We pose whether radical cognitive ontology revision is somehow inevitable as neuroscience progresses. Then, we discuss the tangled knot of empirical and methodological arguments motivating cognitive ontology revision. We identify a problem of mutual interdependence between refining our cognitive ontology, testing our theories of neural functioning, and facilitating terminological and conceptual consistency among different experimental groups and literatures. Many in the cognitive ontology debate hope we can hold our assumptions about the brain's functional organization and our research practices fairly constant, and adjust our cognitive ontology accordingly. Contrary to this hope, we think these problems are so interdependent that fixing one at a time (e.g., revising our cognitive ontology while holding our neurofunctional ontology fixed) or even identifying the main source of error (e.g., our cognitive ontology is what blocks us from achieving one-to-one mappings) is likely impossible.

14.3.1 Cognitive Ontology and Folk Psychology: Is Radical Revision Inevitable?

Philosophers have long pondered the fate of our folk psychological concepts as neuroscience progresses (P. S. Churchland, 1986; Hochstein, 2016; Francken & Slors, 2018; Dewhurst, 2020). Folk psychology is often regarded as a nonscientific ontology that is vague, familiar, and uninformed by cognitive science. The ongoing concern that folk psychology exerts a corrupting influence on cognitive science is reflected in Bunzl and colleagues' (2010) comment that "describing the [inferior parietal lobe] in some familiar and yet vague folk psychological terms creates a hopeless muddle of claims and agendas that get fossilized in the journals and training of graduate students" (p. 54).

The proposal that scientific progress will require replacing our current set of mental concepts with neuroscientific concepts is timely, provocative, and not entirely novel. In the 1980s, philosophers Patricia Churchland and Paul Churchland famously argued for "eliminative materialism," which held that the folk psychological concepts we use to explain human behavior—for example, beliefs, desires, and intentions—will be discarded for neuroscientific concepts (P. M. Churchland, 1981; P. S. Churchland, 1986). For example, the folk concept "memory" lumps together phenomena (e.g., remembering how to ride a bike, remembering that Paris is the capital of France, remembering your third birthday) that cognitive neuroscience splits via multiple memory systems (e.g., procedural memory for motor skills and semantic memory for facts; Michaelian, 2011). Thus, the folk concept of "memory" is ripe for replacement—perhaps all folk psychological concepts are too.

The Churchlands (P. M. Churchland, 1981; P. S. Churchland, 1986) argued that the nature of scientific progress virtually guarantees the replacement of folk psychological concepts by neuroscientific ones. First, they argued that folk psychology is part of a thoroughly prescientific world view, likening "beliefs" and "desires" to "miasmas" or "witchcraft." Second, they argued that the theoretical model of scientific reduction that would preserve folk psychology is wrong.

In the traditional philosophical model of reduction, a higher-level scientific theory "reduces" to a lower-level theory when all of the entities and empirical regularities of the former can be identified with those of the latter. For instance, classical genetics would reduce to molecular genetics if hereditary patterns equate to facts about the cellular transmission of genetic material (Schaffner, 1969). This form of reduction may be folk psychology friendly. Just as identifying molecular genes need not eliminate hereditary concepts, identifying brain systems need not eliminate folk psychological kinds.

P. S. Churchland (1986) thinks this "smooth" sort of inter-theoretic reduction where higher-level entities can be identified with lower-level ones is rare. Instead, the history of science is riddled with cases where the concepts initially used to understand some phenomena were eliminated as science progresses. For instance, the term "caloric" (a hypothetical substance that passes from hotter to colder bodies) was eliminated from chemistry with the kinetic theory of heat (P. S. Churchland, 1986). The same is true of the terms "ether," "phlogiston," and many other once cherished explanatory posits. The Churchlands argued that neuroscientific models of behavior will supplant prescientific folk concepts such as intentions, beliefs, or desires, which will go the way of caloric, ether, and phlogiston (P. M. Churchland, 1981).

However, the Churchlands' arguments are not decisive in retrospect. Not every prescientific concept is discarded as science advances. While the contemporary explanation for Earth's day/night cycle differs from ancient ones (i.e., it invokes the Earth's rotation rather than gods), science has not eliminated the concept of "a day." Second, reductionism is no longer the dominant view of how higher- and lower-level sciences relate. Instead, anti-reductionist views predominate in the philosophy of mind, biology, and even physics (e.g., Fodor, 1974; Batterman, 2000).

Some calls for cognitive ontology revision are Churchlandian in spirit (e.g., Bunzl et al., 2010; Poldrack, 2010). They share with eliminative materialism: (1) worries that the concepts currently governing the science are folksy and ripe for replacement (Poldrack, 2010), and (2) an optimism that neurobiological models will provide grounds for radically revising the existing ontology (Lenartowicz et al., 2010) or starting anew (Yeo et al., 2015). Many in this literature (e.g., Bunzl et al., 2010) argue that folk psychological concepts pervade our cognitive ontology, and that their influence is detrimental to advancing our understanding of cognition.

The idea that folk psychology can and does play a productive role in cognitive science is, like the idea that folk concepts should be eliminated, not novel. Several philosophers have recently argued that folk psychological concepts can be productive participants in cognitive science research (Sullivan, 2014; Hochstein, 2017; Francken & Slors, 2018; Dewhurst, 2020). There are at least three problems with the suggestion that folk psychology corrupts our cognitive ontology. First, human cognitive agents possess internal subjective states, which are what folk concepts aim to capture. Theories that fail to account for these internal states ultimately fail to describe human capacities categorically, and so fail to describe the naturalistic human traits and capacities that cognitive science aims to explain (e.g., Francken & Slors, 2018; Dewhurst, 2020). Second, contrary to the intuition that folk psychology can only corrupt cognitive neuroscience, folk psychology—or a version of it that is open to scientific revision and reconceptualization—remains a productive part of theorizing about human and animal cognition (Sullivan, 2014; Hochstein, 2017). Third, theories that focus on the emergent level of cognition, which is often cashed out in folk psychological terms, can achieve practical outcomes in settings, such as a clinic, where more detailed mechanistic accounts that spell out the chemicals and biological pathways correlated with behavior do not (e.g., Tabb & Schaffner, 2017).

Tension between folk psychology and cognitive science is easy to find. Brain training programs that make bold promises to "change" participants' brains through the power of "neural plasticity," ongoing debates about whether brain scans can serve as lie detectors, or any other click-bait headline describing a finding in cognitive science provide examples of folk terms and scientific terms failing to individuate the same phenomena. It is not just in scientific reporting that these ontological frameworks are at odds. Consider the character Dory from *Finding Nemo*. Dory is a fish described in the film as having a problem with short-term memory. However, as is observed in scene after scene, Dory has no problems with working memory, as she is able to solve problems she faces and to make considered decisions. Instead, speaking scientifically, she has a problem with consolidation in long-term memory. This is a case in which a scientific term, "short-term memory," is colloquially used in a manner that does not correspond at all with scientific research.[7]

Francken and Slors (2018) call this a "translation problem," and locate some of the responsibility for addressing it on the shoulders of cognitive scientists. In particular, they note that a many-to-many relationship holds between common-sense cognitive concepts (i.e., folk concepts) and scientific concepts (i.e., our cognitive ontology). This is not, on their view, grounds to eliminate the folk from cognitive science, but instead grounds for cognitive scientists to reevaluate how their research relates to the everyday concepts of which it purports to advance understanding.

A central goal of cognitive science is to clarify how concepts studied in the laboratory, such as response inhibition and task switching—relate to real-world mental phenomena such as multitasking while driving or impulsive behavior in addiction. Folksy concepts often delineate the phenomena in the real world that cognitive scientists aim to understand better through controlled laboratory experiments. Indeed, research in cognitive science is often motivated by and framed around real-world examples that lean on common sense, or folk psychological, intuitions about cognitive capacities. A talk about response inhibition may open with a colloquial example of impulsive shopping or stopping a car when a child runs out into the road.

These examples are not just rhetorical tools used to prime other scholars to think about a tricky scientific question or to convince funding agencies of the relevance of research. They can, and often do, become entrenched cases around which intuitions, hypotheses, and theories are refined and evaluated. For example, recognition memory, which is the capacity to identify previously encountered objects, people, or situations as such, is often

treated as having two sub-component processes: recollection (or remembering) and familiarity (or knowing). Recollection is described as the capacity to
retrieve details associated with the previous experience of the object, person,
or event in question. Familiarity, on the other hand, merely delineates the
feeling that the target has been previously encountered, and does not include
the additional contextual details that may allow one to "place" the memory.
This distinction, and empirical methods for investigating its biological realizers, is frequently motivated by appeal to the "butcher on the bus" scenario:

> Consider seeing a man on a bus whom you are sure that you have seen before;
> you "know" him in that sense. Such a recognition is usually followed by a search
> process of asking, in effect, Where could I know him from? . . . Eventually the
> search may end with the insight, That's [sic] the butcher from the supermarket!
> (Mandler, 1980, pp. 252–253).

This observation leads us toward the second reason folk psychology
should not be hastily jettisoned from cognitive science: it is a framework
with considerable empirical utility.

Sullivan (2014), through a detailed analysis of the history of the Morris
water maze in rodent memory research, argues that establishing the reliability of an experiment, and developing and refining tasks more generally,
"for individuating a discrete cognitive function requires a consideration
of 'what' an organism trained in the paradigm is learning" (p. 58). She
proposes that this is done by attributing beliefs, desires, and intentions to
that organism. Furthermore, she notes that ascribing internal cognitive
states to research subjects (by adopting the "intentional stance") also assists
researchers with inferring from observations of behavior to claims about
cognitive functions.[8]

In a similar vein, Hochstein (2017) defends the use of folk psychology
(specifically, the ascriptions of propositional attitudes to research subjects)
by arguing that folk psychological concepts are indispensably used to frame
comparative psychology research—for example, disputes about whether
chimpanzees or scrub jays have theory of mind concern whether one animal can *know* what another animal *knows* or what another animal *intends*
(italics denote folk concepts). As Hochstein (2017) notes, the fact that folk
concepts are interwoven into the linguistic practices of researchers is not
evidence that they are indispensable. Nonetheless, "[t]o be dismissive of
such theories simply because of the terminology they choose to invoke
would be to cut ourselves off from decades of psychological research that

we know empirically provides exactly the sort of information that neuroscientists need in order to refine and improve their models" (p. 1142).

If the above arguments are correct, then folk psychological concepts are motivation for, targets of, and supporting scaffolds for research in cognitive science. The scientific and folk ontologies are not in direct conflict, and may even be symbiotically related. When ontologies of the mind are viewed pluralistically, this situation is less surprising. After all, different ontologies serve different aims. Where the ontology of cognitive science has been developed to individuate and explain human cognitive capacities, folk psychology is used to predict and explain the behavior of whole organisms (Dewhurst, 2020). Tabb and Schaffner (2017) note that, at least in the case of schizophrenia research and treatment, different ontologies may be more important in different contexts. They notice that while folk psychiatry categorizes patterns most readily visible in a clinical context (after all, it is through the lens of folk psychiatry that patients view themselves), the aims and circumstances of laboratory research may be better facilitated by different ontological commitments (p. 355).

Even if our cognitive ontology contains folk concepts, nothing guarantees their elimination as neuroscience progresses. Furthermore, the targets of cognitive ontology revision are the empirical posits of cognitive psychology (e.g., response inhibition) rather than true folk concepts (e.g., impulsivity). Though some psychologists study familiar mental concepts—for example, lust, attachment, and romantic love (Fisher et al., 2002)—other terms and concepts—for example, "priming," "semantic memory," or "bottom-up attention"—may be alien to most people. And folksy terms such as "working memory" often refer to elaborate cognitive models (Baddeley & Hitch, 2019) rather than self-reported experiences. Our general point here is that the mere fact that a cognitive concept has folk psychological roots does not corrupt its scientific value. Furthermore, the persistence and utility of folk psychological concepts and categories strongly suggests that instead of calling for the elimination of folk psychology, those concerned with the ontology of cognitive science ought to pay more careful attention to the nuanced relationship between folk cognitive concepts and scientific ones.

14.3.2 Mutual Interdependence and the Case for Revision

Shorn from philosophical arguments about the nature of scientific progress and the explanatory deficiencies of folk psychology, claims that our psychological taxonomy will be replaced with neuroscientific concepts reflect a stance rather than a guarantee. We concede that brain research may spur extensive taxonomic revision. Psychology has made and will continue to make taxonomic progress. Perhaps our current ontology will someday seem as provincial and unscientific as Gall's phrenological faculties strike contemporary readers. Neuroscience will likely inform this revisionary process. Although philosophers usually resist the idea of psychoneural reduction, the mere fact that " evidence from neuroscience is *relevant* to models of cognition is sufficient to license the thought that it could be used to revise psychological constructs" (Anderson, 2015, p. 70, emphasis added). The work reviewed above demonstrates that neuroscience can productively challenge our cognitive theories and models.

So, we agree that cognitive ontology revision is a worthy project and that the existing literature raises foundational issues for improving cognitive neuroscience (Poldrack, 2010; Sullivan, 2016, 2017; Anderson, 2015). Nevertheless, we think the major arguments for cognitive ontology revision—no one-to-one mappings and conceptual woes—fail to establish that our cognitive ontology needs massive revision. For one, there are limitations with existing proposals.

Consider attempts to identify novel cognitive constructs by mining task-based patterns of brain activation (Poldrack, Halchenko, et al., 2009; Yeo et al., 2015). While this bottom-up methodology promises to reveal the brain's functional preferences in a manner unbiased by our existing ontology, it has significant drawbacks. First, the dimensionality reduction techniques these approaches employ do not have unique solutions in terms of the number of constructs identified. Thus, it is premature to reify these mathematical posits as real mental entities (McCaffrey & Machery, 2016). Second, the constructs are (deliberately) uninterpretable from a cognitive standpoint, meaning the constructs that maximally *predict* brain activation patterns may do less work *explaining* what is going on cognitively. But our goal is not to criticize individual studies, as the work is promising and novel, and future iterations may assuage these concerns.

Our major criticism is that the cognitive ontology debate points to such a tangle of issues that we cannot revise our cognitive ontology in a vacuum, or even pinpoint whether our existing ontology is in jeopardy. This literature raises foundational doubts about: (1) our practices for mapping cognitive kinds onto neural structures, (2) our cognitive ontology, and (3) the terminological and conceptual rigor of experimental psychology. The overall situation is knotted indeed. But since these problems are mutually interdependent (resolving one requires simultaneously grappling with the others), we cannot definitively say that our cognitive ontology is the problem. For all we know, our cognitive ontology will emerge fairly unscathed while we massively revise our views of how the brain carries out psychological functions (Anderson, 2014; Burnston, 2016) and/or our practices for linking terms to constructs and constructs to tasks (Figdor, 2011; Burnston, 2016).

Neurofunctional revision or taxonomic revision?   The fact that brain structures are implicated in multiple cognitive processes (and vice versa) is widely considered evidence of deficiencies in our ontology, since existing concepts fail to capture what brain areas are doing (Price & Friston, 2005; Anderson, 2010). But as many authors (e.g., Klein, 2012; McCaffrey, 2015; Viola, 2017) note, neurofunctional revision (revising our views of the brain's functional organization) may undercut the case for ontological revision. For example, Burnston (2016) argues that brain regions perform multiple functions in a context-sensitive manner (i.e., according to neuromodulatory effects and/ or their neural context). But if areas can perform multiple functions, then a region's involvement in multiple functions is not itself an indictment of our existing ontology. The multi-functionality of brain areas only weighs against our ontology if we assume the area performs some unified computation or function (Anderson, 2010; Shine et al., 2016), but this assumption is debatable (Anderson, 2014; Viola, 2017). These problems only worsen if we consider more radical models in which cognitive processes are carried out by different coalitions of brain regions depending on the circumstances (Anderson, 2014; Hutto et al., 2017).

Neurofunctional revision or revising psychological practice?   Thus, neurofunctional revision may provide means for resisting taxonomic revision. It is unclear whether we have failed to find one-to-one mappings between brain regions and cognitive functions because we have the wrong ontology or because such mappings do not exist (Anderson, 2014; Viola, 2017). But the evidence motivating neurofunctional revision is bound in concerns about psychology's conceptual practices. For example, formal meta-analyses play an important role in claiming that cortical regions are typically multifunctional (Anderson, 2010). But these meta-analyses (e.g., showing that brain areas are implicated in many different cognitive domains) compare brain activation patterns to the cognitive terms used in particular studies. And as we have seen, psychologists deploy these terms in a heterogeneous matter. Perhaps if psychologists deployed their concepts, terms, and tasks consistently, the relationship between mental categories and neural entities would not appear so messy (Poldrack, Kittur, et al., 2011; Sullivan, 2016).

Taxonomic revision or revising psychological practice? Efforts to revise our cognitive ontology using brain data likewise depend on psychologists' task analyses and conceptual practices (Sullivan, 2016). For example, Lenartowicz and colleagues (2010) claim that since a classifier applied to brain data cannot distinguish task-switching patterns from response selection or response inhibition ones, we should question the underlying reality of task switching. But as Figdor (2011) notes, this meta-analysis pools studies using different tasks (e.g., go/no-go task vs. stop-signal task) to study their target construct (e.g., response inhibition). These tasks may not be equivalent measures of the same neural process—for example, the go/no-go task and the stop-signal task elicit distinguishable patterns of brain activity (Swick et al., 2011). Therefore, one must address the relationship between tasks and constructs while trying to address the relationship between constructs and patterns of brain activity (Figdor, 2011).

Proponents of large-scale cognitive ontology revision often claim that neuroscience suggests our ontology is mistaken. We agree that the current situation is tangled. However, assuming that fixing our ontology will fix problems such as our inability to find one-to-one mappings between brain structures and cognitive processes is like finding a particularly tangled knot with three ends and believing you know which one to pull. We think the problems with structure-function mappings, our cognitive ontology, and conceptual practices in psychology are mutually interdependent—they cannot be solved independently. The situation is like a knot with three ends that need to be simultaneously worked on. It is likely that progress in the mind–brain sciences will require simultaneously untangling these three issues rather than addressing any individually.

14.4   Atlases, Not Maps: A Case for Taxonomic Pluralism

Aligning our cognitive ontology with brain data will require simultaneously reckoning with three issues: (1) achieving consistency and conceptual clarity about mental kinds and the tasks that engage them, (2) revising our views of the brain's functional neuroanatomy, and (3) adjusting our cognitive ontology through lumping, splitting, discovery, or elimination of members. Will progress on these three fronts achieve a unified cognitive ontology in which mental kinds and brain structures map smoothly onto one another?

The problem actually runs much deeper. Many authors implicitly or explicitly assume that there is a correct cognitive ontology in which every cognitive kind maps onto a particular brain mechanism (Poldrack, 2010; Anderson, 2015; Viola, 2017). In other words, an assumption of taxonomic monism (there is one true taxonomy of human cognition) pervades the cognitive ontology literature. For example, Anderson (2015) describes the issue as a "debate in the cognitive sciences over the *right taxonomy* for understanding cognition—the *right theory of* and *vocabulary for* describing the structure of the mind" (p. 68, emphasis added). Additionally, many researchers think the right cognitive ontology is one where each element maps onto a brain structure. Price and Friston (2005) write, "a systematic ontology for cognition would *facilitate the integration of cognitive and anatomical models* and organise the cognitive components of diverse tasks into a *single framework*" (p. 262, emphasis added).

Here, we challenge this widespread assumption of taxonomic monism, echoing and reinforcing authors advocating taxonomic pluralism in science generally (e.g., Dupré, 1995; Danks, 2015) and in cognitive neuroscience specifically (Hochstein, 2016; Sullivan, 2017). Taxonomic pluralism, applied to these issues, holds that we will need multiple cognitive ontologies to capture the diverse aims of researchers in the mind–brain sciences. Therefore, we disagree that cognitive neuroscience ultimately will, can, or even should aim to, converge on a unified cognitive ontology charting the entire mental domain.[9]

An orthodox view in philosophy holds that there is a single, correct way to classify empirical domains—for example, chemical substances, diseases, and species (Wilkerson, 1988; Boyd, 1991). According to this view, a correct taxonomy identifies a set of "natural kinds"—that is, categories that "carve nature at its joints" independently of researchers' goals and interests.[10] Recent work in philosophy of science challenges this traditional picture, pointing to various ways in which taxonomic categories (1) partly depend on the goals and interests of particular researchers and (2) are not fixed by understanding the domain's metaphysical (e.g., causal or mechanistic) structure (Dupré, 1995; Craver, 2009; Danks, 2015; Hochstein, 2016; Sullivan, 2016, 2017; Plutynski, 2018).[11] Next, we review these developments in the philosophy of science (section 14.4.1) and illustrate with examples how they motivate goal-dependent pluralism about cognitive ontologies (section 14.4.2).

14.4.1   Goal Dependence and Pluralism about Cognitive Ontology

A perennial debate in philosophy concerns the status of natural kinds (Wilkerson, 1988; Boyd, 1991; Khalidi, 2013). Natural kinds are allegedly categories reflecting the world's real structure in a mind-independent way (i.e., the categories do not depend on human thoughts or goals). Paradigmatic cases of natural kinds include biological categories (e.g., mollusks), chemical substances (e.g., gold), and so on as opposed to conventional categories that depend on our goals, judgments, and preferences—for example, weeds are plants we don't like, and jocks are people who privilege athletic prowess. Essentialist views of natural kinds—dating to John Locke (1632–1704)—propose that while attending to superficial qualities leads to merely conventional categories (e.g., gold and silver are "sparkly metals"), discovering hidden essences (e.g., gold and silver each consist of atoms with a specific atomic number) leads to natural kinds sharing innumerable properties and supporting innumerable inferences (e.g., all members of the category "gold" have the same atomic number, melting point, conductivity, etc.; Wilkerson, 1988).

A famous example is jade (Putnam, 1975). Jade was originally classified by observable properties such as color, malleability, and so on. But jade actually consists of two separate minerals: jadeite ($NaAlSi_2O_6$) and nephrite ($Ca_2(MgFe)_5Si_8O_{22}(OH)_2$). Jadeite and nephrite differ in chemical composition and physical properties, including density, hardness, and refractive index (Harlow & Sorensen, 2005). A standard conclusion is that "jade" is a merely conventional category (useful for ceremonial purposes or commerce), while "jadeite" and "nephrite" are natural kinds that license numerous inductive generalizations (generalizations about what will happen if the rock is heated, scratched, subjected to pressure, etc.). The jade example shows how uncovering microstructural differences (Putnam, 1975) can motivate splitting folk taxonomic categories; another case would be dividing "fish" into "bony fishes" versus "cartilaginous ones." One can also lump together once-separate categories after discovering deeper similarities (e.g., lumping whales, dogs, and bats into the category "mammal" based on homologies).

There are many philosophical views about what natural kinds are and how we identify them—for example, whether members of a natural kind share microstructural essences (e.g., Putnam, 1975) or homeostatic property clusters (e.g., Boyd, 1991). What these views share is the commitment that questions of scientific taxonomy—for example, how many species of insects there are—are settled by properly understanding the metaphysical structure of the world. Thus, there is a single, correct taxonomy that carves nature at its joints.

Philosophers and cognitive scientists have employed the natural kinds framework to theorize about whether our mental categories correctly reflect the mind's structure (Hochstein, 2016). Michaelian (2011) questions whether memory can be a natural kind, since there are multiple memory systems (e.g., episodic vs. procedural memory). Barrett (2006) argues that basic emotion categories (e.g., anger and fear) are not natural kinds, given evidence against specific neural and physiological signatures for particular emotions. Machery (2005) claims that the mental kind concept is not a natural kind, since different mental representations (e.g., prototypes, exemplars, theories) housed in different brain systems actually do the work psychologists attribute to concepts.

Some authors recommend lumping or splitting cognitive kinds based on neural structures or representational features. As an example of kind splitting, Michaelian (2011) argues that memory is not a natural kind, since declarative memory and non-declarative memory are computationally distinct (the former involves representing the past, while the latter arguably does not) and have distinct neural implementations (explicit memory involves the hippocampus and associated structures, while implicit memory involves the basal ganglia). Thus, perhaps we can make interesting scientific generalizations about declarative memory but not memory as a whole (Michaelian, 2015). This is a psychological analogue to the jade versus jadeite/nephrite case. As a potential example of lumping, De Brigard (2014b) argues that recalling the past (i.e., episodic memory), envisioning the future, and pondering counterfactual scenarios rely on a common set of brain areas. For De Brigard (2014b, p. 179) this raises (though does not settle) the possibility that these disparate cognitive abilities are manifestations of one general capacity for projecting the self into different hypothetical situations.

The cognitive ontology debate essentially takes these local concerns about whether memory or emotion categories are natural kinds and applies them to the entire taxonomy of psychology. The hope is that we can articulate our cognitive ontology and then use brain data to lump, split, eliminate, or otherwise revise its elements until we carve the mind at its natural joints (Poldrack, 2010; Anderson, 2015). Success would entail completely mapping the cognitive domain, with each concept aligning with an element of our neural ontology (Price & Friston, 2005). This taxonomy could be represented in a database akin to the *Diagnostic and Statistical Manual of Mental Disorders* or the *International Classification of Diseases*. We support these efforts, and agree that brain data are relevant to our cognitive ontology (Michaelian, 2011; Anderson, 2015; Khalidi, 2017). But we caution against the idea that revising our cognitive ontology will lead to a Periodic Table of the Mind—that is, a monolithic database representing the true structure of human cognition.

Recent philosophy of science suggests that while developing taxonomies in biology, chemistry, psychiatry, and so on requires correctly tracking the world's metaphysical structure, understanding that structure does not determine what taxonomy we should have. Instead, scientific ontologies unavoidably depend on researchers' goals and interests (Dupré, 1995; Craver, 2009; Danks, 2015; Hochstein, 2016; Plutynski, 2018). Since these goals and interests are plural, so too are our taxonomies. As Danks (2015) puts it, "the world does not come to us 'carved up' into its constituent objects, and we can 'segment' the world (through our scientific theories) in different, incompatible ways depending on our goals and purposes" (p. 3603). Therefore, "the pursuit of a unified (scientific) ontology is fundamentally misguided" (p. 3603).[12]

Consider questions about biological individuals. Some researchers envision a natural way to type or count biological individuals—for example, there is some definitive answer to whether ant colonies (e.g., with one reproductive queen, some reproductive male drones, and many sterile workers) are one big organism or many little ones (e.g., using genetic, immunological, or evolutionary criteria). Kovaka (2015) argues that instead biologists classify individuals variously based on their idiosyncratic goals and aims. For example, models of ant foraging may construe ant colonies as collections of separate individuals that make and interpret communications, while some evolutionary models consider the whole colony as effectively one individual. Taxonomic pluralism holds that neither way of thinking is strictly correct, and both are warranted in different circumstances.

Philosophers have recently imported these insights to cognitive neuroscience, arguing that taxonomic pluralism is required for meeting the science's diverse aims (Hochstein, 2016; Sullivan, 2017). Sullivan (2017) proposes that cognitive neuroscience needs coordinated pluralism to progress. Sullivan (2017, p. 141) argues that cognitive neuroscience currently exhibits an uncoordinated or self-defeating pluralism in which researchers unwittingly have different cognitive ontologies and preferred tasks for studying their proposed cognitive elements (see section 14.2.1). While this situation hinders research, Sullivan claims that deliberate, coordinated pluralism can facilitate it.

For Sullivan, current practices can give the false appearance that researchers are studying the same cognitive concept—for example, response inhibition. In fact, different laboratories often use different tasks and protocols to study them—for example, the go/no-go task and the stop-signal task. The stop-signal task and the go/no-go task may not recruit the same cognitive process (Swick et al., 2011). Not realizing this fosters miscommunication and stymies research. But we cannot solve this problem by simply curtailing some lines of research and insisting that laboratories use the same task. Researchers want to know, for example, what role response inhibition plays in addiction and other risky behaviors (Morein-Zamir & Robbins, 2015). There are many live possibilities for how the inhibitory processes targeted by these tasks relate to real-world behavior (perhaps they each target a different process relevant to addiction, perhaps only one is relevant to addiction, but the other is relevant to other risky behaviors, etc.). Instead of cutting down lines of research, we need a pluralism in which researchers develop different taxonomies and employ different methods while building the physical structures (e.g., databases) and social structures (e.g., research initiatives) needed to facilitate communication and coordination.

One might reply that taxonomic pluralism may be useful now, but this results from our relative ignorance. Perhaps as our understanding of the brain matures, neuroscience will achieve a unified neurocognitive ontology. Hochstein (2016; see also Craver, 2009) argues that even when we understand the underlying neurobiology, pragmatic considerations necessarily factor into classifying mental kinds. Hochstein argues that classifying mental kinds (e.g., lumping or splitting cognitive concepts) based on neural mechanisms requires answering whether the mechanisms in question are the same or different. The problem is that brain mechanisms—within individuals, between individuals, and between different species—are similar and different in innumerable respects. According to Hochstein (2016), whether we highlight the similarities or the differences "depends largely on how abstract or detailed our descriptions are, and what the interests and goals of individual scientists are when describing them" (p. 754).

Consider the neural circuits governing digestion or swimming/crawling behavior in invertebrates (e.g., crabs, lobsters, and leech). In these circuits, "similar rhythms can arise from different mechanisms" (Marder, Goeritz, & Otopalik, 2015, p. 156). Put differently, from animal to animal, the same circuit (a neural network with a largely similar oscillation pattern) can be built out of neurons with different properties connected in different ways. For most purposes, these subtle differences do not matter. Most of these circuits respond similarly to changes in temperature, neuromodulatory input, lesions, and so on. But for other cases, small differences in circuit composition critically matter. For example, whether these circuits will break down when subjected to unusually high or low temperatures depends on individual circuit properties (Marder et al., 2015, p. 158). Do different crabs have the same neural circuits? The answer is that it depends. If you are interested in whether the circuits produce a generally similar rhythm or whether the rhythm will persist under minor temperature fluctuations, the answer is yes. If you are interested in whether they produce the same exact rhythm or will persist under the same large temperature fluctuations, the answer is no. So, whether these animals have the same circuit seems to depend on what you mean by "the same" and for what purpose you are asking.

If this analysis is correct (even for animals with relatively simple nervous systems), then we should not expect studying brain mechanisms to settle questions about our cognitive ontology. Instead, pragmatic considerations will invariably weigh into our decisions about whether we are dealing with the same cognitive kinds. Next, we present two examples from cognitive neuroscience motivating taxonomic pluralism. We illustrate how neuroscientists can justifiably categorize cognitive processes differently according to their particular aims and interests.

14.4.2 Two Examples of Goal Dependence: Translational Psychiatry and Cognitive Modeling

Here, we give two kinds of examples motivating taxonomic pluralism in discussions of cognitive ontology. The first comes from studies of children with disruptive behavioral disorders (DBD) in translational psychiatry (White et al., 2016; Blair et al., 2018). This case suggests that the same behavioral disposition can arise from distinct neural underpinnings. While this example deals specifically with the neural basis of mental disorders, the general considerations apply to attempts to link personality traits (e.g., impulsiveness or introversion) to their neural bases. The second comes from cognitive models of reading (Seghier et al., 2012; Price, 2018) and working memory (Baddeley & Hitch, 2019).

Divergent neural underpinnings of reactive aggression   White and colleagues (2016) performed behavioral and neuroimaging studies designed to test the neural underpinnings of reactive aggression (aggression in response to perceived threats and provocations) in children with DBD such as conduct disorder and oppositional defiant disorder. Children with DBD tend to exhibit higher levels of reactive aggression than controls. In laboratory settings, they have a higher propensity to retaliate in a social ultimatum game, even when it is economically counterproductive (e.g., if retaliation leads to losses). Among children with DBD, some are also high in callous-unemotional traits (e.g., they have reduced prosocial capacities such as guilt and empathy [DBD/+CU]), while others score low on callous-unemotional traits [DBD/–CU]). Previous studies demonstrated that DBD/+CU children may have heightened emotional responses to perceived threats—for example, they seem to have heightened amygdala response when shown pictures of fearful faces (Viding et al., 2012). This raises the possibility that while DBD/–CU and DBD/+CU children both have heightened rates of reactive aggression, the neural underpinnings of this tendency for reactive aggression (partly) differ between the groups.

This is precisely what White and colleagues (2016) found. During ultimatum game retaliation, DBD/–CU children exhibited heightened responses in brain regions associated with reactive aggression and threat detection in animals (e.g., the amygdala and periaqueductal gray), whereas DBD/+CU children did not. Furthermore, activity levels in these regions were predictive of the propensity to retaliate for DBD/–CU but not DBD/+CU, children. This suggests that reactive aggression arises in DBD/–CU children partly due to a heightened threat response, where DBD/+CU may undertake reactive aggression in a more cold and calculated manner.

White and colleagues (2016) conclude that these results show "differences in the underlying neurobiology of maladaptive reactive aggression" (p. 282). Furthermore, these differences may matter for clinical practice—for example, they argue that interventions designed to reduce emotional responses would only help mitigate reactive aggression in DBD/–CU children. At this juncture, it might be tempting to suggest that DBD/–CU and DBD/+CU children are entirely distinct clinical populations with distinctive neural underpinnings. But there is more to the story. While some of the mechanisms underlying retaliatory behavior differ, others may be the same. White and colleagues (2016) suggest that while the temptation to carry out retaliation may arise by different means in CBD/–CU and CBD/+CU children, the propensity to retaliate is also partly rooted in a shared inability to appreciate the consequences of retaliation fully. They hypothesized that alterations in the ventromedial prefrontal cortex (vmPFC) may render CBD children more likely to escalate a situation without appreciating the ensuing negative consequences. Indeed, they found that connectivity patterns involving the vmPFC corresponded to a propensity to punish in CBD youths. Thus, they claim that while there are clinically important neural differences in reactive aggression in CBD/–CU and CBD/+CU youths, there is also a shared mechanism that is also a worthwhile clinical target.

Plural taxonomies in cognitive models   Early cognitive models of reading involved a single pathway for articulating sounds on the basis of orthography (written characters). However, patients with impaired reading exhibited a puzzling pattern of deficits in which some patients had difficulty reading real words with exceptional phonological features (e.g., pint), while others had trouble reading non-words with typical phonological features (e.g., pord). This led to modeling two pathways from orthography to phonology: a sub-lexical pathway going from orthography to phonological rules, and a lexical pathway going from orthography to a whole-word phonological store. Patients with deficits for atypical words (e.g., pint) were hypothesized to have damage to the lexical pathway, while patients with deficits for non-words were hypothesized to have damage to the sub-lexical pathway (Price, 2018).

This implies that the same (at some level of description) component in a cognitive model can be carried out by different brain structures at different times—that is, retrieving phonology from orthography in healthy patients is done by at least two brain mechanisms (Price, 2018). This represents a broader pattern (called neural degeneracy; see sections 14.2 and 14.3) in which studies of neuropsychological deficits reveal that a cognitive process can be carried out by different brain structures (Price, 2018; Seghier & Price, 2018). Similarly, Seghier and colleagues (2012) found that while the parallel letter processing required for skilled reading (measured via word identification under rapid presentation) is usually associated with the left ventral occipitotemporal cortex, some patients with damage to this area read using an alternate pathway involving the left superior temporal sulcus (STS). Interestingly, Seghier and colleagues (2012) then found that neurotypical individuals also utilize this STS pathway to varying degrees when reading.

The challenge to taxonomic monism is whether we consider retrieving phonology from orthography or parallel letter processing single items in a cognitive ontology or multiple items. On one hand, they achieve the same end in some sense. On the other hand, they are implemented in disparate neural systems (between and even presumably within individuals) that plausibly vary in computational properties. We think that for some purposes (e.g., studying memory with lists of familiar words), it does not matter whether reading involves a lexical and a sub-lexical route. But for other purposes (e.g., explaining the pattern of deficits above), it clearly matters a great deal. Once more, whether we lump or split the cognitive process in question seems to depend on what we are trying to investigate or explain.

A similar picture emerges when looking at models of working memory. Early models of working memory had components such as the central executive, the phonological loop, and the visuospatial sketchpad (e.g., Baddeley & Hitch, 1974). A neurotypical individual can attempt to rehearse short lists of numbers, words, and so on by different means, including conscious or unconscious subvocalization or visualization (Baddeley & Hitch, 2019). Does this mean that working memory is a single cognitive component or multiple ones (e.g., visual working memory, auditory working memory, etc.)? Once again, for some purposes (e.g., studying whether working memory impairments generally affect task switching), it might not matter whether people tend to use the visuospatial sketchpad or phonological loop. For other purposes, it matters very much. For example, there are tasks (e.g., verbal word repetition) that selectively interfere with the phonological loop; the same goes for the visuospatial sketchpad. Thus, questions about what tasks interfere with working memory require dividing the mechanism further (Baddeley & Hitch, 2019). To make matters even worse, there is evidence that speech and music may be processed differently in the phonological loop. Thus, some studies may want to divide the phonological loop further.

We have endeavored to show that there is no definitive answer to questions such as whether working memory is one cognitive kind or many sub-kinds. Instead, whether we lump or split the cognitive kind in question depends on what we want to explain or investigate. If we care about some downstream interactions of working memory, we may want to consider its various implementations as a single entity. If, on the other hand, we care about what processes interfere with working memory online or about individual or clinical differences in working memory, we may need to spit the kind in a more fine-grained way. One knee-jerk reply to this argument is that since every difference makes a difference, we should always split kinds when there is *any* difference in the neural mechanism. But as Hochstein (2016) and Craver (2009) point out, this has the unfavorable consequence of multiplying our ontology endlessly as we garner trivial mechanistic details. If every human hippocampus has a slightly different way of implementing episodic memory, then we need a new entry in our ontology for every brain hemisphere.

To navigate the Scylla of coarse-grained categories that obscure important differences and the Charybdis of multiplying categories in ways that obscure interesting similarities, cognitive ontologies will need to articulate the various investigational purposes for which they are constructed (Hochstein, 2016). The end result will be less like a giant map charting all of human cognition and more like an atlas that tells you what map you need at the moment. Sometimes, you need to know a region's topography and local customs in great detail; sometimes you just need a gas station.

14.5   Conclusion

In this chapter, we examined the growing literature on cognitive ontology revision, highlighting how neuroscientific theories and methods might be used to develop, shape, and refine psychology's taxonomy. Then, we challenged two of the debate's central assumptions. First, we challenged the assumption that failures to achieve one-to-one mappings between brain structures and cognitive functions necessarily imply that our ontology is mistaken. Instead, it is uncertain where the central problem lies. The numerous issues surrounding our cognitive ontology, the brain's functional organization, and conceptual practices in psychology are deeply intertwined and must be addressed jointly. Second, we challenged the assumption that achieving a single, correct cognitive ontology is possible or desirable. Instead, we think researchers should embrace taxonomic pluralism—the need for different ontologies for different purposes (Hochstein, 2016; Sullivan, 2017).

In advocating taxonomic pluralism, we are not claiming that all taxonomies are equally worthy or that our psychological taxonomy should float free of brain science. But proposals to use neuroscience to revise our cognitive ontology are too often bound in the claim that counting brain mechanisms will reveal what categories psychology should have. Psychologists and neuroscientists have a diverse range of goals and interests. Some have clinical aims, while others want to understand typical behavior. Some are interested in individual or cultural differences, while others are interested in commonalities. Some strive to understand unique features of human cognition, while others aim to understand features shared with nonhuman animals. We doubt that a single cognitive ontology can serve these diverse purposes.

It is an exciting time to be thinking about these questions. By making the structure of our cognitive ontology (and its relationship to tasks and brain regions) explicit, initiatives such as the Cognitive Atlas can spur new studies and conversations about the nature of the human mind. We hope that initiatives such as these will increasingly incorporate concepts from diverse theoretical origins—for example, from embodied cognition, ecological psychology, non-Western psychology, and so on—and attend to the diverse predictive, explanatory, and clinical goals of the mind–brain sciences. Doing so may facilitate cross talk between siloed corners of cognitive science, and make our theories of social and moral behavior more neurally plausible.

Notes

1. By debates about the structure of cognition, we mean disputes about broad theoretical frameworks for understanding cognition, such as embodied approaches, predictive processing approaches, etc. vs. standard representationalist views (Janssen et al., 2017).

2. Sigmund Freud (1855–1939), the founder of psychoanalysis, divided the mind into three parts: the id, the ego, and the super-ego. In his cognitive ontology, the id (an instinctual drive for sexual and bodily pleasure) and its repression by the super-ego (an internalization of social rules) explains numerous psychological phenomena from dream symbols to mental disorders. Besides psychoanalysts, psychologists no longer use this concept. Response inhibition—the ability to resist an impulse (e.g., to suppress a desire to look toward a blinking light)—is an element of cognitive control (control over one's thoughts and behaviors) in contemporary cognitive ontologies.

3. Franz Gall is most known for founding phrenology—the belief that studying skull "bumps" reveals someone's intellectual and moral character. Despite his reputation for promulgating pseudoscience, he presciently advocated the ideas that mental functions localize to parts of the cerebral cortex and that studying brain lesions informs theories of mental faculties.

4. For example, Anderson (2010) reports meta-analyses in which every cortical area is associated with numerous cognitive functions (e.g., reading, mathematics, etc.) spanning numerous domains of cognition (vision, perception, memory, etc.).

5. Researchers express this notion differently. Anderson (2010) claims that multi-functional brain regions contribute a common working to different traditionally defined psychological processes. Rathkopf (2013) describes a functional label for a brain area that does not reference psychological task as its intrinsic function. Shine and colleagues (2016) claim that regions contribute the same computation to different cognitive tasks. See McCaffrey (2015) for a detailed discussion.

6. Anderson (2014, chapter 1.4) argues that multi-functionality is observed in small human cortical regions as well as small neural circuits in invertebrates.

7. Thanks to Felipe De Brigard for this delightful example.

8. Dennett (1987) coined the notion of the intentional stance to capture how folk psychology aids in the evaluation and prediction of behavior. To adopt the intentional stance is to ascribe beliefs and desires to a target in order to make predictions about how the target will act or behave (see Dennett, 1987).

9. Our view echoes De Brigard's (2014a) point that "the organizational principles of the brain might not mirror the categories we use to describe the mind's many functions. The brain is not an atlas of the mind" (p. 43).

10. The idea that our best theories "carve nature at its joints" comes from Plato (*Phaedrus*, 265e1).

11. Interested readers should consult Dupré (1995) and Plutynski (2018) for examples of goal-dependent taxonomies in evolutionary biology and medicine.

12. As Danks (2015) notes, taxonomic pluralism does not entail that objects come into and out of existence as we classify them, or that all classification schemes are equal. Instead, it recognizes that dividing the world in different ways can support different but equally useful inductive inferences, depending on the context.

References

Aminoff, E. M., Kveraga, K., & Bar, M. (2013). The role of the parahippocampal cortex in cognition. *Trends in Cognitive Sciences, 17*(8), 379–390.

Anderson, M. L. (2010). Neural reuse: A fundamental organizational principle of the brain. *Behavioral and Brain Sciences, 33*(4), 245–266.

Anderson, M. L. (2014). *After phrenology: Neural reuse and the interactive brain.* Cambridge, MA: MIT Press.

Anderson, M. L. (2015). Mining the brain for a new taxonomy of the mind. *Philosophy Compass, 10*(1), 68–77.

Anderson, M. L., Kinnison, J., & Pessoa, L. (2013). Describing functional diversity of brain regions and brain networks. *NeuroImage, 73*, 50–58.

Baddeley, A. D., & Hitch, G. (1974). Working memory. In *Psychology of Learning and Motivation* (Vol. 8, pp. 47–89). Academic press.

Baddeley, A. D., & Hitch, G. J. (2019). The phonological loop as a buffer store: An update. *Cortex, 112*, 91–106.

Barrett, L. F. (2006). Are emotions natural kinds? *Perspectives on Psychological Science, 1*(1), 28–58.

Batterman, R. W. (2000). Multiple realizability and universality. *The British Journal for the Philosophy of Science, 51*(1), 115–145.

Blair, R. J. R., Meffert, H., Hwang, S., & White, S. F. (2018). Psychopathy and brain function: Insights from neuroimaging research. In C. J. Patrick (Ed.), *Handbook of psychopathy* (pp. 401–421). New York: Guilford Press.

Boyd, R. (1991). Realism, anti-foundationalism and the enthusiasm for natural kinds. *Philosophical Studies, 61*(1), 127–148.

Brooks, G. P. (1976). The faculty psychology of Thomas Reid. *Journal of the History of the Behavioral Sciences, 12*(1), 65–77.

Bunzl, M., Hanson, S. J., & Poldrack, R. A. (2010). An exchange about localism. In S. J. Hanson & M. Bunzl (Eds.), *Foundational issues in human brain mapping* (pp. 49–54). Cambridge, MA: MIT Press.

Burnston, D. C. (2016). A contextualist approach to functional localization in the brain. *Biology and Philosophy, 31*(4), 527–550.

Churchland, P. M. (1981). Eliminative materialism and propositional attitudes. *The Journal of Philosophy, 78*(2), 67–90.

Churchland, P. S. (1986). *Neurophilosophy: Toward a unified science of the mind–brain.* Cambridge, MA: MIT Press.

Craver, C. F. (2009). Mechanisms and natural kinds. *Philosophical Psychology, 22*(5), 575–594.

Danks, D. (2015). Goal-dependence in (scientific) ontology. *Synthese, 192*(11), 3601–3616.

De Brigard, F. (2014a). The anatomy of amnesia. *Scientific American Mind, 25*(3), 39–43.

De Brigard, F. (2014b). Is memory for remembering? Recollection as a form of episodic hypothetical thinking. *Synthese, 191*(2), 155–185.

Dennett, D. (1987). *The intentional stance.* Cambridge, MA: Bradford.

Dewhurst, J. (2020). Folk psychological and neurocognitive ontologies. In F. Calzavarani & M. Viola (Eds.), *Neural mechanisms: New challenges in the philosophy of neuroscience* (pp. 311–334. Heidelberg: Springer.

Duckworth, A. L., Peterson, C., Matthews, M. D., & Kelly, D. R. (2007). Grit: Perseverance and passion for long-term goals. *Journal of Personality and Social Psychology, 92*(6), 1087.

Dupré, J. (1995). *The disorder of things: Metaphysical foundations of the disunity of science.* Cambridge, MA: Harvard University Press.

Feest, U. (2020). Construct validity in psychological tests—The case of implicit social cognition. *European Journal for Philosophy of Science, 10*(1), 4.

Figdor, C. (2011). Semantics and metaphysics in informatics: Toward an ontology of tasks. *Topics in Cognitive Science, 3*(2), 222–226.

Fisher, H. E., Aron, A., Mashek, D., Li, H., & Brown, L. L. (2002). Defining the brain systems of lust, romantic attraction, and attachment. *Archives of Sexual Behavior, 31*(5), 413–419.

Fodor, J. A. (1974). Special sciences (or: The disunity of science as a working hypothesis). *Synthese, 28*(2), 97–115.

Francken, J. C., & Slors, M. (2018). Neuroscience and everyday life: Facing the translation problem. *Brain and Cognition, 120,* 67–74.

Freud, S. (1989). The ego and the id. *TACD Journal, 17*(1), 5–22. (Original work published 1923)

Gall, F. J. (1835). *On the functions of the brain and of each of its parts: With observations on the possibility of determining the instincts, propensities, and talents, or the moral and intellectual dispositions of men and animals, by the configuration of the brain and head* (Vol. 1). Boston, MA: Marsh, Capen & Lyon.

Gentilucci, M., & Volta, R. D. (2008). Spoken language and arm gestures are controlled by the same motor control system. *Quarterly Journal of Experimental Psychology, 61*(6), 944–957.

Harlow, G. E., & Sorensen, S. S. (2005). Jade (nephrite and jadeitite) and serpentinite: Metasomatic connections. *International Geology Review, 47*(2), 113–146.

Hochstein, E. (2016). Categorizing the mental. *The Philosophical Quarterly, 66*(265), 745–759.

Hochstein, E. (2017). When does "folk psychology" count as folk psychological? *The British Journal for the Philosophy of Science, 68*(4), 1125–1147.

Hutto, D. D., Peeters, A., & Segundo-Ortin, M. (2017). Cognitive ontology in flux: The possibility of protean brains. *Philosophical Explorations, 20*(2), 209–223.

Janssen, A., Klein, C., & Slors, M. (2017). What is a cognitive ontology, anyway? *Philosophical Explorations, 20*(2), 123–128.

Khalidi, M. A. (2013). *Natural categories and human kinds: Classification in the natural and social sciences.* Cambridge: Cambridge University Press.

Khalidi, M. A. (2017). Crosscutting psycho-neural taxonomies: The case of episodic memory. *Philosophical Explorations, 20*(2), 191–208.

Klein, C. (2012). Cognitive ontology and region-versus network-oriented analyses. *Philosophy of Science, 79*(5), 952–960.

Kovaka, K. (2015). Biological individuality and scientific practice. *Philosophy of Science, 82*(5), 1092–1103.

Laird, A. R., Eickhoff, S. B., Fox, P. M., Uecker, A. M., Ray, K. L., Saenz, J. J., . . . & Turner, J. A. (2011). The BrainMap strategy for standardization, sharing, and meta-analysis of neuroimaging data. *BMC Research Notes, 4*(1), 1–9.

Lenartowicz, A., Kalar, D. J., Congdon, E., & Poldrack, R. A. (2010). Towards an ontology of cognitive control. *Topics in Cognitive Science, 2*(4), 678–692.

Lindquist, K. A., Wager, T. D., Kober, H., Bliss-Moreau, E., & Barrett, L. F. (2012). The brain basis of emotion: A meta-analytic review. *The Behavioral and Brain Sciences, 35*(3), 121.

Machery, E. (2005). Concepts are not a natural kind. *Philosophy of Science, 72*(3), 444–467.

Mandler, G. (1980). Recognizing: The judgment of previous occurrence. *Psychological Review, 87*(3), 252-271.

Marder, E., Goeritz, M. L., & Otopalik, A. G. (2015). Robust circuit rhythms in small circuits arise from variable circuit components and mechanisms. *Current Opinion in Neurobiology, 31*, 156–163.

McCaffrey, J. B. (2015). The brain's heterogeneous functional landscape. *Philosophy of Science, 82*(5), 1010–1022.

McCaffrey, J. B., & Machery, E. (2016). The reification objection to bottom-up cognitive ontology revision. *Behavioral and Brain Sciences, 39*, E125.

Menon, V., & Uddin, L. Q. (2010). Saliency, switching, attention and control: A network model of insula function. *Brain Structure and Function, 214*(5–6), 655–667.

Michaelian, K. (2011). Is memory a natural kind? *Memory Studies, 4*(2), 170–189.

Michaelian, K. (2015). Opening the doors of memory: Is declarative memory a natural kind? *Wiley Interdisciplinary Reviews: Cognitive Science, 6*(6), 475–482.

Morein-Zamir, S., & Robbins, T. W. (2015). Fronto-striatal circuits in response-inhibition: Relevance to addiction. *Brain Research, 1628*, 117–129.

Pessoa, L. (2014). Understanding brain networks and brain organization. *Physics of Life Reviews, 11*(3), 400–435.

Plutynski, A. (2018). *Explaining cancer: Finding order in disorder*. Oxford: Oxford University Press.

Poldrack, R. A. (2006). Can cognitive processes be inferred from neuroimaging data? *Trends in Cognitive Sciences, 10*(2), 59–63.

Poldrack, R. A. (2010). Mapping mental function to brain structure: How can cognitive neuroimaging succeed? *Perspectives on Psychological Science, 5*(6), 753–761.

Poldrack, R. A., Halchenko, Y. O., & Hanson, S. J. (2009). Decoding the large-scale structure of brain function by classifying mental states across individuals. *Psychological Science, 20*(11), 1364–1372.

Poldrack, R. A., Kittur, A., Kalar, D., Miller, E., Seppa, C., Gil, Y., . . . & Bilder, R. M. (2011). The cognitive atlas: Toward a knowledge foundation for cognitive neuroscience. *Frontiers in Neuroinformatics, 5*, 17.

Poldrack, R. A., & Yarkoni, T. (2016). From brain maps to cognitive ontologies: Informatics and the search for mental structure. *Annual Review of Psychology, 67*, 587–612.

Price, C. J. (2018). The evolution of cognitive models: From neuropsychology to neuroimaging and back. *Cortex, 107*, 37–49.

Price, C. J., & Friston, K. J. (2002). Degeneracy and cognitive anatomy. *Trends in Cognitive Sciences, 6*(10), 416–421.

Price, C. J., & Friston, K. J. (2005). Functional ontologies for cognition: The systematic definition of structure and function. *Cognitive Neuropsychology, 22*(3–4), 262–275.

Putnam, H. (1975). The meaning of "meaning." *Minnesota Studies in the Philosophy of Science Papers, 7* (131–193).

Quesque, F., & Rossetti, Y. (2020). What do theory-of-mind tasks actually measure? Theory and practice. *Perspectives on Psychological Science, 15*(2), 384–396.

Rathkopf, C. A. (2013). Localization and intrinsic function. *Philosophy of Science, 80*(1), 1–21.

Sabb, F. W., Bearden, C. E., Glahn, D. C., Parker, D. S., Freimer, N., & Bilder, R. M. (2008). A collaborative knowledge base for cognitive phenomics. *Molecular Psychiatry, 13*(4), 350–360.

Scarantino, A., & Griffiths, P. (2011). Don't give up on basic emotions. *Emotion Review, 3*(4), 444–454.

Schaffner, K. F. (1969). The Watson–Crick model and reductionism. *The British Journal for the Philosophy of Science, 20*(4), 325–348.

Scholz, J., Triantafyllou, C., Whitfield-Gabrieli, S., Brown, E. N., & Saxe, R. (2009). Distinct regions of right temporo-parietal junction are selective for theory of mind and exogenous attention. *PLoS One, 4*(3), e4869.

Seghier, M. L., Neufeld, N. H., Zeidman, P., Leff, A. P., Mechelli, A., Nagendran, A., . . . & Price, C. J. (2012). Reading without the left ventral occipito-temporal cortex. *Neuropsychologia, 50*(14), 3621–3635.

Seghier, M. L., & Price, C. J. (2018). Interpreting and utilising intersubject variability in brain function. *Trends in Cognitive Sciences, 22*(6), 517–530.

Shine, J. M., Eisenberg, I., & Poldrack, R. A. (2016). Computational specificity in the human brain. *Behavioral and Brain Sciences, 39*, e131.

Sullivan, J. A. (2014). Is the next frontier in neuroscience a "decade of the mind"? In C. Wolfe (Ed.), *Brain theory: Essays in critical neurophilosophy* (pp. 45–67). London: Palgrave Macmillan.

Sullivan, J. A. (2016). Construct stabilization and the unity of the mind–brain sciences. *Philosophy of Science, 83*(5), 662–673.

Sullivan, J. A. (2017). Coordinated pluralism as a means to facilitate integrative taxonomies of cognition. *Philosophical Explorations, 20*(2), 129–145.

Swick, D., Ashley, V., & Turken, U. (2011). Are the neural correlates of stopping and not going identical? Quantitative meta-analysis of two response inhibition tasks. *NeuroImage, 56*(3), 1655–1665.

Tabb, K., & Schaffner, K. (2017). Causal pathways, random walks, and tortuous paths: Moving from the descriptive to the etiological in psychiatry. In K.S. Kendler, & J. Parnas (Eds.), *Philosophical Issues in Psychiatry IV* (pp. 342–360). Oxford: Oxford University Press.

Tettamanti, M., & Weniger, D. (2006). Broca's area: A supramodal hierarchical processor? *Cortex, 42*(4), 491–494.

Tulving, E. (2007). Are there 256 different kinds of memory? In J. S. Nairne (Ed.), *The foundations of remembering: Essays in honor of Henry L. Roediger, III* (pp. 39–52). Hove, UK: Psychology Press.

Vazsonyi, A. T., Ksinan, A. J., Jiskrova, G. K., Mikuška, J., Javakhishvili, M., & Cui, G. (2019). To grit or not to grit, that is the question! *Journal of Research in Personality, 78,* 215–226.

Viding, E., Sebastian, C. L., Dadds, R. M., Lockwood, P. L., Cecil, D. A., De Brito, S. A., & McCrory, E. J. (2012). Amygdala response to preattentive masked fear in children with conduct problems: The role of callous-unemotional traits. *American Journal of Psychiatry, 169*(10), 1109–1116.

Viola, M. (2017). Carving mind at brain's joints. The debate on cognitive ontology. *Phenomenology and Mind,* (12), 162–172.

Viola, M., & Zanin, E. (2017). The standard ontological framework of cognitive neuroscience: Some lessons from Broca's area. *Philosophical Psychology, 30*(7), 945–969.

White, S. F., VanTieghem, M., Brislin, S. J., Sypher, I., Sinclair, S., Pine, D. S., . . . & Blair, R. J. R. (2016). Neural correlates of the propensity for retaliatory behavior in youths with disruptive behavior disorders. *American Journal of Psychiatry, 173*(3), 282–290.

Wilkerson, T. E. (1988). Natural kinds. *Philosophy, 63*(243), 29–42.

Yarkoni, T., Poldrack, R. A., Nichols, T. E., Van Essen, D. C., & Wager, T. D. (2011). Large-scale automated synthesis of human functional neuroimaging data. *Nature Methods, 8*(8), 665–670.

Yeo, B. T., Krienen, F. M., Eickhoff, S. B., Yaakub, S. N., Fox, P. T., Buckner, R. L., . . . & Chee, M. W. (2015). Functional specialization and flexibility in human association cortex. *Cerebral Cortex, 25*(10), 3654–3672.

Glossary

Compiled by Maria Khoudary

Disclaimer: Some of the definitions included in this glossary are controversial and do not reflect universal agreement among researchers. The definitions provided below report what the authors of this volume intend to refer to when they use the term and are meant to serve as a guide for orienting the reader through this volume. We encourage skepticism and curiosity and do not intend to imply that these definitions are indisputably correct.

access consciousness: the functional aspects of being consciously aware of contents as they become accessible to cognitive systems.

action binding: the perceived shift forward of the time of action toward the outcome in voluntary conditions; a kind of intentional binding.

actuarial risk assessment: generating predictions about an offender's future behavior on the basis of algorithms and statistics that assess the offender's degree of fit with a large group of offenders with known histories.

acute sleep deprivation: a situation or environment in which wakefulness extends over twenty-four hours.

agency: the manifestation of a capacity (or capacities) to act; paradigmatic instances of agency include intentional action, where what an agent wants and believes causally produces what she does.

awareness/consciousness: umbrella terms that refer to different phenomena; common distinctions include state versus creature consciousness and phenomenal versus access consciousness.

belief: an attitude with a world-to-mind direction of fit that takes propositions as its content and is such that having a belief disposes one to assert sincerely the content of the belief when prompted in the right circumstances.

blindsight: after damage to the primary visual cortex, the ability to perform above chance in perceptual tasks while reporting no conscious awareness of the relevant stimuli.

catatonia: a behavioral symptom characterized by reduced responsiveness to the environment, immobility or slow movement, rigidity, and/or mutism or difficulty with speech.

ceteris paribus: all else equal, or holding everything else constant.

channel specificity: whether a representation is specific to processing information from a certain sensory modality.

chronic sleep restriction: a situation or environment in which an individual is awake for more than 80 percent of a twenty-four-hour period across more than five days.

clinical risk assessment: generating predictions about an offender's future behavior on the basis of a clinician's professional judgment.

cognitive behavioral therapy (CBT): a class of therapeutic interventions in which a therapist assists a client in reducing negative symptoms by challenging unhelpful thoughts, beliefs, and attitudes and then finding and practicing new, more effective strategies.

cognitive empathy: inferring another's non-affective mental state in order to empathize with them.

cognitive map: a structured mental representation that is organized by dimensions shared with physical space; recently suggested to underlie understanding of both physical and abstract spaces.

commissurotomy (in neurosurgery): the severing of the anterior commissure and usually also the corpus callosum, which thereby disconnects the two hemispheres of the brain; sometimes used as a treatment for severe epilepsy.

concepts: in cognitive science, elements of the mind that allow for tracking observations and forming thoughts about a certain kind of thing; these elements can be used to identify observations, build propositional thoughts, and create and understand linguistic utterances.

confabulate: to construct an explanation of one's behavior that does not accurately represent the true cause of one's behavior (e.g., a patient explains her inability to remember your name in terms of fatigue rather than dementia).

construct validation: the process of determining whether a psychological scale or instrument (e.g., the Stroop task or Implicit Association Test) actually measures the real-world mental construct it is intended to measure (e.g., cognitive control or implicit bias).

content (of consciousness): the object of a conscious state; what the state is about.

contralateral: pertaining to the opposite side.

contrast sensitivity: the responsiveness of a neuron to changes in luminance across space.

cortico-cortical: of or relating to transfers of information from one cortical region to another.

criterion: in signal detection theory, the amount of evidence needed for an observer to decide that a stimulus or feature was present.

d' (d prime): in signal detection theory, the sensitivity of an observer to discriminate signal from noise (measured by the distance between the means of evidence distributions).

deep brain stimulation (DBS): the delivery of electrical current to a set of neurons somewhere in the brain; an invasive technique that requires surgical implantation of electrodes into the brain (i.e., requires opening the skull); is used for therapeutic and research purposes.

deontological: moral reasoning in terms of rules (e.g., do not kill, do not intend harm) that is relatively insensitive to the effects of violating such rules.

effect binding: the perceived shift backward of the time of the outcome toward the action in voluntary conditions; a kind of intentional binding.

electroconvulsive therapy (ECT): a therapeutic intervention used in some cases of severe mental disorder wherein, under anesthesia, a small amount of electric current is applied to induce a brief seizure.

emotional contagion: non-voluntarily adopting another's affective states; in infants, this presents as contagious crying upon hearing sounds of another person's distress.

encode: come to represent or to carry information about; when sensory neurons encode some information, it means that they represent it through their pattern of activity.

endogenous: internally generated, occurring (as in cognitive process), or existing (as in ion channels).

episodic memory: memory for particular past experiences or events (in contrast to memory for general facts or procedures); often thought to involve imagery and other phenomenological features.

exogenous: externally generated, directed, or existing.

experiential scope: the extent to which a representation aggregates over specific experiences rather than representing individual episodes.

experimental philosophy: a subfield that uses quantitative methods to understand or answer philosophical questions; often contrasted with "armchair philosophy," which answers similar questions on the basis of introspection.

folk intuitions: immediate reactions of non-philosophers to moral dilemmas or other philosophical problems.

folk psychiatry: prescientific or commonsense terms and concepts about mental health; folk psychiatric categories include insane, psychotic, irrational, and delusional.

folk psychology: prescientific or commonsense psychological terms and concepts used to understand others' thoughts and behaviors; folk psychological categories include beliefs, desires, and intentions.

functional connectivity: a measure of whether spatially distinct brain regions have correlated mental functions or properties (e.g., memory, moral judgment).

Gaussian: of or relating to a type of probability distribution, which is symmetric about its peak; also called a normal distribution, or more colloquially, a "bell curve."

germline editing: genetic modification of reproductive (egg or sperm) cells, resulting in hereditary alterations.

graphical model: a declarative representation that consists in a set of nodes connected by edges.

grid cells: neurons in the medial entorhinal cortex that fire in multiple regularly spaced locations.

identity theory: in philosophy of mind, the position that the workings of the mind are identical with the workings of the brain.

identity: sameness; can also refer to qualities or characteristics of a person.

instrumental response: a behavior (e.g., work) that an animal performs as a result of learning that the behavior is a means (instrumental) to a reward (e.g., money).

intentional binding: an indirect measure of senses of agency; measured as the time interval between perceived time of action and perceived time of outcome, usually compared between active and passive action initiation conditions (see also "action binding" and "effect binding").

internal response: in signal detection theory, a quantity returned by a detector, indicating how much evidence there is for the presence of a stimulus or feature on a given trial.

interoception: the sensory system that informs us about the internal state of our body (heart rate, arousal, hunger, thermoregulation, etc.).

intrusive thoughts: a psychiatric symptom characterized by the presence of involuntary and unwelcome ideas, images, or thoughts.

ipsilateral: pertaining to the same side.

mask: in psychophysics, a stimulus briefly presented before or after a target stimulus with the goal of rendering the target stimulus unconscious.

material monism: in philosophy of mind, a metaphysical position arguing that all naturally occurring entities and phenomena can be reduced to the same fundamental material or physical components; that is, a rejection of the idea that the mind is composed of different material than the brain/body (as in dualism).

mentalizing: the process of attributing mental states to a target.

mental kinds: categories for classifying human cognition and emotion; for example, attempts to delineate how many kinds of memory there are (e.g., semantic memory, episodic memory, working memory, sensory memory, implicit memory, etc.).

mental representation: any feature of the mind symbolizing or referring to some thing or property, x, and employed in mental activities (e.g., imagining, reasoning, feeling, deciding, controlling behavior, etc.) concerning x.

metacognition: the capacity to monitor and evaluate one's own cognitive processes.

metacontrast mask: a mask presented after a target stimulus that surrounds and abuts it without overlap.

meta-d' (meta d prime): a measure that extends signal detection theory by quantifying how well confidence judgments correlate with accuracy, independently of response bias.

mind wandering: a mode of thinking where the structure of mental content is less constrained by personal-level goals and concerns relative to goal-directed thinking; a typical (though not necessary) feature of mind wandering is shifting between topics without feeling distracted.

mirror neurons: cells that are activate both when one performs an action and when one observes another performing that action.

momentary evidence: in drift diffusion models, an internal response—that is, measured as a quantity indicating the amount of evidence for deciding between two choices that is generated at each moment during stimulus processing.

multi-voxel pattern analysis: a way to analyze brain activity in many voxels (through functional magnetic resonance imaging data) that uses pattern-recognition algorithms and machine learning to predict behavioral responses.

neural resonance: in the case of empathy, the activation of similar populations of neurons in the empathizer and in the individual with whom they empathize.

neuroethics: an interdisciplinary subfield of philosophy that considers the ethical implications of advances in neuroscience, as well as whether and to what extent evidence from neuroscience factors into ethical theory.

neuroprediction: the use of neuroscientific measures to characterize biological markers of human behavior that increase the ability to predict particular behavioral outcomes accurately.

nociception: the sensory system that encodes harmful stimuli (e.g., burning, stinging, cutting, crushing, etc.); conceptually distinct from pain perception, which represents the way the brain interprets nociceptive signals (e.g., football players may experience several nociceptive episodes while playing, but *feel* the pain only after the match).

nosology: the classification of diseases.

ontology: in informatics, an ontology represents knowledge in a particular domain by representing the concepts in that domain and the relationships between them.

opsins: light-sensitive proteins.

optogenetics: in neuroscience, a method that uses light-sensitive proteins to manipulate, with very high precision and specificity, the activity of neurons.

phenomenal consciousness: first-person, subjective experience of some dimension of the world.

place cells: neurons in the hippocampus that fire preferentially in certain locations; originally discovered in rats (O'Keefe, J., & Dostrovsky, J. [1971]. The hippocampus as a spatial map. Preliminary evidence from unit activity in the freely-moving rat. *Brain Research, 34*(1), 171–175); thought to exist in all mammals with similar hippocampal structures; Key component of the idea that memory is based in cognitive maps.

prima facie: "on the face of it"; defeasible epistemic or practical justification.

probabilistic: describes a process or entity that involves chance variation; in neuroscience, a probabilistic representation of some feature of the world is one that reflects the system's uncertainty about that feature.

proprioception: the sensory system that enables us to perceive the location, movement, and sense of effort of parts of the body.

psychological continuity: persistence of psychological characteristics (memories, dispositions, attitudes, preferences, etc.) over time.

psychotropic: substance that affects behavioral states and cognitive functioning via effects on neurotransmitter systems.

readiness potential: a slow, negative, bilateral buildup of precentral and parietal electrical activity preceding the initiation of an action; also termed the Bereitschaftspotential.

response-congruent evidence rule: the finding that confidence judgments depend heavily on evidence congruent with a perceptual decision while ignoring or downweighting evidence that contradicts the decision.

risk assessment: in criminal justice, generating probabilistic predictions about an offender's future behavior on the basis of their attributes and/or circumstances.

saccade: a rapid, jerky eye movement between fixation points that often occurs automatically and without any initiation.

self-report: a method of recording judgments about whether one has (or had) a particular experience or holds (or held) a particular attitude; for example, a researcher might ask a participant whether she feels focused after performing a task.

semantic memory: type of long-term memory that contains concepts and other general knowledge about the world.

sensory field: in essence, the extent of a particular sensory system (e.g., humans have a visual field of ~210°, and an auditory field of 20 Hz to 20 kHz); the concept of a sensory field implies that all sensory objects must occupy a certain space.

sharedness: quality of a memory by which it refers to information likely also known to others.

slow-wave brain activity: electrical brain signals in the delta (0.5–4 Hz) and theta (4–8 Hz) ranges that are markers of restful and recuperative sleep states.

somatic editing: genetic modification of nonreproductive cells resulting in non-hereditary alterations.

somatosensation: any sensory activity originating elsewhere than in the specialized sense organs (eyes and ears) and conveying information about the state of the body and its immediate environment.

striatal: pertaining to the corpus striatum, a nucleus within the basal ganglia located lateral to the thalamus; the dorsal portion is comprised of the caudate nucleus and the putamen, and the ventral portion is comprised of the nucleus accumbens and the olfactory tubercle.

sympathy: a process wherein an agent evaluates the perceived mental/affective state of another and forms a motivational attitude toward it.

tactile: of or relating to touch.

thought experiment: hypothetical scenario designed to probe intuitions about a particular philosophical hypothesis or position.

triangulation: combining results from multiple different experiments or disciplines, each of which addresses a unique dimension of an explanatory target, in order to come to a robust conclusion about that target.

tuning curve: the response profile of a neuron to different values of a particular feature (e.g., orientation, wavelength of light, spatial frequency, etc.).

unicept: a term coined by Millikan: "neural node that helps in storing factual or procedural knowledge through its connections with other unicepts or with behavior controllers. Each unicept is supplied with its own unitracker" (Millikan, R. G. [2017]. *Beyond concepts: Unicepts, language, and natural information* (p. 225). Oxford: Oxford University Press).

unitracker: a term coined by Millikan: "neural network whose function is to recognize information arriving at the sensory surfaces that concerns one particular thing and present it for use or storage by its proprietary unicept" (Millikan, R. G. [2017]. *Beyond concepts: Unicepts, language, and natural information* (p. 225). Oxford: Oxford University Press).

Contributors

Sara Abdulla is a recent graduate in neuroscience and philosophy from Georgia State University. She is interested in economics, ethics, and technology and is pursuing a career in research.

Eyal Aharoni is an Associate Professor of Psychology, Philosophy, and Neuroscience at Georgia State University. His research investigates risk factors for antisocial behavior and the influence of emotion and other extralegal factors on legal decision making.

Corey H. Allen is a doctoral candidate in neuroscience with a concentration in neuroethics at Georgia State University. His work focuses on neurobiologically informed risk assessment of antisocial behaviors and on moral decision making as it pertains to punishment and agency attribution.

Sara Aronowitz is an Assistant Professor in Philosophy and Cognitive Science at the University of Arizona. She studies how rationality unfolds over time and how memory and imagination support learning.

Ned Block is Julius Silver Professor, with appointments in the Departments of Philosophy and Psychology and Center for Neural Science at New York University (NYU). He did his PhD at Harvard and taught at MIT for twenty-five years before moving to NYU. He is a Fellow of the American Academy of Arts and Sciences and the Cognitive Science Society. He has given the William James Lectures at Harvard, the John Locke Lectures at Oxford, and the Immanuel Kant Lectures at Stanford.

Jennifer Blumenthal-Barby is the Cullen Professor of Medical Ethics and Associate Director of the Center for Medical Ethics and Health Policy at Baylor College of Medicine. Dr. Blumenthal-Barby's primary research focuses on medical decision making and the ethics of the use of behavioral economics and decision psychology to shape people's decisions and behaviors. She recently published a book with MIT Press titled *Good Ethics and Bad Choices: The Relevance of Behavioral Economics for Medical Ethics*.

Maj. Allison Brager, PhD, is presently the Director of Human Performance and Outreach Education for the United States Army Recruiting Command. In addition to being author of the popular science book, *Meathead: Unraveling the Athletic Brain*, she has more than thirty publications in flagship journals, including *Science, eLife*, and *Journal of Neuroscience*. She holds several board of directors and editorial board positions for sleep- and neuroscience-focused societies.

Antonio Cataldo is a postdoctoral researcher with the Cognition, Values, and Behavior (CVBE) group at Ludwig Maximilian University of Munich (LMU) and the School of Advanced Study, University of London where he investigates the behavioral and neural correlates of sense of agency for gaze-mediated actions. His research mainly focuses on the study of how the bodily senses shape our perception of the world while simultaneously creating a representation of our own body as a unique object in the world.

Tony Cheng is an Assistant Professor at National Chengchi University (NCCU), Department of Philosophy, Taiwan, where he is also the Director of the Center of Phenomenology and an assistant research fellow of the Research Center for Mind, Brain, and Learning. He primarily works on the subjective, the objective, and the relation between them. The entry points are the senses and consciousness as they mediate varieties of minds and the world.

Felipe De Brigard is Fuchsberg-Levine Family Associate Professor of Philosophy and Associate Professor in the Departments of Psychology and Neuroscience and the Center for Cognitive Neuroscience at Duke University. He is also Principal Investigator of the Imagination and Modal Cognition Laboratory (IMC Lab) within the Duke Institute for Brain Sciences. His research focuses on the nature of memory and its relations to other cognitive faculties, such as perception, imagination, attention, and consciousness, but he is also interested in the philosophy of neuroscience and the cognitive psychology and neuroscience of causal reasoning.

Rachel N. Denison is an Assistant Professor of Psychological and Brain Sciences at Boston University. She studies visual perception, attention, and decision making, with a focus on how the brain generates visual experience in real time. Her research aims to link behavior to neural activity using computational models—integrating approaches from experimental psychology, human cognitive neuroscience, computational neuroscience, and cognitive science.

Jim A. C. Everett is a lecturer (Assistant Professor) at the University of Kent and Research Associate at the Uehiro Centre for Practical Ethics at the University of Oxford. He specializes in moral judgment, perceptions of moral character, and parochial altruism.

Gidon Felsen, Associate Professor, is a neuroscientist and neuroethicist in the Department of Physiology and Biophysics at the University of Colorado School of Medicine. His lab studies how the nervous system makes and acts upon decisions under normal and pathological conditions, using behavioral, electrophysiological, genetic, and computational approaches. Dr. Felsen also examines ethical, legal, and social issues associated with advances in neuroscience.

Julia Haas is an Assistant Professor in the Department of Philosophy and the Neuroscience Program at Rhodes College. Her research is in the philosophy of cognitive science and neuroscience, including moral artificial intelligence.

Hyemin Han is an Assistant Professor in Educational Psychology and Educational Neuroscience at the University of Alabama. His research interests include moral development, moral education, social neuroscience, and computational modeling.

Zachary C. Irving is an Assistant Professor at the Corcoran Department of Philosophy, where he works on philosophy and cognitive science. He defends a philosophical theory of mind wandering as unguided attention and examines general philosophical topics—including mental action, attention, and introspection—through the lens of mind wandering. Zac has two projects outside of mind wandering. One concerns the norms of attention: How do (and should) we evaluate people's attention? Another project uses x-phi methods to illuminate psychological concepts, including self-control, conscientiousness, and mind wandering.

Maria Khoudary is a research assistant in the Imagination and Modal Cognition Lab at Duke University, program coordinator for the Summer Seminars in Neuroscience and Philosophy, and an incoming PhD student in the Cognitive and Neural Computation Lab at the University of California, Irvine.

Kristina Krasich is a postdoctoral fellow in the Center for Cognitive Neuroscience at Duke University. She studies the neurocognitive mechanisms that support visual attention and conscious perception, and how each process impacts how people generate mental simulations and make causal judgments. She uses behavioral, eye tracking, and electroencephalography methods and explores philosophical implications of her work for moral agency and responsibility.

Enoch Lambert is a postdoctoral associate at the Center for Cognitive Studies at Tufts University. He received his PhD in philosophy from Harvard University where he worked on implications of evolutionary theory and species concepts for how we think about human nature. He recently co-edited the book *Becoming Someone New: Essays on Transformative Experience, Choice, and Change* with John Schwenkler.

Cristina Leone graduated from the University of Toronto as a double-major in Neuroscience and Cognitive Science. Her research interests include causal cognition and clinical psychology. She is currently pursuing her PhD in experimental psychology at University College London, studying causal cognition.

Anna Leshinskaya received her PhD from Harvard University in 2015, working with Alfonso Caramazza on the neural organization of semantic memory. She then worked as a postdoc with Sharon Thompson-Schill at the University of Pennsylvania, where she investigated questions regarding how new semantic knowledge is learned from cognitive and neural perspectives. She now works with Charan Ranganath at UC Davis studying how the neural mechanisms for forming new episodic and semantic memory interact.

Jordan L. Livingston is a postdoctoral fellow in the Department of Psychology at the University of Toronto. Her research examines the social psychology and social neuroscience of self and identity.

Brian Maniscalco is currently an Associate Project Scientist at the Cognitive and Neural Computation Lab at UC Irvine, seeking to understand the nature and functioning of consciousness and metacognition. A central premise of his research is that it is methodologically crucial to distinguish consciousness and metacognition from correlated but ultimately dissociable perceptual and cognitive processes. To that end, he has helped develop analytical frameworks and experimental paradigms that can tease apart consciousness and metacognition from confounding factors such as objective task performance.

Joshua May is an Associate Professor of Philosophy at the University of Alabama at Birmingham. He draws on empirical research to understand better the development, improvement, and breakdown of moral knowledge and virtue.

Joseph McCaffrey is an Assistant Professor of Philosophy at the University of Nebraska, Omaha, and a member of the Medical Humanities Faculty. He is a philosopher of neuroscience and cognitive science. His research concerns the nature of structure–function relationships in the human brain and whether and how neuroscience informs our philosophical understanding of the mind.

Jorge Morales is an Assistant Professor of Psychology and Philosophy at Northeastern University. His research program aims to increase our understanding of the subjective character of the mind, in particular the cognitive architecture, the neural implementation, and the mental properties that make subjectivity possible.

Samuel Murray is a postdoctoral associate in the Psychology and Neuroscience Department at Duke University. He earned his PhD in philosophy from the University of Notre Dame in 2019 and was a research fellow at the Universidad de los Andes (Bogotá) from 2017 to 2019. His research focusses on mind wandering, vigilance, and moral responsibility.

Thomas Nadelhoffer is an Associate Professor in the Philosophy Department at the College of Charleston. He is also an affiliate member of the Psychology Department and a roster faculty member in the neuroscience program. Professor Nadelhoffer specializes in the philosophy of mind and action, moral psychology, and the philosophy of law.

Laura Niemi is Assistant Professor of Psychology at Cornell University and the Charles H. Dyson School of Applied Economics and Management at SC Johnson College of Business. Her research combines a variety of methods from experimental social psychology and psychology of language to investigate topics including causal attribution and blame, the psychological sources of moral values, and the role of moral cognition in well-being and conflict.

Brian Odegaard currently works as an Assistant Professor in the Department of Psychology at the University of Florida, where he serves as the Principal Investigator for the Perception, Attention, and Consciousness Laboratory. His primary research focus is to study the neural and computational basis of perceptual decision making, attention, and metacognition in order to further our understanding of how the brain produces conscious experience. His areas of expertise include multisensory integration, peripheral vision, perceptual biases, and the role of the prefrontal cortex in consciousness.

Hannah Read is a PhD candidate in philosophy at Duke University. Before Duke, she completed her MA in philosophy at Tufts University and her BA in philosophy and literary studies at The New School University. Her work falls primarily within moral philosophy and moral psychology. She has additional interests in social and political philosophy, feminist philosophy, and the philosophy of education.

Sarah Robins is an Associate Professor of Philosophy at the University of Kansas (KU) and an Affiliate Faculty Member of KU's Cognitive and Brain Sciences PhD program. Her research is at the intersection of philosophy and psychology, with a primary focus on memory. In her work, she has argued for an expanded taxonomy of memory errors, the philosophical significance of optogenetics, and the causal theory of memory.

Jason Samaha is a faculty member at the University of California, Santa Cruz, where he runs the Cognitive and Computational Neuroscience Lab. The lab uses recordings of electrical brain activity along with psychophysics, computational modeling, and brain stimulation to understand better the neural basis of visual consciousness, top-down processing, decision making, and metacognition.

Walter Sinnott-Armstrong is Stillman Professor of Practical Ethics in the Philosophy Department and Kenan Institute for Ethics at Duke University, with secondary appointments in the Psychology and Neuroscience Department and the Law School. His current work focuses on moral artificial intelligence, arguments and political polarization, free will and moral responsibility, and various topics in moral psychology and brain science, including how narratives and reasons shape our moral judgments and actions.

Joshua August (Gus) Skorburg is Assistant Professor of Philosophy, Faculty Affiliate at the Centre for Advancing Responsible and Ethical Artificial Intelligence (CARE-AI), and Faculty Affiliate at the One Health Institute at the University of Guelph in Ontario, Canada. His research spans topics in applied ethics and moral psychology.

Shannon Spaulding is an Associate Professor of Philosophy at Oklahoma State University. She works on philosophy of psychology, especially social cognition. Her research explores the ways in which our knowledge of social norms, situational contexts, stereotypes, and biases influence how we interpret social interactions. Her book on these topics is titled *How We Understand Others: Philosophy and Social Cognition.*

Arjen Stolk is an assistant professor in psychological and brain sciences at Dartmouth College. His research focuses on the neurocognitive mechanisms supporting human mutual understanding, taking into account the key role of the shared knowledge people build up over even a short time, and considering alterations of that conceptual ability in psychiatric and neurological disease.

Rita Svetlova is Assistant Research Professor of Psychology and Neuroscience at Duke University. She works primarily on social and moral development as well as empathy and prosocial behavior.

Robyn Repko Waller is an Assistant Professor of Philosophy and Neuroscience program faculty member at Iona College in the Greater New York City area. Her research lies in philosophy of mind and philosophy of psychology, with a particular focus on free will and neuroethics. She has published in *The Monist, Philosophia, Ethical Theory and Moral Practice, Philosophical Psychology*, and *Cambridge Quarterly of Healthcare Ethics*, as well as forthcoming work in academic press anthologies.

Natalia Washington is an Assistant Professor of Philosophy in the Philosophy Department at the University of Utah, specializing in philosophy of psychiatry, cognitive science, and mental health. Using the conceptual and critical tools of philosophy, she seeks to understand how human minds are both shaped by and integrated with our physical and social environments, especially with regard to cognitive biases, social cognition, and the construction of diagnostic categories.

Clifford I. Workman is a postdoctoral fellow in the Penn Center for Neuroaesthetics at the University of Pennsylvania. His research examines moral dysfunction in healthy populations (e.g., support for violence, biases relating to beauty) as well as disordered populations (e.g., symptomatic guilt in major depression).

Jessey Wright's last academic appointment was as a postdoc at Stanford University. There, they collaborated as a philosopher in the Poldrack Lab. Their research examined how the methods and means by which researchers obtained and interacted with data affected their understanding of its significance, and ultimately shaped how they reasoned about the claims about phenomena that data are situated as evidence in support of.

Index

Printed in the United States
by Baker & Taylor Publisher Services